Sustainable Agriculture and Climate Change

Special Issue Editors

Suren(dra) Nath Kulshreshtha
Elaine E. Wheaton

MDPI • Basel • Beijing • Wuhan • Barcelona • Belgrade

MDPI

Special Issue Editors
Suren(dra) Nath Kulshreshtha
University of Saskatchewan
Canada

Elaine E. Wheaton
University of Saskatchewan
Canada

Editorial Office
MDPI AG
St. Alban-Anlage 66
Basel, Switzerland

This edition is a reprint of the Special Issue published online in the open access journal *Sustainability* (ISSN 2071-1050) in 2017 (available at: http://www.mdpi.com/journal/sustainability/special_issues/agricultural_climate_change).

For citation purposes, cite each article independently as indicated on the article page online and as indicated below:

Lastname, F.M.; Lastname, F.M. Article title. *Journal Name*. **Year**. *Article number*, page range.

First Edition 2018

ISBN 978-3-03842-725-4 (Pbk)
ISBN 978-3-03842-726-1 (PDF)

Table of Contents

About the Special Issue Editors

Suren(dra) Nath Kulshreshtha is a Professor in the Department of Agricultural and Resource Economics at the University of Saskatchewan. He received his earlier degrees from Agra University, India, and a Ph.D. in agricultural economics from the University of Manitoba. He has taught quantitative methods, and project evaluation, where the incorporation of environmental considerations is a major focus. His research has focused on resource economics and economics of climate change. He has served various professional societies in capacities such as Editor of the Canadian Journal of Agricultural Economics, Associate Editor of the Canadian Water Resources Journal, and a Regional Editor of the journal Impact Assessment. He has participated in several oversees projects in Indonesia, Zambia and India through the Canadian International Development Agency. He has over 600 publications to his credit, with over 150 refereed journal articles. In 2004, based on his contributions, the Canadian Society of Agricultural Economics selected him as a Fellow of the Society.

Elaine E. Wheaton is an Adjunct Professor at the Department of Geography and Planning and the School of Environment and Sustainability, and Emeritus Researcher at Saskatchewan Research Council. She has considerable research experience in the areas of climate change, impacts, adaptations, hazards, risks and vulnerability assessment. Specific sectors of expertise include agriculture, water, health and risk assessment. She has international research experience in several countries, including those in South and Central America, Asia, and Europe. Her awards include the 2007 Nobel Peace Prize certificate for substantial contributions to the work of the Intergovernmental Panel on Climate Change, Wolbeer Award for contributions to water resources research, Emeritus Researcher and Distinguished Scientist appointments at the Saskatchewan Research Council, and the YWCA Science and Technology award. She is widely published in refereed science journals and the author of the award-winning book, But Its a Dry Cold! Weathering the Canadian Prairies.

Preface to "Sustainable Agriculture and Climate Change"

Globally, our food system is not sustainable, does not provide adequate nutrition to everyone on the planet and, at the same time, changes to our climate threaten the future of farming as we know it (a quote from Achieving Food Security in the Face of Climate Change, final report from the Commission on Sustainable Agriculture and Climate Change, 2012). Agriculture is both part of the problem and part of the solution to climate change. With our great pleasure, we present this book based on articles published in the Special Issue of Sustainability—Sustainable Agriculture and Climate Change. International research, including the work by the Intergovernmental Panel on Climate Change (IPCC), has stated that climate change is a reality. Most parts of the world are vulnerable to the impacts of global warming expected under these conditions, as confirmed by international research. Climate change therefore has a strong connection with the sustainability of agriculture, and perhaps of the entire global economic system. This aspect of the relationship between climate change and sustainable agriculture is investigated in this book. Producers facing climate change have the option of adapting to the changed conditions as well partnering in reducing the greenhouse gas (GHG) emissions and reducing the damage inflicted by current and future climate change. This book explores both of these topics.

Resilience is an important feature of sustainability and is addressed by Herrera in Chapter 1. Having resilience is a key characteristic of any production system. Although global warming would affect agriculture in different parts of the world differently, a major part of these impacts would be created through increased frequency of extreme events. These events may be observed in the form of changes in extreme cold or heat, or in terms of precipitation—droughts and floods. Under climate change, the frequencies of such events would increase, as well as their intensities and durations. More droughts, for example, would be back-to-back or even longer in duration. These would affect environmental sustainability (as suggested by Wheaton and Kulshreshtha in Chapter 2) and in turn, sustainability of agriculture. Agricultural production, particularly crop production, depends heavily on water resources. Under climate change, producers may develop surface water systems (discussed in Chapter 3 by Berry et al.), change planting dates (suggested by Wang et al. in Chapter 7), as well as expand irrigation, as suggested by Zhang et al. in Chapter 8.

Adaptation for climate change is not going to be easy and inexpensive. In fact, as Lazurko and Venema indicate in Chapter 4, some methods of financing such costs would be needed. Such financing could be easier for producers if they lead to higher returns from agriculture. Some adaptation measures have been shown to have significant positive impacts (see the case of Chile by Roco et al. in Chapter 10, and the use of conservation farming in Chapter 12 by Mubiri et al.). Changing greenhouse gas (GHG) emissions and climate change would have impacts on crop yields (Chapter 5 by Gao and Bian; Chapter 9 by Lim et al.), and increase social vulnerability (Chapter 6 by Ye et al.), among other changes. Although adaptation to changing climates is needed, efforts can also be focused on reducing GHG emissions through practices such as integrated nutrient management, as suggested by Graham et al. (Chapter 13). Machekano et al. (Chapter 14) indicate that adoption of better technologies may be limited, particularly in the case of small holders.

Although Coutinho et al., in Chapter 11, have applied the Participatory Sustainability Assessment in Brazil, they feel that it is still necessary to develop instruments for studies on sustainability. As the Commission on Sustainable Agriculture and Climate Change has noted: The transition to a global food system that satisfies human needs, reduces its GHG footprint, adapts to climate change and is in balance with the planets resources requires concrete and coordinated actions, implemented at scale, simultaneously and with urgency. In this light, overcoming technical, social, financial and political barriers to adaptation to climate change and adoption of measures for GHG mitigation are needed around the world, both by the public as well as the private sectors in order to maintain the sustainability of agriculture and food systems.

Suren(dra) Nath Kulshreshtha, Elaine E. Wheaton
Special Issue Editors

sustainability

MDPI

Article

Resilience for Whom? The Problem Structuring Process of the Resilience Analysis

Hugo Herrera [1,2]

1   Geography Department—System Dynamics Group, Bergen University, Fosswinckelsgate 6, 5007 Bergen, Norway; hugojhdl@gmail.com
2   Department DEMS, University of Palermo, 90100 Palermo, Italy

Received: 6 May 2017; Accepted: 3 July 2017; Published: 11 July 2017

Abstract: Resilience is a flexible concept open to many different interpretations. The openness of resilience implies that while talking about resilience, stakeholders risk talking past each other. The plurality of the interpretations has practical implications in the analysis and planning of resilience. This paper reflects on these implications that have so far not explicitly been addressed in the literature, by discussing the problem structuring process (PSP) of a modelling-based resilience analysis. The discussion is based on the analysis of food security resilience to climate change in Huehuetenango, Guatemala, jointly undertaken by the author, governmental authorities, small-scale farmers and academics of the national university. The aim of this discussion is to highlight the underestimated challenges and practical implications of the resilience concept ambiguity and potential avenues to address them. The contributions of the results presented in this paper are twofold. First, they show that, in practice, the resilience concept is constructed and subjective. Second, there remains a need for a participatory and contested framework for the PSP of resilience.

Keywords: food security; resilience; power; system dynamics; problem structuring process

1. Introduction

Climate change effects start to be recognised as threats to food system sustainability and food security [1]. Sustainability involves maintaining the functionality of the system without compromising its capacity to do so in the future [2]. However, undergoing effects of climate change compromises food system functionality by contributing to water scarcity and pest exacerbation [3]. Resilience is understood as the system adaptive ability of maintaining its functionality even when the system is being affected by a disturbance [4–6]. For this reason, resilience is a compelling framework for researchers and policymakers seeking to understand how socio-ecological systems (SESs) adapt and transform to withstand changes in the environment. In practice, resilience is often used as a measure of a SES's capability to respond and adapt to new conditions (e.g., climate change). Like Tendall et al. [2] (p. 18) describe, "sustainability is the measure of system performance, whereas resilience can be seen as a means to achieve it". Resilience has the potential to contribute to food security by enhancing farmers, and other stakeholders, capacity "for foreseeing and adapting to possible changes" [5] (p. 270). For instance, in the food systems literature, a number of studies have used resilience as framework for understanding how systems can adapt and transform in the presence of disturbances in the environment while still providing required amounts and quality of food [2,7].

Applications of resilience can be found in numerous disciplines, ranging from engineering to psychology to disaster risk management [8]. The increased popularity of resilience is due, at least partially, to the flexible meaning of the concept [8,9]. Resilience definitions have often been characterised as vague and unprecise in practical terms [2,9]. While the flexibility of resilience has moved it to the category of mainstream concepts and buzzwords, the same ambiguity represents a challenge to its application in prescriptive and normative settings. These challenges manifest when practitioners need to operationalise the concepts described in the literature to the context in which resilience will be applied. Unsurprisingly, different stakeholders of the analysed system have different and sometimes conflicting interpretations of what resilience means in practical terms.

Since each stakeholder interprets resilience differently, the scope of the analysis to be undertaken is not a given but is constructed through a problem structuring process (PSP). The term PSP is used in this paper to describe the "process by which a presented set of conditions is translated into a set of problems, issues sufficiently well-defined to allow specific research action" [10]. During the PSP, stakeholders interpret the available information in light of their values and knowledge and negotiate what is the purpose and the boundaries of the study to commence (referred to from now on as the "scope of the resilience analysis") [11,12]. The cognitive, social and political components, involved in the construction of the scope of analysis, condition its development and outcomes. The social and political nature of the PSP make it impossible to separate the conclusions and recommendations produced from the context in which they were produced. When talking about resilience, we cannot avoid the question: resilience for whom?

Literature has recently started to recognise some of the practical challenges of resilience ambiguity [2,9,13]; however, it still lags behind on recognising the political implications of resilience ambiguity in the analysis and its outcomes [8,14]. While some progress has been made by operationalising the definition of resilience (see for example [2,13]), resilience frequently continues to be presented as a "politically neutral approach" [9] (p. 134). The influence of stakeholders' agendas and power relationships are often overseen by practitioners [8,14]. Although these dimensions of the PSP have been discussed for a long time in the literature regarding problem structuring methods (PSMs), their implications for the resilience analysis are still unexplored.

This paper contributes to closing these gaps by discussing the political and social implications of resilience ambiguity in the PSP. To this purpose, this paper looks at the PSP of a modelling-based analysis of food security resilience to climate change. This case is used to discuss some of the cognitive and political challenges of resilience. This discussion is informed by the personal construct theory [15] and enriched by a post-normal science epistemology [16] for managing a wide range of perspectives. The aim of this discussion is to reflect on (a) the implications of having a diversity of resilience interpretations in the PSP and (b) the potential avenues to mediate stakeholder engagement and mitigate the challenges this diversity entails.

2. Case Study: Analysing the Resilience of Food Security to Climate Change in Guatemala

This research was conducted within the qualitative paradigm of case study research [17,18] and is part of an independent modelling-based discussion for the analysis of and planning for food security resilience to climate change in Guatemala. Specifically, this case study describes the PSP followed to define the scope of the resilience analysis undertaken in the district of Huehuetenango. As part of this PSP, the author conducted a series of semi-structured interviews among relevant stakeholders in the local maize production system.

2.1. Background

Guatemala, similar to other developing countries, faces food security challenges that will only increase as climate change affects small-scale farmers' capabilities to produce food. Guatemala's chronic malnutrition, an accepted measure of food insecurity, is one of the highest in the world [19], reaching 55% in rural areas [20]. Climate change effects, such as severe droughts and increased average temperatures, already compromise the food production in Guatemala, especially among small-scale farmers [21].

Recognising this as problematic, some studies that explore potential means to mitigate climate change effects have been commenced separately by academics, nongovernmental organisations (NGOs) and the local and central government in Guatemala. This research is part of these initiatives, independently conducted by the author with the cooperation of numerous stakeholders in the district of Huehuetenango.

Huehuetenango is located in the Northwest region of Guatemala, on the border with the South of Mexico. Huehuetenango is one of the poorest, most vulnerable districts in Guatemala. In 2014, its population was estimated at 1,150,000 people, with 67.6% of these people under the line of poverty [22]. Huehuetenango's main economic activities are the mining industry of silver and gold and the production of coffee [23]. Nevertheless, the production of maize is an important activity for self-consumption. The majority of the population is indigenous, from the ethnics of Mam and Quechi, with a cultural dependence on maize as the main source of calories. Among indigenous groups, maize represents a 71.2% of share in basic grains consumption).

2.2. Methodology

The intention of the study was to discuss potential policies to enhance food security resilience and to explore in an operational manner the impacts of these policies on different parts of the system. The author, with the support of two academics from the Universidad de San Carlos de Guatemala (national university in Guatemala), started by identifying (mapping) and engaging relevant stakeholders as early as possible and throughout the PSP. The following stakeholder groups accepted the invitation to participate in the PSP: (i) the central government; (ii) NGOs; (iii) farmers from Huehuetenango and (iv) academics and agronomists from the University. The number of delegates from each group and their backgrounds are presented in Table 1.

Table 1. Stakeholders' group representatives.

Stakeholder Group	Number of Delegates Participating	Background
Central Government (CG)	4	Agronomists Policymakers
Non-Governmental Organization (NGO)	3	Agronomist Project Managers
Farmers (F)	6	Maize Farmers
Academics (AC)	2	Agronomist Professor Researcher

During the PSP, the author conducted semi-structured interviews to gather stakeholders' perspectives about the food security resilience of the small-scale maize production system of the region. In the first part of the interviews, the author asked the delegates of the different stakeholder groups about the agendas they have for the local food system. Subsequently, causal loop diagrams (CLDs) were used to capture stakeholders' broad understanding of the underlying causes of system vulnerability (the extent to which the system will be affected by) climate change. Finally, the delegates were also asked to rank the stakeholders in the system in terms of influence on and interest in the local food system.

The elicitation of stakeholders' agendas for the local food system was done by discussing the following general questions with the delegates of each stakeholder group:

- What would you like to get from the small-scale maize production system?
- In this context, what does resilience of food security to climate change mean?
- What are the critical success factors of policies enhancing food security?

After the interviews, the author compiled and summarised the different answers. Similar answers were grouped in the same variable or short statement to simplify further analysis. The resulting statements were discussed in further interviews with each delegate to ensure they reflected their own perspectives. When needed, changes were made and again discussed with the specific delegate requesting the change.

Beside the narratives provided by the delegates, this paper uses CLDs as a means for capturing stakeholders' assumptions. CLDs are diagrams representing, in a simple manner, a possible set of causal relationships between different variables of the systems [24,25]. CLDs are particularly useful for identifying circular relationships known in the systems' literature as feedback loops. The rigor of diagramming forces the participants to "carefully and consistently" make their assumptions explicit and to "put their problem definition to test" [26] (p. 384). Thus, CLDs are a suitable way to represent and compare different interpretations of the problem and the causal explanations held by the stakeholder groups participating in the PSP.

CLDs might be employed in the PSP (also known as the conceptualisation stage of the modelling process) [27,28] to elicit participants' understanding of the problem. During the conceptualisation, the modeller focuses on "a verbal description of the feedback loops that are assumed to have caused the reference mode" [19] (p. 119). Namely, in this paper, the CLDs were used to diagrammatically represent the causal explanations for the lack of resilience of food security in the region. This elicitation might be done, as it was in the case of this paper, during one to one interviews with experts in the field, in our case an agronomist from the university, and stakeholders of the problem at hand.

During the semi-structured interviews, the author drafted CLDs representing what the delegates were describing. The author started by asking the delegates what were the main causes of the decrease and fluctuations of the affordability of maize (as a measure of food security [29]) experienced in the past 10 years in the region of Huehuetenango (see Figure 1). The causes stated by the delegates were summarized by the author in relevant variables while transcribing them to the diagram. Then, the author asked delegates to explain how those variables influenced each other. These causal links between different variables were represented in the diagram by arrows connecting the cause with its effects. When needed, new variables were added to the diagram.

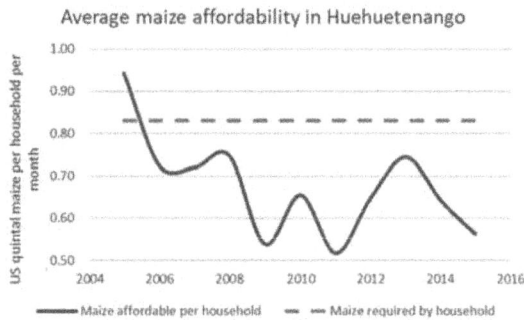

Figure 1. Maize affordability in Huehuetenango.

At the end of the interview, the delegates were asked to complete the CLDs drafted by the author by adding variables, causal relationships or any elements missing in the diagram. Later, the author worked on his own by summarising all the CLDs produced by each delegate into a single CLD per stakeholder group. The single CLDs were validated and discussed with the delegates of each stakeholder group in separate interviews to ensure all of their views were appropriately captured in the diagrams. If participants found important issues missing in the diagram, those issues were added to the final version.

Finally, delegates were asked to characterise the different stakeholders in the system. To be precise, participants were asked to rank from 1 (low) to 5 (high) the level of influence each stakeholder group has on the local food system. During this characterisation, participants were invited to consider in their assessment what resources each stakeholder can allocate for this purpose and the level of organisation and the reputation of each. Similarly, participants were asked to rank the stakeholders from 1 (low) to 5 (high) according to their interest in the problem (i.e., resilience of food security). The author tabulated the results into a single chart showing the average level of influence of each stakeholder group.

Analytical Framework

The results were analysed in light of personal construct theory (PCT) [15]. PCT is based on the assumption that a person needs to make sense of the problem to address it: "a person's processes are psychologically channelized by the ways in which he anticipates events" [30] (p. 7). Thus, to analyse resilience, stakeholders need first to make sense of what resilience means. To illustrate how this cognitive process unfolds, this paper adapts the simplified model proposed by Eden [31] to examine how stakeholders construct their own interpretations of resilience (see Figure 2). According to Eden's [31] model, stakeholders make sense of the concept of resilience by selecting particular elements that are applicable to the problem at hand and its context. This perception is then filtered through the individual system of values and beliefs to articulate its own interpretation of what resilience means in practical terms. This separation of selective perception and construal follows the personal construct theory of Kelly [15].

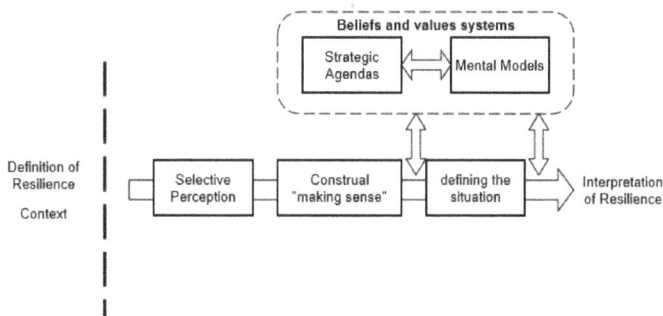

Figure 2. Construction of stakeholders' interpretation of resilience. Note: Adapted from Eden [31].

There is no clear distinction between values and beliefs, as they are closely interconnected [31]. However, for analysis purposes, this paper explores two separate interconnected aspects of the beliefs and values systems: strategic agendas and mental models. The term strategic agenda is used here to describe the set of goals each stakeholder has for the system. Similarly, the term mental model is used to describe the conceptual representations each stakeholder has about how the system works [32]. Strategic agendas and mental models are not separate entities. They support each other, and together, they are supported by wider individual value systems [31].

In policymaking settings, closely linked to the understanding of what resilience means in practical terms, is the concept of adaptability or the "the capacity of actors in the system to influence resilience" [33] (p. 5). Stakeholders' adaptive actions depend on how they perceive the disturbance is changing the conditions of their system. Since timing, magnitude and origin of the disturbance are, at least to some extent, unpredictable, the nature of the change that the disturbance produces deviates from the normal system-near-equilibrium analysis [34]. In these conditions of high uncertainty, identifying the mechanisms driving adaptation is not straight forward but depends on the stakeholder's mental models about how the system works.

To analyse how stakeholders' understand the system, this paper uses the reflections of Mayumi and Giampietro [35] about self-modifying systems and the theories of Funtowicz and Ravetz [36] on emergent systems. According to the aforementioned sources, the explanations each stakeholder group gave to the system behaviour were classified into:

(a) endogenously driven: the observed effects of disturbances affecting the system are the result of the functional links between its different elements. Adaptation emerges from the mechanisms the system has to regulate itself and can only be enhanced by strengthening them [35]. The solution to the problem is within the system boundaries.

(b) exogenously driven: the disturbance affecting the system comes from outside the system and, to adapt to the new conditions introduced, the system needs of external interventions that "push" it back to its equilibrium state. The solution is outside the system boundaries.

(c) chaos: the uncertainty about the disturbance affecting the system and complexity of the system itself are perceived so high that it is impossible to identify links between actions (outside or within the system) and their consequences. The solution is unknown.

This classification offers a helpful analytical framework to explain how delegates from different stakeholder groups understand the system and the differences in the policies they will propose in further stages.

Nonetheless, the agendas and mental models used by stakeholders to construct their own interpretation of resilience are only some of the ingredients for the scope of the resilience analysis. The manifestation of power in the PSP is indeed critical analytical lens to understand complications of resilience ambiguity. In fact, power effect on resilience is one of the most unexplored but most contested characteristics of resilience [14].

Case study research shows that in prescriptive settings, the PSP of resilience is predominantly a negotiation endeavour. For instance, Lebel et al. [37] describe that in many case studies undertaken by the Resilience Alliance, the scope of resilience analysis reflects, to a large extent, the interest of powerful stakeholders, undermining perspectives of ethnic minorities and small-villages (powerless stakeholders). Similarly, Larsen et al. [38] highlight the tensions regarding roles, control and ownerships between powerful stakeholders during the process of building resilience in Thailand tourism-dependent communities.

These cases studied in the literature show that during the PSP of resilience, stakeholders will try to persuade the others to join or accept their own interpretation of resilience and to articulate the scope of the resilience analysis accordingly. As illustrated in Figure 3, the scope of analysis is a negotiated outcome of the PSP that reflects not only the interpretations of each stakeholder in the system but also the power relationships between them.

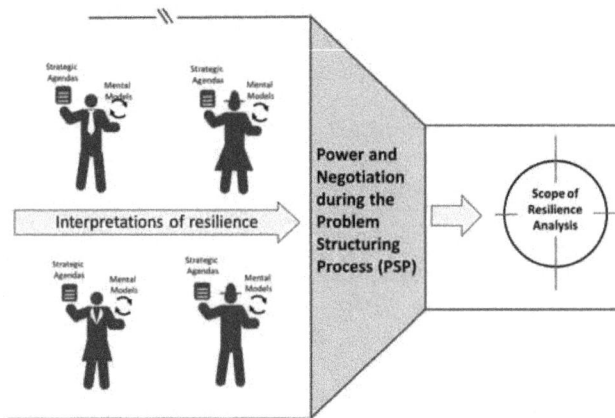

Figure 3. Simplified representation of the problem structuring process (PSP) of resilience analysis.

2.3. Results

2.3.1. Strategic Agendas

Tabulated results from the interview show that delegates from the same group coincide to a large extent in the answers they provided about their agendas. Table 2 summarizes these tabulated answers. In Table 2, it is noticeable that most of the delegates of the same group agreed on a similar answer.

Table 2. Summarized answers to the semistructured interviews.

Delegate Code	CG1	CG2	CG3	CG4	NGO1	NGO2	NGO3	AC1	AC2	F1	F2	F3	F4	F5	F6
What would you like to get out from the small-scale maize production system?															
Increase households' wealth	X	X		X										X	
Produce revenues			X		X	X	X	X	X	X		X		X	
Produce food															
Produce food for locals						X			X		X	X	X		X
In this context, what resilience of food security to climate change means?															
Being able to afford food even when droughts	X	X	X		X	X	X	X						X	
Produce food constantly in despite of the droughts	X		X	X	X	X	X	X	X		X				
Don't starve during the bad years								X							
Have always enough food										X	X	X	X	X	X
What are the critical success factors of policies enhancing food security?															
Money available for purchasing food	X	X	X		X		X	X							
Crop productivity		X					X								
Maize Yield		X		X		X									
Maize reserve		X		X		X			X	X	X	X	X	X	X

Note: CG: Central Government; NGO: Non-Governmental Organization; AC: Academics; F: Farmers.

Based on the interviews results, the strategic agenda held by each stakeholder group can be summarized as follow:

Central Government (CG): The purpose of the analysis is to identify how to increase the household's wealth and particularly the money available to buy food so that households can afford enough food even when droughts reduce the yields of maize in the region.

Non-Governmental Organization (NGO): The purpose of the analysis is to identify how to enhance crop productivity so that households can produce food and revenues constantly despite the droughts. Note that in the words of the NGO delegates, crop productivity is understood as the amount of crop (not exclusively maize) produced from each Guatemalan Quetzal invested by the farmers.

Academics (AC): The purpose of the analysis is to identify how to increase maize yields and reserves as a mean to prevent starvation by increasing farmers revenues and food supply to the region.

Farmers (F): The purpose of the analysis is to identify how to increase food production (not limited to maize or crops in general) and maize reserves to have food year round.

2.3.2. Causal Loop Diagrams

Figure 4 presents the CLD's prepared jointly by the author and delegates of each group. In general, diagrams are relatively simple and focused (with the exception of diagram in Figure 4c) on one or two main causal explanations of the problem to address (decrease and fluctuations of maize affordability in the region).

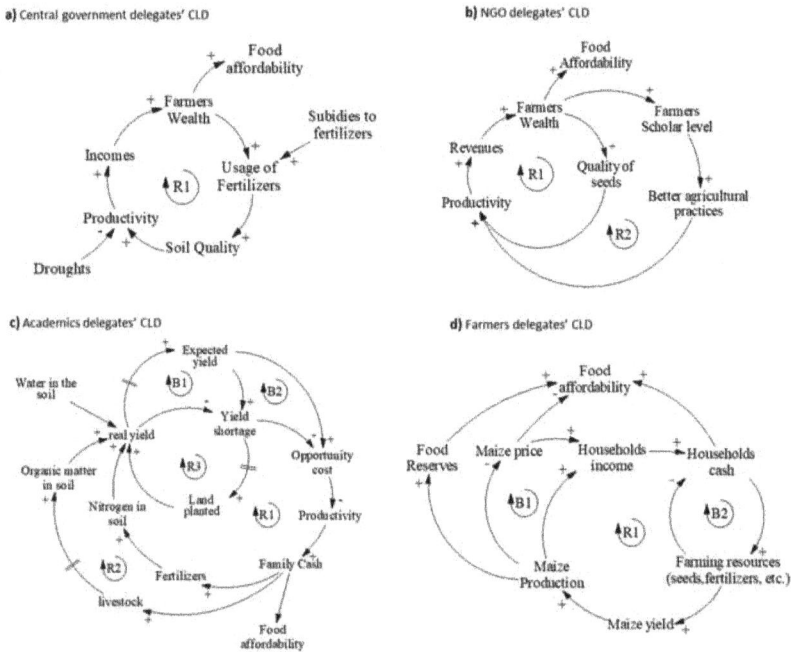

Figure 4. Causal loop diagrams (CLDs) explaining the decrease and fluctuations of maize affordability in Huehuetenango. CLDs were produced by (**a**) central government delegates; (**b**) NGO delegates; (**c**) academics delegates and (**d**) farmers delegates; during one-to-one semi structured interviews.

Next, there is a brief explanation of each diagram.

Central Government (CG): Farmers productivity increases the incomes and, therefore, the wealth of the farmers. Higher wealth increases farmers' capacity to use fertilizers (fertilizers are more affordable). Usage of fertilizers is directly related to the productivity and, therefore, the more fertilizers the farmers use the more productive they become in a virtuous cycle represented by the R1 feedback loop in Figure 4a. This loop, however, is perturbed by droughts (disturbances of the system) that reduce farmers productivity, reducing their overall wealth and hence their capacity to acquire food (food affordability).

Non-Governmental Organization (NGO): Farmers productivity increases the incomes and therefore the wealth of the farmers. Higher wealth increases farmers' capacity to access better seeds and formal education. Seeds of improved varieties, the ones that require less water, are assumed to increase crop productivities, especially during drought seasons, compared to seeds coming from informal sources (on farm save seeds for example). Better seeds increase wealth in the virtuous cycle represented by R1 in Figure 4b. Access to formal education is assumed to be linked to better agriculture practices (e.g., appropriate usage of fertilizers and land planning). Better agriculture practices increase revenues and wealth in the virtuous cycle represented by R2 in Figure 4b.

Academics (AC): The causal explanation represented in Figure 4c focuses on the variation of the real yield against the expected one (yield shortage in the diagram). Yield shortage results into lower productivities and opportunity costs that reduce families' cash and their capacity to invest in fertilizers and livestock (see feedback loops R1 and R2 in Figure 4c). Higher yield shortage also translates into a reduction of the land planted each season (see R3 in Figure 4c), because farmers need to spend more time on other activities (e.g., working on coffee plantations) and less time farming. The expected yield eventually gets adjusted, decreasing the yield shortage and opportunity costs (see loops B1 and B2 in Figure 4c). The increase in droughts occurrence increases yield shortage by affecting the maize system and its real yield, reducing at the same time the land planted and the cash available for the next season's harvest.

Farmers (F): Maize production increases incomes and households' cash, allowing farmers to acquire more resources needed in farming activities (seeds, fertilizers, etc.). This eventually increases the maize production. Higher production results in a) higher food reserves and b) higher incomes (see feedback loop R1 in Figure 4d). However, there are two drawbacks from the feedback loop R1. First, the acquisition of resources decreases households' cash (see feedback loop B1 in Figure 4d) thereby reducing the food affordability. Second, higher production will eventually translate into lower maize prices, reducing farmers' income and profit margins (see feedback loop B2 in Figure 4d).

2.3.3. Influence-Interest Grid

Figure 4 presents the stakeholders' grid produced by the delegates. The four stakeholder groups participating in this case study were consistently identified by all the delegates as those with the highest interest in the problem (see Figure 5). The central government and NGO working in the area were described as being the stakeholders in a better position to solve the problem or those with higher influence on the problem (see quadrant I in Figure 5). Other stakeholders, like the large-scale farmers producing food in the region and traders, were also recognised as highly influential. However, there was an agreement among delegates of all the stakeholder groups participating that, unfortunately, large-scale farmers and maize traders have no interest in enhancing food security in the region (quadrant IV in Figure 5). While recognized as those with higher interest in the problem, small-scale farmers were portraited as the group with the lowest influence on it (see quadrant II in Figure 5). Academics and stakeholders not participating in the PSP (local government) were also portrayed as interested parties with low influence.

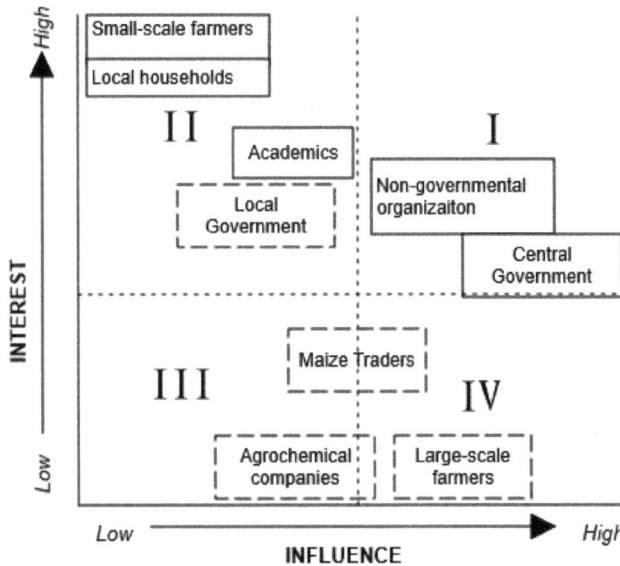

Figure 5. Influence/interest diagram summarizing stakeholders rank in the small-scale maize production system in Huehuetenango. Note: stakeholders in dotted lines did not take part in this research.

3. Complications of the PSP in the Analysis of Resilience

The results presented in Section 2 offer relevant evidence to discuss the ambiguity of resilience and its complications. The ambiguity of resilience, in this case, does not arise from the differences between many definitions of resilience [13], but from the way in which stakeholders interpret it for their specific context and problem. The differences that emerged during the PSP might already be noticeable for the reader, but the analytical lenses proposed in this paper offer a perspective of the deeper and more conflicting differences in the agendas and mental models held by each stakeholder group.

These cognitive differences set the scene for analysing the conflict that could unfold during the negotiation of a single scope of resilience. The more mutually exclusive agendas and mental models are, the harder it is to reach a scope of resilience that satisfies all the stakeholders. As Eriksen et al. [39] pointed out, adaptation will change social, political and economic relationships between stakeholders, "yet not all these changes are desirable for everybody".

This section concludes by discussing the practical implications of resilience ambiguity in the policymaking process. These implications are not only political but also methodological and require thoughtful planning of the PSP. While it might be possible to mitigate some drawbacks, more research is needed before outlining a comprehensive framework for addressing the political challenges that resilience entails.

3.1. Constructing an Interpretation of Resilience

The experience in the district of Huehuetenango in Guatemala shows that different stakeholders have different interpretations of resilience. These interpretations of resilience are context specific [40,41] and reflect the values and beliefs of the stakeholders involved. In other words, stakeholders make sense of what resilience means in their particular context and frame the analysis process accordingly. In this case study, different interpretations of resilience are reflected in (a) the different goals and desired outcomes (strategic agenda) stated during the interviews (see Table 2) and (b) the different descriptions of the causes of the problem (mental models) captured in the CLDs (see Figure 1).

When looking at the strategic agenda, stakeholders see the maize production system at different levels of aggregation (household level vs. regional level). As presented in the results section, delegates from the same stakeholder group share similar perspectives about the purpose of the system (see Table 2). With the exception of the farmers, the groups also share some alignment among themselves. The answers in Table 2 and summarized strategic agendas in the Results section show that most of the delegates have local/regional goals for the system, namely to promote local economic development. Alternatively, farmers focus on their current urgent problem of living in insecure food conditions.

In other words, there are two main strategic agendas for the system. One agenda (shared by many stakeholders) is seeking to use the system as a tool for local and/or regional economic development. The other agenda, held by the farmers, is to have food all year round. While there might be different arguments in favour of one agenda over the other one, it is unlikely that regional solutions will have any impact unless urgent issues challenging the farmers' own subsistence are addressed. Similarly, small-scale solutions, addressing farmers immediate needs, might prove to be unsustainable in the mid-term if the wider problem is not tackled.

Wider differences are found when looking at stakeholders' mental models reflected in the CLDs developed. Academics and NGO delegates describe the system in endogenous terms. This endogenous perspective is reflected in the feedback loops identified in the CLD they drafted (see Figure 4b,c). They look at the problem in a systemic way and try to find solutions within the system boundaries. They have, however, a different understanding of the vicious circles constraining food security. On the one hand, academics focus on the management of the water resources and reservoirs as a potential leverage point.

> *"The obvious cause of the problem is the deficiencies the communities face to access water . . . . This is why that, now that droughts are becoming more common, farmers face more problems."* (Academic delegate 1)

On the other hand, NGO delegates blame farmers' lack of technical skills and training as the cause of their poor productivity and, hence, food insecurity. The solution they propose is to increase training and to provide farmers with better seeds to increase their productivity in a sustainable way.

> *"You see, there are several complications in the situation of these poor people because their culture doesn't let them move forward. They use the same techniques they have been using since pre-colonial times. They have no formal education. You know that most of them cannot read. It is really difficult to teach them and change their minds. We need to make an effort to provide them with the right seeds and the proper instruction to use them well."* (NGO delegate 2)

The government delegates describe the system as exogenous driven. These delegates think the way to influence the system is through the artificial enhancement of farmers' productivity (see Figure 4a). Even though they identified a feedback loop in the system, their proposed solution focuses on ways to quickly boost the system performance, namely by using more fertilizers to increase productivity.

> *"The government is committed to provide a sustainable and plausible solution by providing the fertilizers they (farmers) need to increase their productivity and become more competitive . . . . Once they (farmers) level up with the market, the food affordability should be a natural condition."* (Central government delegate 2)

Farmers perceive the problem in a very different way. In their perspective, the increasing uncertainty about rainfall is transforming the system into a chaotic one. From their perspective, using more expensive seeds or more fertilisers will be useless if the weather conditions are not good. Farmers do not feel in control of the system. They feel they are victims of the uncertainty of the yields that they will get at the end of the season.

"The problem is you don't know if the yield is going to be good or not . . . . Now you never know . . . . If the yield goes bad, we lost the money we spent on seeds and fertilizers." (Farmer delegate 4)

"The weather now cannot be predicted . . . . You gamble every time you plant." (Farmer delegate 1)

Furthermore, the farmers do not see higher production as a means to increase their revenues but only as a means to increase their food reserves (see Figure 4d). In their view, the region is isolated, and they do not have access to other markets to trade. The benefit they perceive from higher production is in having more maize to build food reserves for the future.

Understanding and acknowledging different goals and mental models about the system will lead to a wider scope of analysis and might result in a more balanced decision-making process [34]. Short-term solutions and systemic interventions could provide a balanced view between achieving short-term outcomes and their long-term consequences. Farmers' chaotic view of the world challenges the mechanistic understanding other stakeholders might have and balances their deterministic view by the acknowledgement of uncertainty. The system cannot be assumed mechanistically following economic rules since human behaviour under stressful situations adapts in sometimes unexpected ways [7]. An oversimplified understanding about how different groups will react during a crisis might lead to policy failure [42]. For instance, while most of the stakeholders expect farmers to use a potential production surplus to increase their revenues, farmers will use it to increase their food reserves, affecting the policy's effectiveness.

3.2. Negotiating the Scope of Analysis

The power to influence the final outcome is not symmetrical among stakeholders, with those holding key resources being in an advantageous position to impose their own interpretations in the final scope. System adaptation will "influence social relations, governance and distribution of resources in any given population or place" [39] (p. 2). However, as shown in this case, there is not always agreement about the changes and the scale at which those changes should be made. Those with higher level of influence in the scope of analysis might not be those directly affected by its outcomes. For instance, the small-scale farmers in Huehuetenango are the stakeholders directly affected by potential decisions about how to enhance resilience, but they are also those with the least influence on the decision-making process (see Figure 5).

Power differences have contentious repercussions considering that those with a higher level of influence have different strategic agendas than those suffering the larger impacts of the policies implemented. This is particularly relevant since there is a clear difference between the farmers' interpretations and those held by the rest of the stakeholders. Considering the different interpretations of resilience, the power to set agendas about what issues are to be addressed needs to be an important consideration during the PSP.

Competitive agendas and mental models set the scenario for a game of power where different stakeholders seek to impose their own agendas on the scope of the analysis that will follow. The allocation and distribution of the access to natural resources have been, historically, an expression of power tension between different groups [14,39]. While building the resilience of the system outcomes, the resilience of the institutions and relationships defining those outcomes are also enhanced [43]. Many stakeholders perceive the resilience analysis as an opportunity to gain power or to influence the system towards their own interests [14]. This power might be exercised in many ways. For instance, stakeholders might scope the problem in isolation, ensuring their interpretations are the only ones represented. Alternatively, some groups could try to undermine those with competitive or opposite views by diminishing their credibility as shown in this case. For instance, note the comment above from NGO delegate 2 in which the delegate undermines farmers' practices because they have no formal education. Any analysis that does not account for these tensions would result in an incomplete understanding of the scope of potential responses [14,44].

In short, recognising that there might be different interpretations of resilience implies accepting the PSP as a negotiation and political process. Seeing the PSP as a negotiation forum means that practitioners need to acknowledge the social and political factors (e.g., inequality and legitimacy) shaping the scope of analysis and need to be transparent about the implications of these factors on their recommendations. Otherwise, the resilience analysis risks being used, possibly inadvertently, as a way to legitimise the power of particular groups and to impose particular means to manage natural resources [43].

3.3. What Are the Potential Implications?

There are at least two implications resulting from the flexibility of resilience to interpretation. First, it seems unlikely that a proper analysis would result in a PSP that does not account for the many different interpretations of resilience in each particular context. If the scope of analysis has been defined by only a few groups, it risks being too narrow, excluding important elements from the analysis and reducing the range of solutions explored. For instance, the analysis might focus on short-term solutions, ignoring important feedback loop mechanisms of the system. Alternatively, a pure systemic view of the problem might fail to recognise uncertainty and might oversimplify decision rules and human behaviours.

Second, stakeholders who have a different understanding of the problem will rarely support or get actively engaged in the implementation of a solution that is not addressing their initial understanding of the problem [45]. The contribution of any solution is null if those ultimately responsible for implementing them are not willing to do so [46]. For instance, stakeholders might sabotage the policies proposed at the end of the analysis by refusing to participate in the implementation (e.g., training and the introduction of new practices) or, even worse, by explicitly opposing them (e.g., demonstrations against the introduction of new seeds).

3.4. Potential Avenues for Mitigation

Recommendations are not conclusive, but it is possible to outline avenues for further development with the aim of reducing the potential drawback of power in the PSP. A possible avenue is to advocate for more participatory settings. So far, the SES literature has extensively discussed stakeholders' participation as a requirement for the enhancement of resilience in the SES. However, very little has been elaborated on the role of participation in the formulation of the problem as such. Facilitated modelling approaches, such as Group Model Building [47] or Cognitive Mapping [48], might contribute to mediating this process (e.g., by introducing the CLD as a transitional object that helps to leverage power differences) [49,50]. These methods contribute in leveraging the power between groups by forcing participants to make their assumptions explicit in a diagram that is challenged by the group [47,51,52]. In this case, the diagram is used to jointly represent the problem definition shared by and agreed upon by different stakeholders through a process of negotiation and dialogue [49].

Alternatively, another option is to aim for a broader perspective in the analysis of resilience and to consider possible trade-offs and asymmetries in resilience between different groups and communities within the system. A broader perspective might be particularly useful when there is a conflict between long-term and short-term goals or when the boundaries of the system are not clear [53]. By using computer simulations, for example, it is possible to uncover long-term unintended consequences that might result from short-term perspectives. Uncovering unintended effects is possible because computer simulations are especially useful when the delays between the policies and their results are too large to allow for assessment by simple intuition. Simulations might also uncover unexpected and unintended consequences of policies that are beneficial to one group but negative for others.

The latter is particularly important when analysing climate change problems because there are time lags or delays between policy measures (or non-action), and effects often extend beyond the normal period of analysis [54,55]. When important consequences of current policies materialise several years later (in some cases decades later), significant future stakeholders will not be present to voice their concerns and weigh in when preferences are aggregated into policy decisions. Present stakeholders might be willing to compromise the overall future detriment of the system for short-term benefits. Namely, in the resilience analysis, present stakeholders might favour policies that yield more efficiency in the short term but diminish the capability of the system to continue providing the desired outputs in the long term. The benefits for the few who are defining the problem now might be preferred over the benefits for the many tomorrow.

4. Conclusions

The ambiguity of resilience is a challenge for practitioners that want to implement it as an analytical and policymaking framework in real life problems. This paper addresses the ambiguity of resilience from a cognitive and political perspective by focusing on how resilience is interpreted in practice instead of its theoretical definition. This paper argues that the interpretation of what resilience means in a specific context (resilience of what?) and the ways to achieve it are results of the values and beliefs of those with a stake in the system. In this light, the case study presented methods to identify and highlight some of the challenges and practical implications of resilience ambiguity. Specifically, this paper focuses on strategic agendas and mental models as observable expressions of stakeholders' values, beliefs and knowledge about the system. The results discussed in this paper show that, in practice, different agendas and mental models compete during the PSP to be part of the scope of resilience analysis. The question of what outcome of the system needs to be resilient has many answers (revenues, yield, food supply).

The results presented in this paper show that stakeholders have different understandings of how the system works. For instance, while academics and delegates from the NGO participating in the study focused on enhancing virtuous cycles within the system, the central government delegates proposed solutions outsides the system's boundaries. All of these solutions, however, ignored the bounded rationality of the farmers and the premises of their decision-making process. Including only a few stakeholders in the process risks leaving many important aspects out of the scope of the analysis and therefore undermining its results.

It is also necessary to acknowledge the role of power shaping and filtering different interpretations of resilience into a formal scope of analysis. It is expected that those with more power will attempt to influence the PSP to reflect their views and agendas. In the case presented in this paper, farmers have little influence in the PSP and their agendas might, intentionally or accidentally, be bypassed by experts (e.g., academics and researchers) and policymakers. For instance, as discussed in this paper, farmers bounded rationality and socioeconomic position might be used as an argument for disregarding their knowledge and their claims.

In short, results show that the practical meaning of resilience is socially constructed by those participating in the PSP and the way this process is conducted will affect the result of the analysis. There are at least two practical implications of underestimating resilience ambiguity while structuring the scope of the resilience analysis. First, including only a few stakeholders in the process risks leaving many important aspects of the system out of the scope of the analysis to be undertaken. Second, poor stakeholder management also risks obstructing the implementation of proposed policies and, in the worst case, unintentionally harming those in more vulnerable positions. While literature starts to acknowledge the challenges and contentious implications of power in the resilience analysis (see for instance [7,14,39]), more research is needed toward defining a framework of how to facilitate negotiation during the PSP.

If resilience is to play a significant role in climate change adaptation, policymakers should be careful when structuring the scope of the resilience analysis and should seek for broader participation. Such broadening is not a simple case of bringing more perspectives. Instead, it is a "fundamental shift in how knowledge is understood to operate and consequences of this for the kinds of questions we formulate prior to our analyses" [14] (p. 484). Increasing participation is not a normatively uncontroversial route either, but at least it acknowledges that resilience-based policy solutions and institutions will have distributional and, thereby, moral consequences (as most other forms of public policy do).

Acknowledgments: To the stakeholders in Huehuetenango who participated in this research and kindly agreed to share their perspectives in this paper.

Conflicts of Interest: The authors declare no conflict of interest.

References

1. Food and Agriculture Organization of the United Nations (FAO). *2016 The State of Food and Agriculture. Climate Change, Agriculture and Food Security*; FAO: Rome, Italy, 2016.
2. Tendall, D.M.; Joerin, J.; Kopainsky, B.; Edwards, P.; Shreck, A.; Le, Q.B.; Kruetli, P.; Grant, M.; Six, J. Food system resilience: De fining the concept Resilience Sustainability. *Glob. Food Secur.* **2015**, *6*, 17–23. [CrossRef]
3. Campbell, B.M.; Vermeulen, S.J.; Aggarwal, P.K.; Corner-Dolloff, C.; Girvetz, E.; Loboguerrero, A.M.; Ramirez-Villegas, J.; Rosenstock, T.; Sebastian, L.; Thornton, P.K.; et al. Reducing risks to food security from climate change. *Glob. Food Secur.* **2016**, *11*, 34–43. [CrossRef]
4. Walker, B.; Carpenter, S.; Anderies, J.; Abel, N.; Cumming, G.; Janssen, M.; Norberg, J.; Peterson, G.D. Pritchard, R. Resilience Management in Social-ecological Systems: A Working Hypothesis for a Participatory Approach. *Conserv. Ecol.* **2002**, *6*, 14. [CrossRef]
5. Holling, C.S.; Gunderson, L.H. *Panarchy: Understanding Transformations in Human and Natural Systems*; Island Press: Washington, DC, USA, 2002.
6. Gallopín, G.C. Linkages between vulnerability, resilience, and adaptive capacity. *Glob. Environ. Chang.* **2006**, *16*, 293–303. [CrossRef]
7. Maleksaeidi, H.; Karami, E. Social-ecological resilience and sustainable agriculture under water scarcity. *Agroecol. Sustain. Food Syst.* **2013**, *37*, 262–290. [CrossRef]
8. Duit, A. Resilience Thinking: Lessons for Public Administration. *Public Adm.* **2015**, *92*, 364–380. [CrossRef]
9. Pizzo, B. Problematizing resilience: Implications for planning theory and practice. *Cities* **2015**, *43*, 133–140. [CrossRef]
10. Woolley, A.R.N.; Pidd, M. Problem Structuring? A Literature Review. *J. Oper. Res. Soc.* **1981**, *32*, 197–206.
11. Shaw, D.; Westcombe, M.; Hodgkin, J.; Montibeller, G. Problem structuring methods for large group interventions. *J. Oper. Res. Soc.* **2004**, *55*, 453–463. [CrossRef]
12. Weingart, L.R.; Bennett, R.J.; Brett, J.M. The impact of consideration of issues and motivational orientation on group negotiation process and outcome. *J. Appl. Psychol.* **1993**, *78*, 504–517. [CrossRef]
13. Quinlan, A.E.; Berbés-Blázquez, M.; Haider, L.J.; Peterson, G.D. Measuring and assessing resilience: Broadening understanding through multiple disciplinary perspectives. *J. Appl. Ecol.* **2016**, *53*, 677–687. [CrossRef]
14. Cote, M.; Nightingale, A.J. Resilience thinking meets social theory, Situating social change in socio-ecological systems (SES) research. *Prog. Hum. Geogr.* **2012**, *36*, 475–489. [CrossRef]
15. Kelly, G. The nature of personal constructs. *Psychol. Pers. Constr.* **1955**, *1*, 105–183.
16. Funtowicz, S.; Ravetz, J.R. Uncertainty, complexity and post-normal science. *Environ. Toxicol. Chem.* **1994**, *13*, 1881–1885. [CrossRef]
17. Merriam, S.B. Introduction to qualitative research. In *Qualitative Research in Practice: Examples for Discussion and Analysis*; John Wiley and Sons: New York, NY, USA, 2002.
18. Stake, R.E. *The Art of Case Study Research*; Sage: London, UK, 1995.
19. World Food Programme. *Countries-Guatemala*; World Food Programme: Guatemala, Guatemala, 2016.
20. Guardiola, J.; Cano, V.G.; Pol, J.L.V. *La Seguridad Alimentaria: Estimación de Índices de Vulnerabilidad en Guatemala*; VIII Reunión de Economía Mundial; Universidad de Alicante: Alicante, Spain, 2006.
21. Bouroncle, C.; Imbach, P.; Läderach, P.; Rodirguez, B.; Medellin, C.; Fung, E.; Martinez-Rodriguez, R.; Donatti, C.I. *La Agricultura de Guatemala y el Cambio Climático: ¿Dónde Están Lasprioridades Para la Adaptación?* CGIAR Research Program on Climate Change, Agriculture and Food Security (CCAFS): Copenhague, Dinamarca, 2015.

22. Instituto Nacional de Estadistica (INE). *Caracterización Estadística República de Guatemala 2012*; INE: Guatemala, Guatemala, 2012.

23. Camposeco, M.; Thomas, M.; Kreynmar, W. *Huehuetenango en Cifras*; Centro de Estudios y Documentación de la Frontera Occidental de Guatemala (CEDFOG): Guatemala, Guatemala, 2008.

24. Lane, D.C. The Emergence and Use of Diagramming in System Dynamics: A Critical Account. *Syst. Res. Behav. Sci.* **2008**, *25*, 3–23. [CrossRef]

25. Richardson, G.P. Problems with causal-loop diagrams. *Syst. Dyn. Rev.* **1986**, *2*, 158–170. [CrossRef]

26. Vennix, J.A.M. Group model-building: Tackling messy problems. *Syst. Dyn. Rev.* **1999**, *15*, 379–401. [CrossRef]

27. Luna-Reyes, L.F.; Andersen, D.L. Collecting and analyzing qualitative data for system dynamics: Methods and models. *Syst. Dyn. Rev.* **2003**, *19*, 271–296. [CrossRef]

28. Randers, J. Guidelines for model conceptualization. In *Elements of the System Dynamics Method*; MIT Press: Cambridge, MA, USA, 1980; pp. 117–139.

29. Ingram, J.; Ericksen, P.; Liverman, D. *Food Security and Global Environmental Change*; Earthscan: London, UK, 2009.

30. Kelly, G. A Brief Introduction to Personal Construct Theory. In *International Handbook of Personal Construct Psychology*; John Wiley & Sons: Chichester, UK, 2003; pp. 3–20.

31. Eden, C. Cognitive mapping and problem structuring for system dynamics model building. *Syst. Dyn. Rev.* **1994**, *10*, 257–276. [CrossRef]

32. Doyle, J.K.; Ford, D.N. Mental models concepts revisited: Some clarifications and a reply to Lane. *Syst. Dyn. Rev.* **1999**, *15*, 411–415. [CrossRef]

33. Walker, B.; Holling, C.S.; Carpenter, S.R.; Kinzig, A. Resilience, adaptability and transformability in social-ecological systems. *Ecol. Soc.* **2004**, *9*, 5. [CrossRef]

34. Darnhofer, I.; Bellon, S.; Dedieu, B.; Milestad, R. Adaptiveness to enhance the sustainability of farming systems. A review. *Sustain. Agric.* **2011**, *2*, 45–58.

35. Mayumi, K.; Giampietro, M. The epistemological challenge of self-modifying systems: Governance and sustainability in the post-normal science era. *Ecol. Econ.* **2006**, *57*, 382–399. [CrossRef]

36. Funtowicz, S.; Ravetz, J.R. Emergent complex systems. *Futures* **1994**, *26*, 568–582. [CrossRef]

37. Lebel, L.; Anderies, J.M.; Campbell, B.; Folke, C. Governance and the Capacity to Manage Resilience in Regional Social-Ecological systems. *Ecol. Soc.* **2006**, *11*, 1.

38. Larsen, R.K.; Calgaro, E.; Thomalla, F. Governing resilience building in Thailand's tourism-dependent coastal communities: Conceptualising stakeholder agency in social-ecological systems. *Glob. Environ. Chang.* **2011**, *21*, 481–491. [CrossRef]

39. Eriksen, S.H.; Nightingale, A.J.; Eakin, H. Reframing adaptation: The political nature of climate change adaptation. *Glob. Environ. Chang.* **2015**, *35*, 523–533. [CrossRef]

40. Carpenter, S.; Walker, B.; Anderies, M.J.; Abel, N. From metaphor to measurement: Resilience of what to what? *Ecosystems* **2001**, *4*, 765–781. [CrossRef]

41. Marshall, N.A.; Marshall, P.A. Conceptualizing and Operationalizing Social Resiliance within Commercial Fisheries in Northern Australia. *Ecol. Soc.* **2007**, *12*, 1. [CrossRef]

42. Nightingale, A.J. "The experts taught us all we know": Professionalisation and knowledge in Nepalese community forestry. *Antipode* **2005**, *37*, 581–603. [CrossRef]

43. Peterson, G. Political ecology and ecological resilience: An integration of human and ecological dynamics. *Ecol. Econ.* **2000**, *35*, 323–336. [CrossRef]

44. Adger, W.N.; Arnell, N.W.; Tompkins, E.L. Successful adaptation to climate change across scales. *Glob. Environ. Chang.* **2005**, *15*, 77–86. [CrossRef]

45. Größler, A. System Dynamics projects that failed to make an impact. *Syst. Dyn. Rev.* **2007**, *23*, 437–452. [CrossRef]

46. Ackermann, F. Problem structuring methods 'in the Dock': Arguing the case for Soft OR. *Eur. J. Oper. Res.* **2012**, *219*, 652–658. [CrossRef]

47. Vennix, J.A.M. *Group Model Building*; John Willey and Sons Ltd.: Chichester, UK, 1996.

48. Eden, C.; Ackermann, F. Group Decision and Negotiation in Strategy Making. *Gr. Decis. Negot.* **2001**, *10*, 119–140. [CrossRef]

Sustainability **2017**, *9*, 1196

49. Franco, L.A.; Montibeller, G. Facilitated modelling in operational research. *Eur. J. Oper. Res.* **2010**, *205*, 489–500. [CrossRef]

50. Davies, K.K.; Fisher, K.T.; Dickson, M.E.; Thrush, S.F.; Le Heron, R. Improving ecosystem service frameworks to address wicked problems. *Ecol. Soc.* **2015**, *20*, 2. [CrossRef]

51. Rouwette, E.A.J.A.; Vennix, J.A.M.; Felling, A.J.A. On evaluating the performance of problem structuring methods: An attempt at formulating a conceptual model. *Gr. Decis. Negot.* **2009**, *18*, 567–587. [CrossRef]

52. Akkermans, H.A.; Vennix, J.A.M. Clients' opinions on group model-building: An exploratory study. *Syst. Dyn. Rev.* **1997**, *13*, 3–31. [CrossRef]

53. Duit, A.; Galaz, V.; Eckerberg, K.; Ebbesson, J. Governance, complexity, and resilience. *Glob. Environ. Chang.* **2010**, *20*, 363–368. [CrossRef]

54. Young, O.R. Institutional dynamics: Resilience, vulnerability and adaptation in environmental and resource regimes. *Glob. Environ. Chang.* **2010**, *20*, 378–385. [CrossRef]

55. Warner, K. Global environmental change and migration: Governance challenges. *Glob. Environ. Chang.* **2010**, *20*, 402–413. [CrossRef]

sustainability

MDPI

Article

Environmental Sustainability of Agriculture Stressed by Changing Extremes of Drought and Excess Moisture: A Conceptual Review

Elaine Wheaton [1] and Suren Kulshreshtha [2,*]

[1] Department of Geography and Planning, University of Saskatchewan, Saskatoon, SK S7N 5A8, Canada; elainewheaton@sasktel.net
[2] Department of Agricultural and Resource Economics, University of Saskatchewan, Saskatoon, SK S7N 5A8, Canada
* Correspondence: suren.kulshreshtha@usask.ca; Tel.: +1-306-966-4014; Fax: +1-306-966-8413

Academic Editor: Iain Gordon
Received: 14 February 2017; Accepted: 29 May 2017; Published: 6 June 2017

Abstract: As the climate changes, the effects of agriculture on the environment may change. In the future, an increasing frequency of climate extremes, such as droughts, heat waves, and excess moisture, is expected. Past research on the interaction between environment and resources has focused on climate change effects on various sectors, including agricultural production (especially crop production), but research on the effects of climate change using agri-environmental indicators (AEI) of environmental sustainability of agriculture is limited. The aim of this paper was to begin to address this knowledge gap by exploring the effects of future drought and excess moisture on environmental sustainability of agriculture. Methods included the use of a conceptual framework, literature reviews, and an examination of the climate sensitivies of the AEI models. The AEIs assessed were those for the themes of soil and water quality, and farmland management as developed by Agriculture and Agri-Food Canada. Additional indicators included one for desertification and another for water supply and demand. The study area was the agricultural region of the Canadian Prairie Provinces. We found that the performance of several indicators would likely decrease in a warming climate with more extremes. These indicators with declining performances included risks for soil erosion, soil salinization, desertification, water quality and quantity, and soil contamination. Preliminary trends of other indicators such as farmland management were not clear. AEIs are important tools for measuring climate impacts on the environmental sustainability of agriculture. They also indicate the success of adaptation measures and suggest areas of operational and policy development. Therefore, continued reporting and enhancement of these indicators is recommended.

Keywords: environmental sustainability; agricultural sustainability; environmental indicators; climate change; climate extremes; drought; excess moisture; Canadian Prairie Provinces

1. Introduction

Considerable changes in climate variables relevant to agriculture and the environment have already occurred and have been documented [1–3]. For the agricultural portion of the Canadian Prairie Provinces, where this study was conducted, these agro-climatic changes include longer growing seasons, more crop heat units, decreasing snow-cover area, changes in precipitation from snow to rain during winter months, and warmer winters. Future changes for the prairie agricultural region indicate continued and perhaps accelerated trends in these variables and many others [1,4]. An increase in climate extremes, including droughts, excess moisture, and heat waves is expected for the Canadian Prairies [5,6]. These extremes can often have adverse effects on the environmental sustainability of agriculture. More recent work also confirms that future drought characteristics (frequency of droughts, duration, and intensity) show increases over the southern prairies [7]. Increases in such extremes would have adverse effects on the environmental sustainability of agriculture. The effects of droughts and excessive moisture on environmental sustainability are of special concern, and are the subject of this paper.

Agriculture is an important part of the economy of the Canadian Prairie Provinces of Alberta, Saskatchewan, and Manitoba. The agriculture and agri-food system of these provinces consists of several industries including primary agriculture, farm input and service providers, food and beverage processing, food distribution, as well as retail, wholesale, and food service industries. In Canada, this sector contributed CAD$108 billion (or 6.6% of Canadian gross domestic product) and employed 2.3 million workers [8]. Much of this production activity occurs in the Prairie Provinces. Primary production is a key part of this system as it affects the other components of the regional economy [9].

Although agriculture is important to the economy, environmental impacts must also be considered for achieving sustainability. Agriculture has many effects on the environment and these effects determine the environmental sustainability. Environmental sustainability is defined as sustainability of ecological services that are provided by the ecosystems [10]. Humans depend upon these services directly or indirectly. A strong environmental sustainability would label any practice unsustainable if the natural ecosystems are put to alternative uses, such as conversion of forest ecosystems to agricultural ecosystems. A more practical definition of environmental sustainability requires that those ecosystems and ecosystem services that are essential to humans be conserved to the point of a minimum safe standard. Examples of effects include those on soil and air quality by the use of different tillage and cropping systems, and those on water quality related to the use of fertilizer and pesticides. Climate trends and extremes are expected to affect air, land, and water resources, and knowledge of these effects are crucial to achieving sustainable agricultural production and food and water security. The effects of excess moisture and drought are especially important, as they can have more pronounced impacts on the environmental sustainability of agriculture than gradual increases in temperature. For example, droughts reduce the protection of soil moisture and vegetation, and erosion can result. Excess moisture and flooding can result in water run-off leading to erosion of soil and damage to vegetative cover that protects the soil. Flooding can also damage the storage areas of fertilizer, manure, and pesticides, releasing them as contaminants into the environment. In this paper, we explore the effects of climate change on the environmental sustainability of agricultural systems, with emphasis on the extreme events of droughts and excess moisture. Therefore, our main objective is specifically to assess the effects of future drought and excess moisture on selected agri-environmental indicators. No other investigations have addressed this topic, to our knowledge.

By 1999, the member countries of the Organization for Economic Cooperation and Development (OECD), including Canada, noted that establishing a key set of agri-environmental indicators (AEIs) that could be useful for member countries was important [11]. In Canada, Agriculture and Agri-Food Canada (AAFC) reports on a set of science-based AEIs using mathematical models showing the interactions between agriculture and the environment [9,12]. These two reports are the latest in the series of Canada's agri-environmental reporting. Therefore, they are the basis for the AEIs we have selected for use, as well as associated trend information. This reporting series and their AEIs are not intended for use with climate change scenarios, but their use may be for strategic adaptation to drought and excess moisture.

Wall and Smit [13] noted that agricultural sustainability and climate change adaptation strategies support one another and that ecosystem integrity is needed for sustaining agricultural production. However, Wheaton et al. [14–16] were first, according to the authors' knowledge, to assess the possible changes in agricultural sustainability (using AEIs) as expected under climate change, and this study builds upon and expands that work. They found several of the AEIs to be sensitive to climate change and reported a possible decline in the performance of AEIs with climate change for soil erosion, contamination, soil salinization, and water quality categories.

2. Data and Methods

Environmental sustainability indicators of agriculture considered here were based on changes in the set of science-based AEIs developed by Agriculture and Agri-Food Canada [9,12]. The AEIs report agri-environmental performance under four main categories: soil quality, water quality, air quality, and farmland management. Each category has set of indicators addressing sub-themes within the category (e.g., from the soil health theme, sub-indices include soil erosion, soil organic matter, trace elements, and salinity). Many of the indicators can be integrated within climate change studies either directly or indirectly. Directly, these indices can be calculated using climate change scenario data and compared with values obtained under the observed climate record.

Our study methods included literature reviews, development of a conceptual framework, examination of the possible relationships, sensitivities and responses of selected AEIs to climate by examining their mathematical structures. These approaches were used to suggest possible directions of future trends in AEIs with increases of drought and excess moisture under continued climate change. In the remaining sections, the selected AEIs are described, along with an assessment of their future status.

The AEIs selected for this study were those for the soil quality, water quality, and farmland management themes. We added two more indicator types because of their relevance, one for desertification and another for water supply and demand. AEIs were selected for their utility in assessing the possible effects of current and future drought and excess moisture. We examined the mathematical models (factors affecting the relationship among stimulus that causes a change in the AEI level) of each AEI as the first step in choosing them [15]. The AEIs that contain climate variables in their mathematical models are the most clearly sensitive to climate, and therefore either are directly driven by and may have strong relationships with climate change. We determined the nature of the relationships by the types of climate variables used (e.g., temperature, precipitation) in the indicator and whether the relationship with climate was linear or more complex and direct or inverse. Some AEIs for the category of soil quality are good candidates for exploring the direct effects of drought and excess moisture on the environment. Examples include the wind and water erosion, salinity, and particulate emission models as they include climate variables. Where it was not possible to assess indices due to a lack of direct use of climate variables in the models, climatic effects were indirectly implied from ecosystem assessments of changes in vegetation, insects, and diseases, for example.

Although drought and excess moisture affect most aspects of environmental sustainability, we focused on AEI categories and their indicators for soil quality, water quality, farmland management, and water supply and demand, as guided by our conceptual framework (Figure 1), expertise and available literature. From the conceptual framework, we analyzed how the four AEI themes would be affected by changing climate extremes starting with the knowledge of the main characteristics of future possible drought and excess moisture events, as summarized from the literature.

Drought and excess moisture events are expected to become more common in the future on the Canadian Prairies [5,7]. The frequency, intensity, and extent of moderate to extreme droughts are projected to increase. At the other extreme, the review also found agreement that the frequency of severe storms and unusually wet periods is also projected to increase, leading to the conclusion that wet times will become wetter and dry times will become drier, with several driving forces supporting this finding.

Regarding drought, four main characteristics of future possible droughts in the Canadian Prairies were found: (1) increased intensity of dryness, driven by increased evaporation potential with higher temperatures and longer warm seasons; (2) droughts of 6–10 months and longer become more frequent by the 2050s; (3) the frequency of long duration droughts of five years and longer more than doubles in the future to 2100; and (4) decade-long and longer droughts increase by triple in frequency to 2100 [5,6]. The finding of future possible increase in droughts is confirmed by other work that finds increases in drought characteristics in the Canadian Prairie Provinces, especially over the southern study region [7].

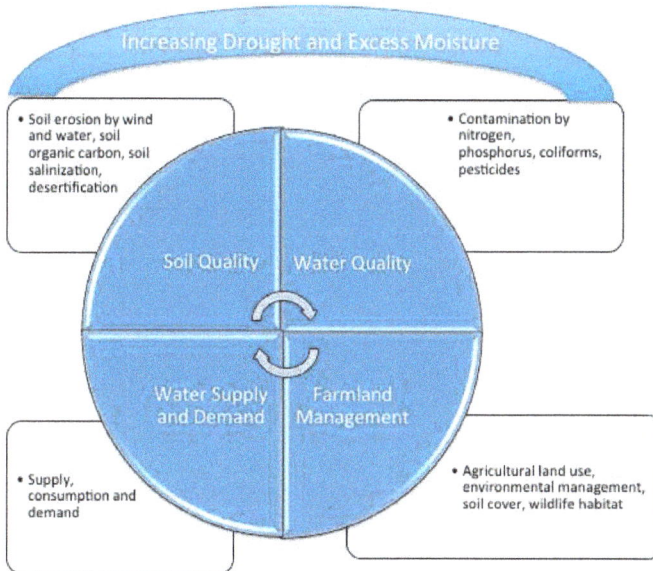

Figure 1. Framework for integrating the effects of changing climate extremes increasing droughts and excess moisture with selected main categories and sub-components of environmental sustainability of agriculture.

Shifting of climate zones poleward with higher temperatures also indicates the occurrence of drought in areas farther north of their usual positions in the study area. Worst-case scenarios should also be considered because of the severe and multiple effects of droughts. Mega droughts have occurred in the past in the Canadian Prairies [17], and it is therefore expected that droughts will be pushed to greater severity with climate warming.

In the context of future trends in the AEIs, it is important to ask: what are the future projections of extreme precipitation events and associated excessive moisture conditions? The IPCC (Intergovernmental Panel on Climate Change) [18] has reported on managing the risks of extreme climate events globally. The report indicates that the frequency of heavy precipitation will increase in the 21st century over many parts of the world. They gave this projection a 66–100% chance of occurring and found that this trend is particularly the case in the high latitudes.

The Canadian Prairie agricultural area has experienced extremely wet conditions in the past and these are projected to increase. Saskatchewan holds Canada's record wettest hour under the current climatic conditions when 250 mm rainfall occurred at Buffalo Gap in the south central area [19]. The largest area eight-hour event in the Canadian Prairies was the rainstorm of 3 July 2000 around Vanguard in southwest Saskatchewan. This storm brought about 375 mm of rainfall, exceeding the average annual precipitation of 360 mm, and caused severe flooding [20]. The projected changes to precipitation amounts in Canada for 2041–2070 show an increase in maximum precipitation in the range of about 10–20% for the prairies for the 20-year return period of one-day precipitation [21]. This means an increase from 40–60 mm (1941–1970) to 48–72 mm for the 20% increase.

Although the work of [17] for the prairies focused on droughts, the climate indices (i.e., Palmer Drought Severity Index and Standardized Precipitation Index) over the future period to 2100 show some very high values, indicating wet periods for a range of Global Climate Model results. For example, some of the future wet periods appear to be as excessive as the wet period of the 1970s. The review of future possible extremes suggested that the overall prairie climate would become drier, but with substantial year-to-year variability, including an increased chance of heavy precipitation and very wet periods [5,6]. The next section provides an assessment of the possible changes in AEIs with projected increases in drought and excess moisture.

3. Results

Descriptions of the environmental sustainability of agriculture as affected by drought and excess moisture are provided in this section for several AEIs. The indicators are in four main categories, soil quality, water quality, water supply and demand, and farmland management (Figure 1).

3.1. Possible Future Trends in AEIs for Soil Quality

Four main AEIs for the soil quality theme are discussed here, namely soil erosion, soil organic carbon, soil salinization, and desertification.

3.1.1. Soil Erosion by Wind and Water

Soil erosion occurs through the action of wind and water, as well as tillage. The soil erosion AEIs had overall improved performances in recent decades in the prairies, indicating reductions in erosion risk between 1981 and 2011 [12]. This trend is mainly due to improved land management, such as adoption of minimum to no-tillage practices, reduced use of summer fallow, and increased forage and cover crops [9]. Recent decades, however, have had severe and extensive droughts, such as in 1999 to 2004 [22], 2008 to 2010 [23], and 2015 [24]. The 2015 drought was found to be likely an outcome of human-influenced warm spring conditions and naturally forced dry weather from May to July. Droughts can result in considerable soil erosion by wind. At least 32 incidents of blowing dust were documented between April and September 2001. This number of incidents was high as it was exceeded only once during the 1977–1988 period of dust storms. Although the wind erosion was severe, it would have been much worse without the increase in soil conservation practices [22]. These events make it clear that drought can result in soil erosion even with the adoption of improved land management practices.

The greater evaporation rates, lower soil moisture, and decreased vegetation cover under droughts result in increased risk of soil erosion (Table 1). This means that management practices to reduce the soil erosion risk would become even more important in the future. Descriptions of future wind speed changes are rare, but Price et al. [25] project little change in wind speed, on average for the prairie semiarid region, with slight reductions in mean summer wind speed of 0.14 m/s for the medium emissions scenario for the 2040–2069 period. However, they did find increases of mean spring wind speed of 0.11 m/s for this scenario and time. Spring is an important time for increased wind erosion risk, as the vegetation cover is not yet well established and the soil is more exposed. Wind speed was a very important factor in the wind erosion component of the AEI as the relationship is direct and cubic [15,26]. The risk of future wind erosion in the province of Saskatchewan, Canada, was estimated to continue to increase with rising temperature and potential evapotranspiration [26].

Agriculture and Agri-Food Canada's assessment of the environmental sustainability of Canadian agriculture found that higher rainfall in eastern provinces, such as Ontario and Quebec, contributed to the lower performances of soil quality indicators [9]. This relationship between higher rainfall and soil quality is useful in assessing future effects in the prairies. Here, drought conditions can shift very quickly to wet conditions. Recent years have shown intense rainfall and severe flooding in several areas of the Prairie Provinces [27]. There were very wet conditions, especially in some parts, in 2010, 2011, and 2012. Spring 2010 was the wettest among the 1948–2012 period, at 64% greater than the areal average for the prairie climate region. Spring 2012 was the third wettest spring, at 52% higher than average. Summer 2010 was also very wet, with a total precipitation amounting to the fourth highest on record at 40% above average. Summer 2012 had the sixth highest areal average precipitation [28]. The heavy rainstorms and high amounts of accumulated precipitation resulted in many excess moisture problems, including more agricultural land being under water than ever recorded. Problems of excess moisture were persistent, lasting from October 2010 to July 2011 for many areas [27]. Intense rainfall events tend to contribute to runoff and increasing soil erosion (Table 1) and do not ease drought conditions as much as gentler rains.

Table 1. Potential effects of increasing droughts and excessive moisture on trends of soil quality indicators.

Soil Health Indicator	Climate Linkage (Direct and Indirect)	Effects of Increased Droughts	Effects of Increased Excess Moisture	Comments Regarding Other Factors
Soil erosion by wind	Wind, temperature and precipitation, soil moisture, vegetation cover	Reduced soil moisture and vegetation cover which increase erosion risk	Increased precipitation intensity can destabilize soil particles	Decreasing snow cover increases exposure to erosion
Soil erosion by water	Precipitation intensity, vegetation cover	Water erosion risk decreases	Increased heavy rainfall increases potential for soil erosion	Heavy rainfall on frozen soil increases erosion risk
Soil organic carbon	Temperature, precipitation, vegetation cover	Reduced vegetation production reduces carbon	Run-off increases carbon losses	Temperature increases tend to increase carbon losses
Soil salinization	Aridity (temperature and precipitation balance), vegetation cover	Evaporation concentrates salts. Reduced vegetation cover can increase salinization	Elevated water tables can increase salinization	Increased variability with drought/wet shifts increases salinization risk
Contamination by trace elements	Precipitation intensity	Possible increased concentrations may occur	Increases	Climate effects estimations require further investigation

The summer of 2011 in Southeastern Saskatchewan provided a good example of several heavy rainfall events [29]. April–June in 2011 had 150 to over 200% of normal precipitation amounts. Multiple rainfall events of 20 mm or greater occurred, and a severe 1:100 year rainfall event occurred on 17 June 2011. These events resulted in unprecedented floods in the Souris River Watershed, causing state of emergency declarations in communities of Weyburn and Estevan. The community of Roche Percee had to evacuate almost every home [30]. Therefore, these extreme precipitation times not only resulted in the flooding of agricultural land with several implications for environmental sustainability, but they also resulted in a loss of homes and other infrastructure (e.g., roads, culverts, and bridges). The damage to infrastructure meant that soil erosion also occurred, though this was more difficult to assess. The impacts to environmental sustainability are discussed with the specific indicator addressed in the following sections. Examples include soil quality (e.g., water erosion of soil) and water quality (e.g., run-off contaminants).

Further changes to snow cover are also expected and have already occurred. Snow cover protects the soil from erosion risk. Northern Hemisphere spring snow cover extent has significantly decreased over the past 90 years and the rate of decrease has accelerated over the past 40 years. An 11% decrease in April snow cover extent has occurred for the 1970–2010 period compared with pre-1970 values. These trends are mainly a result of increasing temperatures [31]. This means that the soil was exposed to wind and water erosion for an increasing length of time in the recent past, and this trend is projected to continue with further warming. Alternatively, snow-melt contributes to overland run-off and can result in water erosion of soil. Recent work finds extensive decreasing trends of snow-water equivalent in Canada related to increasing temperature. The mean size of the decreasing trend for December–April is -0.4 to -0.5 mm/y [32]. Estimates of future possible snow cover changes are challenging because of the complex response of snow cover to warming, but widespread decreases in snow cover duration are projected across the Northern Hemisphere [33]. Implications for the soil erosion AEI are very uncertain because of this complexity, but effects of the continued trend of decreasing snow cover extent should be considered in measures to protect the soil against wind and water erosion.

Intense rainfall events and higher total precipitation accumulations result in greater run-off, with eroded and flooded land (Table 2). The increased water erosion and the flooded land can result in many problems for environmental sustainability of agriculture, including contamination by pollutants of various types resulting in water quality problems.

Table 2. Selected agri-environmental indicator (AEI) categories, indicators, their relationships with climate, and possible future climatic effects related to drought and excess moisture.

Group	Indicator	Measure	Sensitivity to Climate	Links with Climate-Related Changes
Soil Quality	Risk of soil erosion by water	Surface run-off	Strong	Climate change may result in aridity in some parts of the prairies which would increase the probability of surface run-off
				Higher variability in precipitation and incidence of wet events would lead to higher incident of soil erosion
	Risk of wind erosion	Soil loss through wind events	Strong	Future increases are expected with simulated increases in spring wind speed
	Soil organic carbon	Organic carbon level in soil	Medium	Future changes with climate change are not clear because of the interacting effects of management practices
	Risk of soil salinization	Degree of soil salinity	Strong	Climate change may increase salinity from variations of precipitation and dry events
	Contamination by trace elements		Strong	Increased wet and dry periods affect contamination
Water Quality and Quantity	Risk of water contamination by nitrogen	Nitrogen level released by farms into water bodies	Weak	Water run-off containing nitrogen associated with soil erosion is affected by variable precipitation
	Risk of water contamination by phosphorus	Phosphorus level released by farms into water bodies	Weak	Water run-off containing phosphorus associated with soil erosion is affected by variable precipitation
	Water supply and use	Water availability and use	Strong	Climate change would likely impart a reduction in supply, but an increase its demand
Farmland management	Soil cover by crops and residue	Duration of exposed soil	Strong	Vegetative cover is affected by climate change
	Management of farm nutrients and pesticide inputs	Application of organic and inorganic nutrients and pesticides	Medium	Favorable wetter conditions may lead to increased nutrient use. Climate change may lead to increased pest and diseases and the need for their management

3.1.2. Soil Organic Carbon

Other AEI components of soil quality include the tillage erosion risk indicator, soil organic carbon change (SOCC) indicator, the risk of soil salinization indicator, and the risk of soil contamination by trace elements. As indicated earlier, only selected indicators can be considered. The SOCC indicator is affected by land management changes, including the effects of tillage practices, summer fallow frequency, cropping types, and land-use changes. The current trend of the SOCC indicator showed an improved performance from average to a good status as most of the cropland had increasing soil organic carbon from 1981 to 2011. Spatial patterns over the Prairie Provinces to 2011 ranged from no change to large increases in Alberta, large increases over much of Saskatchewan, and mostly moderate increases in Manitoba. The use of reduced tillage practices and reduced summer fallow area was an important influence in this change in the prairies. The Century model was used to predict the rate of change in organic carbon content in soils [9,12].

An important aspect of soil is its carbon storage capacity, which can affect the atmospheric concentrations of carbon dioxide. The level of soil organic carbon would be susceptible to climate change extremes. For example, under a drought period, organic biomass is low which would affect the level of soil organic carbon. Similarly, if land management under climate extremes includes more permanent cover, soil organic carbon would tend to increase because of plant-derived inputs to soils. Vegetation cover is adversely affected by both droughts and flooding and may result in at least short-term decreases in soil organic carbon. Moreover, depending on other constraints, temperature increases would tend to increase soil decomposition loss of carbon [34].

One of the developments needed for the SOCC indicator is to include soil erosion aspects in the model. Even low rates of soil erosion can decrease soil organic carbon [9]. Soil erosion risk increases during droughts and heavy rainfall, so the effects of these extremes should be incorporated into their modeling. Another limitation of the SOCC indicator is that the effects of past (and future) temperature increases do not appear to be assessed and discussed in the reporting by [9]. The Century model does use monthly temperature and precipitation data, so this assessment is possible and is recommended.

3.1.3. Soil Salinization

The risk of soil salinization in the Prairie Provinces decreased from 1981 to 2011, and over this period the land area in the very high-risk class decreased by 2%. The spatial patterns of risk of soil salinization (RSS) on the prairies showed no change to decreased risk in Alberta and Manitoba and large areas of decreased risk in Saskatchewan. The improvements were mostly related to land management, including decreased summer fallow area, and increased area under permanent cover [35]. Again, as for the SOCC, the impact of changing land use practices on the risk of soil salinization dominated, and climate change did not appear to be considered in the modeling for this AEI and results. However, growing season moisture deficits were a factor in the calculations, but the significant yearly variation in the risk of soil salinization was not considered in the indicator [35]. Such sensitivity of RSS to changes in moisture deficits could be determined. Our early estimate of the possible effects of future increased droughts and excessive moisture is the reduced performance of the RSS indicator (Table 2).

3.1.4. Desertification

A risk of desertification indicator is under development [36], and is included here because of its relevance to the topic. This indicator is not included in the most recent AEI reporting [12]. Desertification is the degradation of land in arid to dry sub-humid regions. The preliminary results indicate that average soil erosion rates were usually below the soil tolerance level, meaning that desertification risk due to erosion was low as of 2006 in the Prairie Provinces [36].

Desertification risk increases with soil erosion, losses of soil organic matter, and fluctuating soil salinity [36]. Climatic extremes have the potential to increase all of these factors as discussed previously, and therefore can increase the risk of desertification. Research indicates that the area of land at risk of desertification in the Prairies could increase by about 50% between conditions of 1961–1990 and the 2050s [37]). The World Meteorological Organization [38] states that climate change may exacerbate desertification and soil salinization through alteration of spatial and temporal patterns in temperature, rainfall, solar radiation, and winds. The threat means that the indicator development is recommended to be completed (including climate drivers) and implemented as led by Agriculture and Agri-Food Canada.

The agricultural area of the Canadian Prairies has a large semi-arid climate zone in southwest to west central Saskatchewan and corresponding regions in Alberta. The remainder of the area is mostly classified as dry sub-humid. These climate classifications are based on the Thornthwaite method using a moisture index with the input of monthly mean temperature and precipitation data [39]. These are the climate zones targeted in the desertification definition [37]. A warming climate is expected to expand these dry zones northward in the prairies to cover even greater areas.

3.2. Possible Future Trends for Water Quality

Climate change can be a major instrumental factor affecting water quality, both for surface as well as ground water (Table 2). These changes would occur as a result of two types of developments, both related to climate change. (1) Climate change would likely reduce water quantity (as described later), which would result in changes in flow regimes influencing the chemistry, hydro-morphology and ecology of regulated water bodies [40]. (2) Agricultural activity would face longer growing seasons combined with reduced water availability, with new crops suited to drier [41], warmer conditions. In addition, wetlands that play an important role in water purification may also dry up during such heat events and longer evaporative seasons. The longer growing season and cropping changes could increase the use of fertilizers with subsequent leaching to watercourses, rivers, and lakes, increasing the risk of eutrophication and loss of biodiversity [42]. Many information gaps exist regarding the effects of climate change (e.g., cyclical variability between wet and dry periods) and these are important to quantify to meet the needs of flood control and water quality improvements, for example [43].

Changes in water quality during storms, snowmelt, and periods of elevated air temperature or drought can cause conditions that exceed thresholds of ecosystem tolerance and, thus, lead to water quality degradation [44]. Such precipitation extremes can pose significant risks to water quality outcomes, resulting in a degradation trend of drinking water quality and potential health impacts [45]. At the same time, the impacts of drought and excess moisture are superimposed onto other pressures on water resources [46] and can exacerbate the other pressures. Such pressures may include market pressures, pest, and disease infestations, and effects on producer incomes from other bottlenecks in the agricultural and food complex [8].

3.3. Possible Future Trends in Water Supply and Demand

Water supply and demand are considered (even though they are not included in Canada's AEI reporting series) because as they are critical for environmental sustainability of agriculture. Water demand for agricultural purposes is expected to increase in the future unless conservation measures are in place. Facing periods of frequent droughts, more farmers would lean towards having irrigation on their farms. However, whether this demand would be met or not depends on water availability and its competing uses [47–49].

Water quantity under climate extremes would be affected through reductions in the water stored in glaciers and snow cover. These water sources are currently declining and this trend is projected to continue, e.g., [50]. This trend reduces water availability especially during warm and dry periods (through a seasonal shift in streamflow, an increase in the ratio of winter to annual flows, and reductions in low flows) in regions supplied by this source [51]. Where storage capacities are not sufficient, much of the winter runoff will be lost to the oceans, and this will create regional water shortages.

In addition to surface water, future changes in climate extremes could affect groundwater. Longer droughts may be interspersed with more frequent and intense rainfall events. These changes in climate may affect groundwater through changes in their recharge and discharge [52]. The aquifers where water withdrawal is already higher than their respective recharge amounts would be even more vulnerable to climate change. Such high levels of withdrawals would reduce available quantities considerably [51].

3.4. Farmland Management

Under climate extremes, land management would be affected through changes in soil moisture, which is directly related to climate extremes. Management of soil moisture and water harvesting would be significant adaptation measures to cope during climate extremes. However, many producers may not anticipate and react appropriately to the occurrence of climate extremes and make appropriate adaptations. A survey by [53] shows that, even during serious drought and flood years, only one third of farmers in China were able to use farm management measures to cope with the extreme weather events. In the Prairie Provinces, a survey of producers regarding the 2001–2002 drought indicated that no producer had made any changes in their cultural practices in anticipation of the drought [54]. However, prairie producers are adaptable, and much adaptation occurred during and after the 2001–2002 drought. In this region, many producers have switched from intensive tillage practices to conservation tillage practices. In 1991, only a third of the cropped area was under conservation tillage methods, but by 2011 this area rose by 157% of the 1991 area, thus constituting 85% of the total area prepared for seeding [55].

The four main AEI categories considered here, along with their indicators, measures, estimates of their sensitivities, and relationships with a changing climate are summarized in Table 2. Several of the indicators are estimated to be fairly sensitive to climate extremes, including soil quality, water supply and demand, and portions of farmland management. The reasons for the indication of weaker relationships with climate may be somewhat related to the lack of understanding of the relation of climate variables with the key parameters describing environmental health.

4. Discussion and Conclusions

This paper was an attempt to explore the possible effects of future drought and excess moisture on the environmental sustainability of agriculture. Methods included examining the possible relationships and responses of AEIs to climate drought and excess moisture using the conceptual framework of Figure 1, by evaluating the relationship of AEI models with climate variables, and by using literature reviews. These approaches were used to suggest possible directions of future trends in AEIs with increased drought and excess moisture. The AEIs assessed were those for soil and water quality, and farmland management as developed by Agriculture and Agri-Food Canada [9], with additions of water supply and demand categories.

The estimation of any future occurrence is difficult with many limitations because of several unknowns. However, the projections using several different methods, including climate indices, climate models, and emission scenarios provide strong agreement of the findings of increased intensity and frequency of both future droughts and extreme precipitation (e.g., [6,7,17]). Measuring, monitoring, modeling, projecting, and communicating the characteristics of wet and dry climate extremes are becoming even more critical as the climate shifts and becomes less stable. Sufficient information is needed to guide planning for and implementation of effective actions to adapt to the impacts of climate extremes.

The critical issues of the effects of climate extremes on environmental sustainability of agriculture include effects on natural resources and their ecosystems, including soil quality, water quality, and water supply and demand. Results indicated the nature of future possible changes in AEIs as affected by trends in climate change and extremes. In order to meet the goal of environmental sustainability of agriculture, climate trends and extremes need to be carefully considered. Much better use of climate information and services are required to meet the goal of environmental sustainability of agriculture. The lack of consideration of climate change reduces the capability to adapt and increases vulnerability.

The possible future effects of climate change extremes examined here are conceptual, but are plausible based on the current data from climate science. Actual results may be lower, but they also might be much higher in terms of worst-case scenarios. Solutions for effects of climate extremes should also considered, especially those with the most serious consequences.

Soil quality, as measured by AEIs in the agricultural region of the Prairie Provinces, has showed an improving trend for the 1981–2011 period [12]. However, these AEIs have strong land management drivers, and the effects of climate trends and extremes are not clear. Results regarding the effects of climate change indicate possible declining performances for soil erosion, salinization, and desertification. Results regarding the effects of climate change for other soil AEIs such as soil organic carbon, contamination by trace elements, and farmland management have even less information. All of these AEIs require more work to fully assess the effects of climate change, especially extremes, such as drought and excessive moisture.

The AEIs are numerous with four main categories containing several indicators apiece as described in [12]. Therefore, many could not be addressed here, including air quality and biodiversity indicators. Alternatively, a critical indicator, that of water supply and demand, is not a part of the AEI indicator series by Agriculture and Agri-Food Canada [9,12]. However, water supply and demand was discussed here as a possible indicator, and we recommend it to be included in the AEI series. Next steps in the AEI assessments are recommended to include additional indicators and their relationships with climate change.

Results indicate that the performance of several indicators would likely decrease in a warming climate with more extremes of droughts and extreme moisture. These indicators include risks of soil erosion, soil salinization, water quality and quantity, and soil contamination. Thresholds of climate extremes, however, may be reached and result in accelerated negative performances of such indicators. The impacts of climate change are more difficult to assess for several indicators because of the effect of other factors, such as land management. AEIs are important tools to measure climate impacts on environmental sustainability of agriculture. They also indicate the success of adaptation measures and of required policy development. The climate change risks to environmental sustainability of agriculture require much more attention.

Acknowledgments: We thank the three anonymous reviewers and the Journal editors for their useful comments for the improved version of this manuscript. We thank the Organization for Economic Cooperation and Development (OECD) for the impetus of our earlier work towards assessing the implications of climate change for Agri-Environmental Indicators (AEIs).

Author Contributions: The authors cooperated on all parts of the manuscript.

Conflicts of Interest: The authors have no conflict of interest for this manuscript.

References

1. Kulshreshtha, S.; Wheaton, E. Climate change and Canadian agriculture: Some knowledge gaps. *Int. J. Clim. Chang. Impacts Responses* **2013**, *4*, 127–148. [CrossRef]
2. Qian, B.; Gameda, S.; Zhang, X.; De Jong, R. Changing growing season observed in Canada. *Clim. Chang.* **2012**, *112*, 339–353. [CrossRef]
3. Nyirfa, W.; Harron, B. *Assessment of Climate Change on the Agricultural Resources of the Canadian Prairies*; The Prairie Adaptation Research Collaborative, University of Regina: Regina, SK, Canada, 2004; 27p.
4. Qian, B.; De Jong, R.; Gameda, S.; Huffman, T.; Neilsen, D.; Desjardins, R.; Whang, H.; McConkey, B. Impacts of climate change scenarios on Canadian agroclimatic indices. *Can. J. Soil Sci.* **2013**, *93*, 243–259. [CrossRef]
5. Wheaton, E.; Bonsal, B.; Wittrock, V. *Possible Future Dry and Wet Extremes in Saskatchewan, Canada*; The Water Security Agency, Saskatchewan Research Council: Saskatoon, SK, Canada, 2013.
6. Wheaton, E.; Sauchyn, D.; Bonsal, B. Future Possible Droughts. In *Vulnerability and Adaptation to Drought: The Canadian Prairies and South America*; Diaz, H., Hurlbert, M., Warren, J., Eds.; University of Calgary Press: Calgary, AB, Canada, 2016.
7. Masud, M.; Khaliq, M.; Wheater, H. Future changes to drought characteristics over the Canadian Prairie Provinces based on NARCCAP multi-RCM ensemble. *Clim. Dyn.* **2016**, *48*, 2685–2705. [CrossRef]
8. Agriculture and Agri-Food Canada. *An Overview of the Canadian Agriculture and Agri-Food System 2016*; Agriculture and Agri-Food Canada: Ottawa, ON, Canada, 2017.
9. Eilers, W.; MacKay, R.; Graham, L.; Lefebvre, A. *Environmental Sustainability of Canadian Agriculture: Agri-Environmental Indicator Report Series*; Report #3; Agriculture and Agri-Food Canada: Ottawa, ON, Canada, 2010; 235p.
10. Markandya, A.; Perelet, R.; Mason, P.; Taylor, T. *Dictionary of Environmental Economics*; Earthscan: London, UK, 2002.

11. Organization for Economic Cooperation and Development (OECD). *Environmental Indicators for Agriculture: Concepts and Framework*; OECD: Paris, France, 1999; Volume 1.
12. Clearwater, R.; Martin, T.; Hoppe, T. (Eds.) *Environmental Sustainability of Canadian Agriculture: Agri-Environmental Indicators Report Series—Report #4*; Agriculture and Agri-Food Canada: Ottawa, ON, Canada, 2016; 239p.
13. Wall, E.; Smit, B. Climate change adaptation in light of sustainable agriculture. *J. Sustain. Agric.* **2005**, *27*, 113–123. [CrossRef]
14. Wheaton, E.; Kulshreshtha, S.; Eilers, W.; Wittrock, V. *Trends in the Environmental Performance of Agriculture in Canada under Climate Change*; The Organization for Economic Cooperation and Development (OECD), Saskatchewan Research Council: Saskatoon, SK, Canada, 2010; 10p.
15. Wheaton, E.; Eilers, W.; Kulshreshtha, S.; MacGregor, R.; Wittrock, V. *Assessing Agri-environmental Implications of Climate Change and Agricultural Adaptation to Climate Change*; SRC Publication No. 10432-1E11; The Organization for Economic Cooperation and Development (OECD), Saskatchewan Research Council: Saskatoon, SK, Canada, 2011; 31p.
16. Wheaton, E.; Kulshreshtha, S. Agriculture and climate change: Implications for environmental sustainability indicators. In Proceedings of the Ninth International Conference on Ecosystems and Sustainable Development, Bucharest, Romania, 18–20 June 2013; Marinov, A.M., Bebbia, C.A.B., Eds.; Wessex Institute of Technology, WIT Press: Southampton, UK, 2013; pp. 99–110.
17. Bonsal, B.; Aider, R.; Gachon, P.; Lapp, S. An assessment of Canadian prairie drought: Past, present, and future. *Clim. Dyn.* **2013**, *41*, 501–516. [CrossRef]
18. IPCC (Intergovernmental Panel on Climate Change). Summary for Policymakers. In *Managing the Risks of Extreme Events and Disasters to Advance Climate Change Adaptation*; A Special Report of Working Groups I and II of the IPCC; Cambridge University Press: Cambridge, UK, 2012.
19. Phillips, D. *The Day Niagara Falls Ran Dry! Canadian Geographic*; Key Porter Books: Toronto, ON, Canada, 1993; 226p.
20. Hunter, F.; Donald, D.; Johnson, B.; Hyde, W.; Hanesiak, J.; Kellerhals, M.; Hopkinson, R.; Oegema, B. The vanguard torrential storm. *Can. Water Res. J.* **2002**, *27*, 213–227. [CrossRef]
21. Mladjic, B.; Sushama, L.; Khaliq, M.; Laprise, R.; Caya, D.; Roy, R. Canadian RCM projected changes to extreme precipitation characteristics over Canada. *J. Clim.* **2011**, *24*, 2566–2584. [CrossRef]
22. Wheaton, E.; Kulshreshtha, S.; Wittrock, V.; Koshida, G. Dry times: Lessons from the Canadian drought of 2001 and 2002. *Can. Geogr.* **2008**, *52*, 241–262. [CrossRef]
23. Wittrock, V.; Wheaton, E.; Siemens, E. *More than a Close Call: A Preliminary Assessment of the Characteristics, Impacts of and Adaptations to the Drought of 2008–2010 in the Canadian Prairies*; Saskatchewan Research Council: Saskatoon, SK, Canada, 2010; 124p.
24. Szeto, K.; Zhang, X.; White, R.; Brimelow, J. The 2015 Extreme Drought in Western Canada. In *Explaining Extreme Events of 2015 from a Climate Perspective*; Herring, S., Hoell, A., Hoerling, M., Kossing, J., Schreck, C., III, Stott, P., Eds.; Bulletin of the American Meteorological Society; American Meteorological Society: Boston, MA, USA, 2016; Volume 97, pp. S42–S45.
25. Price, D.; McKenney, D.; Joyce, L.; Siltanen, R.; Papadopol, P.; Lawrence, K. *High-Resolution Interpolation of Climate Scenarios for Canada Derived from General Circulation Model Simulations*; Information Report NOR-X-421; Northern Forestry Center, Canadian Forest Service: Edmonton, AB, Canada, 2011.
26. Williams, G.; Wheaton, E. Estimating biomass and wind erosion impacts for several climatic scenarios: A Saskatchewan case study. *Prairie Forum* **1998**, *23*, 49–66.
27. Phillips, D. Canada's Top Ten Weather Stories for 2011. Available online: http://www.ec.gc.ca/meteo-weather/default.asp?lang=En&n=0397DE72-1 (accessed on 5 March 2013).
28. Environment Canada. Climate Trends and Variations Bulletin, Summer 2012, Spring 2012. Available online: http://www.ec.gc.ca/adsc-cmda/default.asp?lang=En&n=30EDCA67-1 (accessed on 5 March 2013).
29. Hopkinson, R. *Anomalously High Rainfall over Southeast Saskatchewan—2011*; Custom Climate Services; The Saskatchewan Watershed Authority: Regina, SK, Canada, 2011.
30. United States Army Corps of Engineers. 2011 Post-Flood Report for the Souris River Basin. Submitted to The International Souris River Board and The United States Department of the Interior. Available online: http://swc.nd.gov/4dlink9/4dcgi/GetSubContentPDF/PB-2794/Souris%202011%20Post%20Flood%20Report.pdf (accessed on 5 March 2013).

31. Brown, R.D.; Robinson, D.A. Northern Hemisphere spring snow cover variability and change over 1922–2010 including an assessment of uncertainty. *Cryosphere* **2011**, *5*, 219–229. [CrossRef]
32. Gan, T.; Barry, R.; Gizaw, M.; Gobena, A.; Balaji, R. Changes in North American Snowpacks for 1979–2007 detected from the Snow Water Equivalent Data of SMMR and SSM/I Passive Microwave and Related Climatic Factors. *J. Geophys. Res. Atmos.* **2013**, *118*, 7682–7697. [CrossRef]
33. Brown, R.; Mote, P. The response of Northern Hemisphere snow cover to a changing climate. *J. Clim.* **2009**, *22*, 2124–2145. [CrossRef]
34. Davidson, E.; Janssens, I. Temperature sensitivity of soil carbon decomposition and feedbacks to climate change. *Nature* **2006**, *440*, 165–173. [CrossRef] [PubMed]
35. Wiebe, B.; Eilers, W.; Brierley, J. Soil Salinity. In *Environmental Sustainability of Canadian Agriculture: Agri-Environmental Indicator Report Series*; Report #3; Eilers, W., MacKay, R., Graham, L., Lefebvre, A., Eds.; Agriculture and Agri-Food Canada: Ottawa, Ontario, Canada, 2010; p. 66.
36. Townley Smith, L.; Black, M. Desertification. Sidebar. In *Environmental Sustainability of Canadian Agriculture: Agri-Environmental Indicator Report Series*; Report #3; Eilers, W., MacKay, R., Graham, L., Lefebvre, A., Eds.; Agriculture and Agri-Food Canada: Ottawa, ON, Canada, 2010; p. 235.
37. Sauchyn, D.; Wuschke, B.; Kennedy, S.; Nykolyak, M. *A Scoping Study to Evaluate Approaches to Developing Desertification Indicators*; Agriculture and Agri-Food Canada, Prairie Adaptation Research Collaborative: Regina, SK, Canada, 2003; p. 109.
38. World Meteorological Organization (WMO). Climate Change and Desertification. Available online: http: //www.wmo.int/pages/prog/wcp/agm/publications/documents/wmo_cc_desertif_foldout_en.pdf (accessed on 11 April 2016).
39. Fung, K.; Barry, B.; Wilson, M.; Martz, L. *Atlas of Saskatchewan*; University of Saskatchewan: Saskatoon, SK, Canada, 1999.
40. Waggoner, P.; Revelle, R. Summary. In *Climate Change and U.S. Water Resources*; Waggoner, P.E., Ed.; John Wiley and Sons: Toronto, ON, Canada, 1990.
41. Whitehead, P.G.; Wilby, R.L.; Battarbee, R.W.; Kernan, M.; Wade, A.J. A review of the potential impacts of climate change on surface water quality. *Hydrol. Sci. J.* **2009**, *54*, 101–123. [CrossRef]
42. Moss, B.; Stephen, D.; Balayla, D.; Bécares, E.; Collings, S.; Fernandez-Alaez, C.; Fernandez-Alaez, C.; Ferriol, C.; Garcia, P.; Goma, J.; et al. Continental-scale patterns of nutrient and fish effects on shallow lakes: Synthesis of a pan-European mesocosm experiment. *Freshw. Biol.* **2004**, *49*, 1633–1649. [CrossRef]
43. Anteau, M.; Wiltermuth, M.; van der Burg, M.P.; Pearse, A. Prerequisites for understanding climate-change impacts on northern prairie wetlands. *Wetlands* **2016**, *36*, 299–307. [CrossRef]
44. Murdoch, P.S.; Baron, J.S.; Miller, T.L. Potential effects of climate change on surface water quality in North America. *J. Am. Water Resour. Assoc.* **2000**, *36*, 347–366. [CrossRef]
45. Delpla, I.; Jung, A.-V.; Baures, E.; Clement, M.; Thomas, O. Impacts of climate change on surface water quality in relation to drinking water production. *Environ. Int.* **2009**, *35*, 1225–1233. [CrossRef] [PubMed]
46. Kundzewicz, Z.W.; Mata, l.J.; Arnell, N.W.; Döll, P.; Jimenez, B.; Miller, K.; Oki, T.; Şen, D.; Shiklomanov, I. The implications of projected climate change for freshwater resources and their management. *Hydrol. Sci. J.* **2008**, *53*, 3–10. [CrossRef]
47. Medellin-Azuara, J.; Harou, L.; Olivares, M.; Madani, K.; Lund, J.; Howitt, R.; Tanaka, S.; Jenkins, M.; Zhu, T. Adaptability and adaptations of California's water supply system to dry climate warming. *Clim. Chang.* **2008**, *87*, S75–S90. [CrossRef]
48. Piao, S.; Ciai, P.; Huang, Y.; Shen, Z.; Peng, S.; Li, J.; Zhou, L.; Liu, H.; YihuiDing, Y.; Friedlingstein, P.; et al. The impacts of climate change on water resources and agriculture in China. *Nature* **2010**, *467*, 43–51. [CrossRef] [PubMed]
49. Bates, B.; Kundzewicz, Z.; Wu, S. *Climate Change and Water*; Intergovernmental Panel on Climate Change; Cambridge University Press: Cambridge, UK, 2008.
50. Barnett, T.; Adam, J.; Lettenmaier, D. Potential impacts of a warming climate on water availability in snow-dominated regions. *Nature* **2005**, *438*, 303–309. [CrossRef] [PubMed]
51. Taylor, R.; Scanlon, B.; Döll, P.; Rodell, M.; van Beek, R.; Wada, Y.; Longuevergne, L.; Leblanc, M.; Famiglietti, J.; Edmunds, M.; et al. Ground water and climate change. *Nat. Clim. Chang.* **2013**, *3*, 322–329. [CrossRef]

52. Rosenberg, N.; Epstein, D.; Wang, D.; Vail, L.; Srinivasan, R.; Arnold, J. Possible impacts of global warming on the hydrology of the Ogallala aquifer region. *Clim. Chang.* **1999**, *42*, 677–692. [CrossRef]

53. Huang, J.; Wang, Y.; Wang, J. Farmer's Adaptation to Extreme Weather Events through Farm Management and Its Impacts on the Mean and Risk of Rice Yield in China. In Proceedings of the Agricultural & Applied Economics Association's 2014 Annual Meeting, Minneapolis, MN, USA, 27–29 July 2014.

54. Kulshreshtha, S.N.; Marleau, R. *Canadian Droughts of 2001 and 2002: Economic Impacts on Crop Production in Western Canada*; Publication No. 11602-34E03; SRC Saskatchewan Research Council: Saskatoon, SK, Canada, 2003.

55. Statistics Canada. Table 004-0010-Census of Agriculture, Selected Land Management Practices and Tillage Practices Used to Prepare Land for Seeding, Canada and Provinces, Every 5 Years (Number Unless Otherwise Noted). CANSIM (Database), 2012. Available online: http://www5.statcan.gc.ca/cansim/a47 (accessed on 3 February 2017).

sustainability

MDPI

Article

An Economic Assessment of Local Farm Multi-Purpose Surface Water Retention Systems under Future Climate Uncertainty

Pamela Berry [1,*], Fuad Yassin [1], Kenneth Belcher [2] and Karl-Erich Lindenschmidt [1]

[1] School of Environment and Sustainability, University of Saskatchewan, 11 Innovation Boulevard, Saskatoon, SK S7N 3H5, Canada; fuad.yassin@usask.ca (F.Y.); karl-erich.lindenschmidt@usask.ca (K.-E.L.)

[2] Department of Agricultural and Resource Economics, University of Saskatchewan, Saskatoon, SK S7N 5A8, Canada; ken.belcher@usask.ca

* Correspondence: pamela.berry@usask.ca; Tel.: +1-306-261-0460

Academic Editor: Hossein Azadi
Received: 21 November 2016; Accepted: 16 March 2017; Published: 19 March 2017

Abstract: Regions dependent on agricultural production are concerned about the uncertainty associated with climate change. Extreme drought and flooding events are predicted to occur with greater frequency, requiring mitigation strategies to reduce their negative impacts. Multi-purpose local farm water retention systems can reduce water stress during drought periods by supporting irrigation. The retention systems' capture of excess spring runoff and extreme rainfall events also reduces flood potential downstream. Retention systems may also be used for biomass production and nutrient retention. A sub-watershed scale retention system was analysed using a dynamic simulation model to predict the economic advantages in the future. Irrigated crops using water from the downstream reservoir at Pelly's Lake, Manitoba, Canada, experienced a net decrease in gross margin in the future due to the associated irrigation and reservoir infrastructure costs. However, the multi-purpose benefits of the retention system at Pelly's Lake of avoided flood damages, nutrient retention, carbon sequestration, and biomass production provide an economic benefit of $25,507.00/hectare of retention system/year. Multi-purpose retention systems under future climate uncertainty provide economic and environmental gains when used to avoid flood damages, for nutrient retention and carbon sequestration, and biomass production. The revenue gained from these functions can support farmers willing to invest in irrigation while providing economic and environmental benefits to the region.

Keywords: climate change; multi-purpose retention systems; agriculture; irrigation

1. Introduction

Across Canada, annual mean temperature has increased since the 1950s. As of 2010, Canada has experienced an annual average surface air temperature warming of 1.5 °C. Stronger warming trends have been found for the north and west of Canada, with the greatest warming occurring in winter and spring [1,2]. These climatic changes are expected to increase potential evapotranspiration and lead to moisture deficits [3]. Changes in precipitation timing are also expected, resulting in less snow-cover, shorter snow-cover duration, increased winter river flows, and decreased summer flood events [3–5]. If the timing of seasonal precipitation begins to change and temperatures continue to rise under climate change, there will be severe effects on agriculture, ecosystems, water runoff rates and quantities, as well as groundwater storage. On the Canadian Prairies, where 80 percent of Canada's farmland is situated, strategic water management solutions are needed to deal with the uncertainty associated with climate change and its impact on agricultural production [4,6,7]. Changes in temperature and precipitation expected with climate change will impact a wide range of variables, affecting the productivity of annual crops. Alterations in the growing season length, frost timing, heat waves, precipitation, and moisture availability will be witnessed with temperature and precipitation increases. This will require farmers to be ready to adapt to new climatic patterns [8]. The uncertainty associated with climate also increases risk for farmers, requiring water management solutions that will provide benefits to farmers under all conditions, while reducing agricultural risk [8,9]. Economic consequences of drought or flooding events will depend on the agricultural and water management sectors success in preparing and adapting for climatic extremes [4]. The chosen water management strategy needs to be economically viable, benefitting the farmer and the provincial economy, while reducing risk in agricultural production systems [10].

Climate change is predicted to increase the frequency of extreme drought and flooding events which will impact agricultural sectors, such as southern Manitoba, Canada [4,7,9,11]. Historically, the trend within Manitoba has been to remove excess water from agricultural land as quickly as possible in spring using a series of ditches and drains [3,12]. This strategy to deal with flooding in the province is common throughout the landscape [3,13]. While drainage systems are meant to remove excess water from inundated land quickly, they can actually increase the negative effects of floods by amplifying flood peaks, which then have greater force to cause damage [3]. Drainage also increases the amount of nutrients being removed from the landscape, subsequently impacting water quality as they flow into Manitoba's water bodies [3]. Quickly removing water from the landscape may also leave agricultural lands vulnerable to soil moisture deficits under future climate uncertainty as evapotranspiration quickly removes summer precipitation from the soil [3]. The province of Manitoba aims to increase their adaptive capacity to prepare for future climate uncertainties [9,14,15]. Increasing the adaptive capacity of southern Manitoba communities will require the development of techniques allowing farmers to drought proof their crops as well as to limit damages caused by floods in non-drought years. Strategies should also allow for sustainable water management by providing multiple benefits when possible, such as bio production and nutrient retention [14]. Strategies currently being used on the Prairies include crop insurance, soil and water conservation, improved irrigation (where applicable), exploration of groundwater supplies, as well as introduction of new infrastructure [4,7,9]. Infrastructure implementations range from new wells and pipelines to dugouts [4]. Dryland farmers look to decrease drought risk by conserving soil moisture and nutrients through crop rotation and minimizing tillage practices [9].

The creation of multi-purpose local farm water retention systems, designed to capture and store surface water, may be a viable option that would reduce water stress during droughts by providing water for irrigation [16]. Additionally, the retention systems would serve to capture excess runoff in spring and during extreme rainfall events to help mitigate flood events. The stored water can be used for biomass production and nutrient retention [14,17]. Retention systems also serve to reduce downstream peak flow and aid in retaining flood waters which reduces associated flood risks downstream [16,18]. If water is released from reservoirs, they serve to replenish groundwater stores downstream [16]. Researchers have found these systems to be effective for increasing and stabilizing crop yields via irrigation in locations such as Texas, Kansas, Kentucky, India, and Thailand [19]. Berry [20] reported an average increase in crop yield of $13.00/hectare/year when irrigation was applied to canola, alfalfa, barley, and spring wheat using water from a local farm retention system in Manitoba. However, there is limited pre-existing irrigation infrastructure in Manitoba, so adopting irrigation as a management strategy would require the costly installation of irrigation infrastructure. This would require farmers to earn incremental gross margins of $147.00/hectare of crop land/year to cover the cost of irrigation infrastructure and a surface water retention system as a water source [20]. While irrigation did not provide an increase in gross margins under present day conditions, future climate change scenario predictions of increased temperatures and precipitations may result in more economically advantageous conditions in the future.

Surface water retention systems additionally aid in improving water quality. Retention systems have shown success in reducing nutrient and sediment loading in various locations worldwide. Several examples from the literature provided by [21] on retention systems in America and Europe have shown effectively reduced nutrient loading. A runoff detention pond in Oklahoma, USA reduced sediment discharge downstream by 82%, total nitrogen by 56%, and total phosphorus by 60% [22]. A reduction in total nitrogen of 38%, and 56% in total phosphorus loading from a constructed wetland (a detention basin formed by berms adjacent to a stream) in Illinois, USA was reported by [23]. A shallow predam, a small reservoir aimed at improving water quality of a larger main reservoir downstream, in Luxembourg, was found to retain total phosphorus up to 60%, and a deep predam retained up to 82% [21,24]. A small dam in Spain reduced total phosphorus loads downstream by over 25% [25]. Small ponds in Finland and Sweden reduced total phosphorus loading by 17% and constructed wetlands in Norway and Finland reduced total phosphorus loading by 41% [21,26].

In Manitoba, small on-farm surface water retention systems are scattered throughout agricultural watersheds. The South Tobacco Creek Watershed, in south central Manitoba, is home to twenty-six dams providing management of almost 30% of the watersheds drainage area [18]. The watershed is now home to five dry dams, six back-flood dams, and fifteen multipurpose dams. Each dam was designed to retain 20–25 mm of runoff at full capacity from their catchment area [21]. These dams' capacity to reduce flood risk has been under study since the 1990s. The Watershed Evaluation of Beneficial Management Practices (WEBs) program, an Agriculture and Agri-Food Canada (AAFC) research program, began in 2004 to expand the research on the South Tobacco Creek Watershed dams to include sediment, nitrogen, and phosphorus loadings downstream [18]. Both the dry flood control dam and the multi-purpose dam were effective in reducing total suspended sediment (65%–85% reduction), particulate nitrogen (41%–43% reduction during snowmelt, 7%–11% reduction from summer rainfall events), and particulate phosphorus (27%–38% reduction in snowmelt runoff) [18]. The entire system of dams within the watershed provided a reduction in peak flow of 9%–19% from spring snowmelt runoff and 13%–25% from rainfall runoff [18]. Another on-farm retention pond in Saint-Samuel, Quebec was found to reduce peak flows by 38%, on average, from rainfall runoff events [27]. The pond was also effective at removing total suspended sediment, total nitrogen, and total phosphorus with mean removal efficiency ratios of 50%–56%, 42%–52%, and 48%–59%, respectively.

The ecosystem benefits of reduced nutrient and sediment loading downstream, carbon sequestration, and avoided flood damages downstream have been monetized for surface water retention systems on the Prairie landscape. Berry [20] reported retention system's ability to reduce downstream phosphorus and nutrient loading, sequester carbon, and contribute to avoided flood damages downstream can provide $17,600/hectare of retention basin/year. This estimate used a carbon credit value of $25.00/tonne of carbon dioxide equivalent, which the Manitoba Liquor and Lotteries Corporation is currently offering in Manitoba [28]. Benefit transfer methodology was used to estimate monetary values for reductions in nutrient loading and avoided flood damages downstream [20,29–33]. Cattail harvest from surface water retention systems as a biomass crop has recently been commercialized in Manitoba. Multi-purpose surface water retention systems gain additional ecosystem benefits of biomass production, increased nutrient management, and carbon dioxide emission offsets when cattail harvest is introduced [14,34]. Berry [20] reported that harvesting cattail for biomass production, nutrient removal, and carbon credits from multi-purpose retention systems can provide an additional $7657.00/hectare of retention basin/year. This estimate included monetary compensation for nitrogen and phosphorus removal from the ecosystem from cattail harvest, carbon credit production from biomass replacing coal with a value of $25.00/tonne of carbon dioxide equivalent, and a net value of dry cattail biomass of $16.59/tonne [17,28,34,35].

The purpose of this study is to determine the economic feasibility of the adoption of a multi-purpose local farm surface water retention system as a water management strategy in southern Manitoba to reduce the increased agricultural risks to farmers under future climate uncertainty [9,16,36]. Future climatic conditions will be modelled for the study area to determine the economic and environmental benefits water retention systems can offer as a water management strategy under climate change.

Study Site

The study site for this analysis is a pre-existing water retention system at Pelly's Lake in southern Manitoba (Figure 1). Prime agricultural lands surround the lake, with the lake's watershed currently used to produce a range of crops [6]. The four main crops in the watershed are canola, spring wheat, alfalfa, and barley [37]. Landowners in the area had historically attempted to drain the area of Pelly's Lake to increase hay production. However, these efforts ultimately failed due to the presence of an underground spring, poor drainage, poor water retention potential, and widespread flooding during times of excess water [6,13]. This led to an agreement with the La Salle Redboine Conservation District (LSRCD) to create a back flood system offering multiple benefits [38]. Conservation districts in Manitoba promote sustainable development in working to protect their districts natural resources [39]. The frequently saturated land at Pelly's Lake, MB now provides flood water retention, capturing the spring freshet along with rain runoff, and nutrient retention [17,40,41]. The retained water is released in mid-June to maintain baseflow downstream. Hay production can then occur on the drained land. A water storage capacity of 2,100,000 m^3 additionally provides a large water source if irrigation development in the area is pursued.

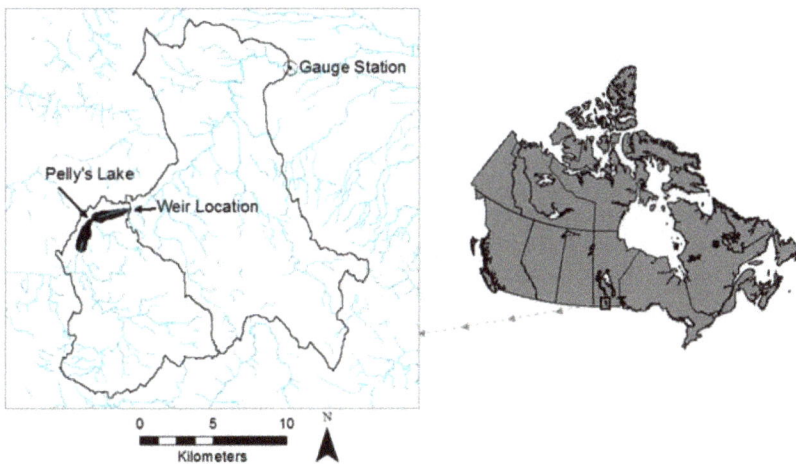

Figure 1. Pelly's Lake, Manitoba, Canada, situated within its watershed. To the right of this is the downstream gauge station and adjacent watershed boundary.

2. Materials and Methods

In order to explore the potential economic advantages of retention ponds under future climate conditions, a hydrologic model was parameterized using future precipitation and temperature data at a fine resolution for the study site. The output from the hydrologic model was input into a modeling system to provide projections of how water storage at the Pelly's Lake multi-purpose retention system would be affected by climate change. Stella modeling software, a program designed specifically for modeling complex system dynamics, allowed for an integrated hydrologic, reservoir, irrigation, plant growth, and economic model to be created [42]. The modeling system was developed based on a daily time step using a growing season simulation period running from April through September. The daily time step captured short-term components of the system while allowing for multiple years to be analyzed for a long-term analysis of the problem [10]. Spatially, study simulations were confined to the

Pelly's Lake watershed as the multi-purpose retention system would collect runoff and precipitation within this boundary (Figure 1).

2.1. Modeling System

2.1.1. Hydrologic Model and Reservoir Module

To model the hydrologic component of the target watershed, the Environment and Climate Change Canada environmental modeling system Modélisation Environnementale Communautaire—Surface and Hydrology (MESH) was chosen. MESH is a distributed land surface model commonly used in Canada for medium to large scale simulations [43,44]. Environment and Climate Change Canada uses MESH as part of an operational forecasting tool and the modeling system is currently being used within research projects such as the Drought Research Initiative (DRI) [45]. MESH requires multiple inputs to provide a complete distributed land surface model. The energy and water balance requirements for the model were determined using the Canadian Land Surface Scheme (CLASS) 1 [44] and CLASS 2 [46]. CLASS 1 is a physically based land surface model which calculates heat and moisture transfer at the surface, while CLASS 2 calculates energy and moisture fluxes at the canopy level [43,46]. Precipitation data for MESH were taken from the Canadian Precipitation Analysis (CaPA) project which produces rainfall accumulations at a six-hour time step and resolution of 15 km over North America in real-time [47]. Further required climatic data such as long wave and short wave radiation, humidity, pressure, and wind speed was acquired from the Global Environmental Multiscale (GEM) Model [43,48]. Routing of water within the study area was performed within the MESH model using a storage-routing technique which applies the continuity equation as outlined in [49]. Optimization of the MESH model was performed. Streamflow outputs from the MESH model were summed for 1 January to 14 April each year to provide an initial reservoir volume from spring freshet. Streamflow values for 15 April to 15 September each year were input into the model (Figure 2). Reservoir outflow considered the height of the emergency spillway, outflow over a rectangular weir, evaporation rates, and withdrawals taken for irrigation purposes (Figure 2). Table 1 provides detailed input parameters and equations for the modeling system.

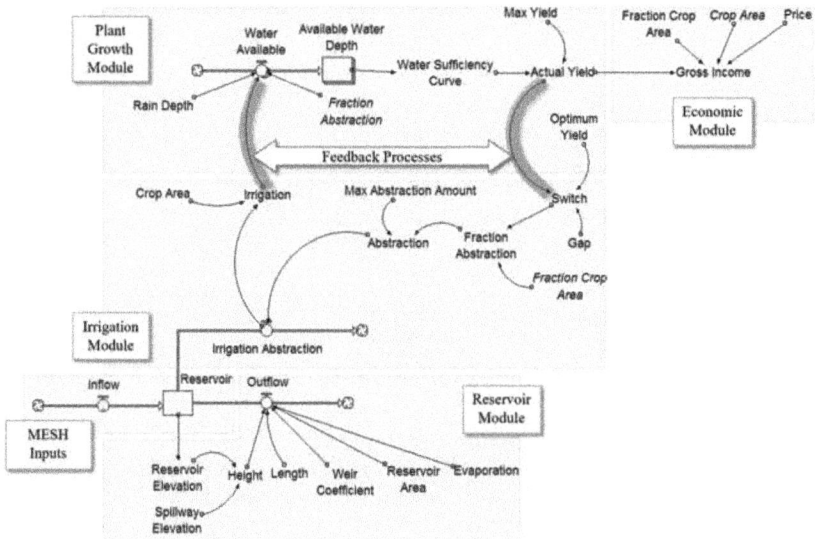

Figure 2. A stock-flow diagram of the modeling system, visually divided into its five component modules.

2.1.2. Irrigation and Plant Growth Module

The irrigation module consisted of irrigation withdrawals and precipitation during the growing season informing soil water volume available for crops (Figure 2). The four most prevalent crops in the study watershed as of 2011, canola, spring wheat, alfalfa, and barley, were modelled with a crop area of 6697 hectares [40]. On average within the study area, there is some initial spring soil moisture associated with snowmelt, however for the purposes of this model, soil moisture was recharged with precipitation and/or irrigation water. Soil moisture was recharged with precipitation and/or irrigation water. With the primary focus of the analysis on water as a crop production input and to maintain the tractability of the model, we assumed that all other production inputs, including nitrogen and phosphorus fertilizer and pesticides, were applied at rates which met crop growth requirements such that water was the only limiting factor to crop yield. This assumption was supported by our use of Government of Manitoba estimated, region specific, production costs including input and insurance cost. These costs are estimated based on the assumption that production inputs are used at rates to meet all crop growth requirements. The resultant crop yields were used in combination with crop prices to determine gross income (Table 1). Crop production costs and input costs were subtracted from gross income in Microsoft Excel to estimate gross margins under irrigation.

2.1.3. Economic Module

Crop prices, production costs, and insurance costs used in the economic module were 2015 values, provided by the Government of Manitoba, and were held constant for all future simulations (Figure 2, Table 1) [50]. Production costs refer to costs for seed and seed treatment, fertilizer, fungicide, herbicide, and insecticide application, machinery operation, fuel, leases, land taxes, and interest costs. Agricultural input costs and crop prices will fluctuate over time due to changes in food demand, changes in crop varieties and production technology, as well as changes in energy prices and climate. However, analysis of Canadian agriculture suggests that the ratios between farm expenses and receipt have been relatively stable over time [51]. As a result, we adopt the relatively strong assumption that input costs including reservoir and irrigation installation and upkeep costs as well as production costs and output prices associated with each crop type were constant for all simulations with the model. The total adjusted cost of the retention system at Pelly's Lake, which included upkeep and accrued interest for a twenty-year time horizon, was \$45,167/year (\$5.26/ha/year) [35,52]. A twenty-year time horizon was chosen as this represents reservoir infrastructure's typical serviceable life [52]. Centre-pivot irrigation infrastructure installation, labour, and maintenance over a twenty-year time horizon totalled \$966,010 (\$112.50/ha/year) [53].

Table 1. Parameter values and equation inputs for each module within the modeling system.

Parameters/Units	Inputs and Equations with Descriptions
Reservoir Module	
Inflow (m^3/day)	Input graphically using output data from the MESH hydrologic model on a daily time step.
Reservoir (m^3)	Initial reservoir volume calculated based on cumulative output from the MESH hydrologic model from 1 January–14 April.
Outflow (m^3/day)	=(3 × Weir_Coefficient × Length × Height$^{1.5}$ × 86,400) − (Evaporation × Reservoir_Area)
	The established engineering equation for discharge over a rectangular weir was multiplied by 86,400 to convert from m^3/s to m^3/day. Evaporation over the reservoir area was subtracted from the discharge equation [54].
Weir Coefficient (dimensionless)	=0.6, An established engineering value was used.
Length (m)	=12, Spillway length was taken from the engineering drawings for the Pelly's Lake weir.
Height (m)	=IF (Reservoir Elevation − Spillway Elevation) > 0 THEN (Reservoir Elevation − Spillway Elevation) ELSE 0, An established engineering equation.

Table 1. *Cont.*

Parameters/Units	Inputs and Equations with Descriptions	
Reservoir Elevation (m)	$=9 \times 10^{-7} \times$ Reservoir + 378.23	
	This equation was determined from the engineering storage rating curve for the Pelly's Lake weir.	
Spillway Elevation (m)	=379.1, This value was provided on the engineering drawings for the Pelly's Lake weir.	
Evaporation (m^3/day)	=0.00182 (April)	=0.00454 (July)
	=0.00422 (May)	=0.00469 (August)
	=0.00460 (June)	=0.00346 (September)
	Mean monthly evaporation values from 1981–2010 were converted to daily values at Brandon, MB [54]. These values were used due to insufficient data available to calculate evaporation at the study site.	
Reservoir Area (m^2)	=85,867,480, calculated in ArcGIS.	
Irrigation Module		
Irrigation Abstraction (m^3/day)	=Abstraction[Canola] + Abstraction[Wheat] + Abstraction[Barley] + Abstraction[Alfalfa], This variable calculated the total water abstraction volume abstracted from the reservoir.	
Abstraction [Crop]	=Max Abstraction Amount × Fraction Abstraction This equation calculated irrigation withdrawal volumes for each crop.	
Max Abstraction Amount (m^3)	=15,000, this amount was calibrated to allow the reservoir to drain at a rate to provide sufficient water for irrigation for the entire growing season.	
Fraction Abstraction [Crop](dimensionless)	=IF ((Switch[Canola] + Switch[Wheat] + Switch[Barley] + Switch[Alfalfa]) > 0) THEN(Switch[Crop] × Fraction_Crop_Area[Crop]/(Switch[Canola] × Fraction_Crop_Area[Canola] + Switch[Wheat] × Fraction_Crop_Area[Wheat] + Switch[Barley] × Fraction_Crop_Area[Barley] + Switch[Alfalfa] × Fraction_Crop_Area[Alfalfa])) ELSE (0)	
	When water requirements were not being met by a specific crop, this algorithm directed water withdrawals to the crop requiring irrigation. It also ensured water was not applied unless required to optimize crop growth.	
Fraction Crop Area	=0.46 (Canola)	=0.11 (Alfalfa)
	=0.38 (Wheat)	=0.05 (Barley)
	Historical patterns of crop production in Manitoba were used to determine the fraction of total crop area each crop comprised [37,55].	
Switch[Crop]	=IF (Actual Yield < Gap × Optimum Yield AND TIME > 30 THEN 1 ELSE 0	
	Irrigation was triggered for a specific crop if the crop's actual yield fell below 80% of optimum yield on day 30 (15 May).	
Gap	=0.8, Irrigation application occurred when available water only allowed for 80% or less of optimum yield growth. Yield reductions due to water stress occur when available water falls below 60% of optimum plant requirements. The threshold of 80% ensured there was always sufficient water available to the plant [53].	
Irrigation (mm/day)	=(Irrigation Abstraction/Crop Area) × 1000	
	This equation served to convert irrigation from a volume to a depth.	
Crop Area (m^2)	=66,976,634, This was calculated in ArcGIS using data on land use downloaded from the Manitoba Land Initiative [55].	
Plant Growth Module		
Rain Depth (mm/day)	Input graphically using values from Environment and Climate Change Canada for Holland, MB [56].	
Water Available [Crop] (mm/day)	=Rain Depth + Irrigation × Fraction Abstraction	
	Daily water available for crop growth [57].	
Available Water Depth [Crop] (mm)	Initial value set at 0. Snowmelt would contribute to initial spring soil moisture, however for the purposes of the model soil moisture was assumed to be recharged solely from precipitation and/or irrigation water.	
Water Sufficiency Curve [Crop]	These curves were input graphically and represented the unique optimal water requirements of each crop. They allowed for crop yield to be calculated based on water availability [57].	
Max Yield [Crop](tonnes/m^2)	= 0.000224124 (Canola)	=0.000376588 (Barley)
	= 0.000336063 (Wheat)	=0.000672126 (Alfalfa)
	Max yield values were held constant for all simulations [50].	

Table 1. *Cont.*

Parameters/Units	Inputs and Equations with Descriptions	
Actual Yield [Crop] (tonnes/hectare)	=Max Yield × Water Sufficiency Curve × 10,000	
	Crop yield was calculated based on water availability. Each crop's water sufficiency curve provided a proportion of maximum growth based on water availability. This proportion was multiplied by max yield and 10,000 to convert from m^2 to hectares.	
Optimum Yield [Crop]	For each crop, values were input graphically. The variable represented the maximum yield of each crop over time when its water requirements were being met. Values were constant for all simulations.	
Economic Module		
Gross Income [Crop] ($)	=Actual Yield × Price × Crop Area × Fraction Crop Area/10,000 Calculated landscape level gross income.	
Price [Crop] ($/tonne)	=418.87 (Canola) =238.83 (Wheat)	=173.23 (Barley) =132.28 (Alfalfa)
	Crop prices were not available for years before 2015. Thus, 2015 crop prices were used in all simulations [50].	

2.2. Climate Change

Statistically downscaled climate data for the study area were acquired from the Pacific Climate Impacts Consortium (PCIC) for the present study [58,59]. Climate scenarios are available across Canada from PCIC. Data are produced at a gridded resolution of roughly 10 km or 300 arc-seconds for 1950–2100. Three output variables on a daily time step are available from PCIC: precipitation, minimum temperature, and maximum temperature [59]. Scenarios for all four representative concentration pathways (RCPs) (Table 2) are available and multi-model ensemble tables are provided to aid the researcher in climate model selection with the widest breadth of future climate simulations. Due to constraints on time and resources, the model ensemble list was narrowed down to contain only four models for this study (Table 3). Historical daily gridded climate data for Canada were used in combination with General Circulation Model (GCM) projections from the Coupled Model Intercomparison Project Phase 5 (CMIP5) [59]. The chosen GCMs have been studied and shown to model climate change most effectively for regional applications over North America [58,60,61]. Of the two downscaling methods provided by PCIC [59], Bias Corrected Spatial Disaggregation (BSCD) was chosen for this study due to its extensive application in previous hydrologic modeling research across North America, its ease of use, daily time series output of gridded temperature and precipitation, and its ability to capture emission scenario transience effectively [58,61–64]. The BSCD method bias-corrects monthly GCM data against GCM gridded observed data. BSCD also downscales monthly GCM data to allow for regional analysis at a daily time step [58,62,63]. Further description of the application of BSCD to PCIC scenarios can be found in Werner [58]. The second downscaling option, Bias Correction/Constructed Analogues with Quantile mapping reordering (BCCAQ), is a recently developed method. As such, at the time of this study its application and accuracy in modeling climate change effects had not been extensively tested and was thus not suitable for this application [63].

Table 2. Representative concentration pathways (RCPs) overview.

RCP	Description
RCP2.6	Radiative forcing will peak at approximately 3 W/m^2 before 2100 and then levels will decline.
RCP4.5	Radiative forcing will stabilize at 4.5 W/m^2 after 2100.
RCP6	Radiative forcing will stabilize at 6 W/m^2 after 2100.
RCP8.5	Radiative forcing will rise resulting in 8.5 W/m^2 in 2100.

Table 3. Selected models used in multi-model ensemble of future climate scenarios.

Modeling Center	Institute ID	Model Name
Canadian Centre for Climate Modeling and Analysis	CCCMA	CanESM2
Meteorological Office Hadley Centre	MOHC	HadGEM2-ES
Max Planck Institute for Meteorology	MPI-M	MPI-ESM-LR
NOAA Geophysical Fluid Dynamics Laboratory	NOAA GFDL	GFDL-ESM2G

Emission scenarios RCP2.6, RCP4.5, and RCP8.5 were used for this study. The emission scenario RCP6 was excluded due to time constraints. The RCP4.5 emission scenario represented a median radiative forcing scenario in place of simulating RCP4.5 and RCP6 as median scenarios. Outputs from the four chosen climate models were downloaded for the study area. As the downscaled models have a grid resolution of approximately 10 km, outputs were spatially constrained over the study area between latitudes 49° N and 50° N and longitudes 98° W and 97° W.

The multi-model ensemble mean historical climate data were validated against observed climate data for the 2005–2014 (present day) study period to confirm its representation of observed values. As shown in Figure 3 there is some discrepancy between modelled and observed precipitation. This is due to the downscaling method application to the GCMs introducing bias. However, it does appear that the multi-model ensemble mean is capturing the annual and monthly cycle of precipitation. Figure 4 illustrated the multi-model ensemble mean's capacity to simulate observed temperature at the study location. These figures provided validation that the GCMs are suitable for simulating future climate change at this location.

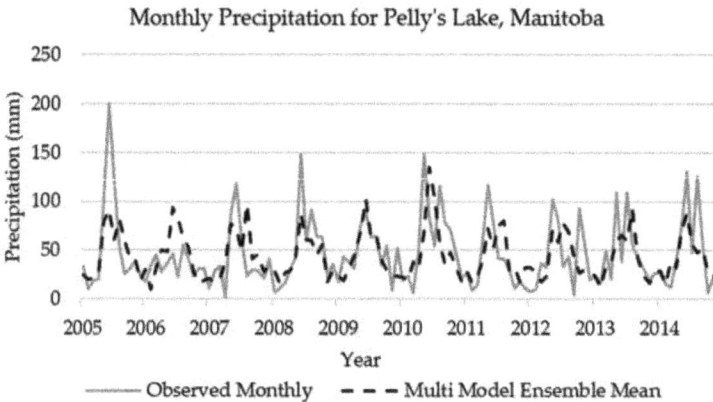

Figure 3. Comparison of observed and modelled monthly precipitation at Pelly's Lake, Manitoba for 2005–2014.

Climate trends at Pelly's Lake for the median emission scenario, RCP4.5, along with the two extreme emission scenarios, RCP2.6 and RCP8.5, were graphed to provide a comparison (Figures 5 and 6). The data were divided into summer and winter to determine multi-model ensemble mean increment changes in precipitation and temperature for the two seasons. These plots indicate a consensus in future climate precipitation trends between the four different models, increasing confidence that the climate models are performing as desired. Confidence in the model's abilities to simulate future climate conditions was further increased by the clear trends between models for each RCP. Outputs from the PCIC downscaled models support the findings published by Warren and Lemmon [1] that precipitation is projected to increase for all seasons across Canada in the future.

Future precipitation simulation outputs in the same report also indicated precipitation increases will be greater in the winter than the summer [1].

Monthly Mean Temperature for Pelly's Lake, Manitoba

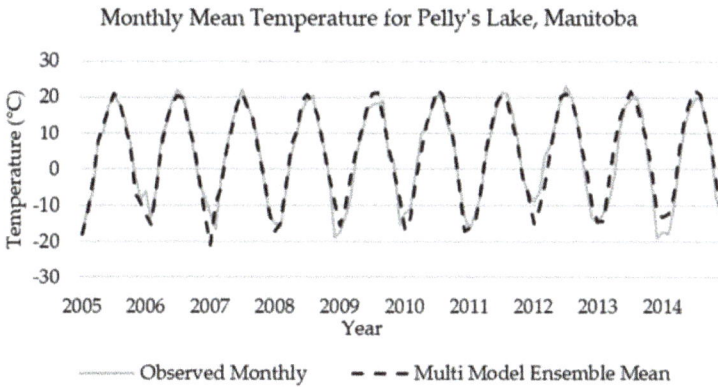

Figure 4. Comparison of observed and modelled monthly mean temperature at Pelly's Lake, Manitoba for 2005–2014.

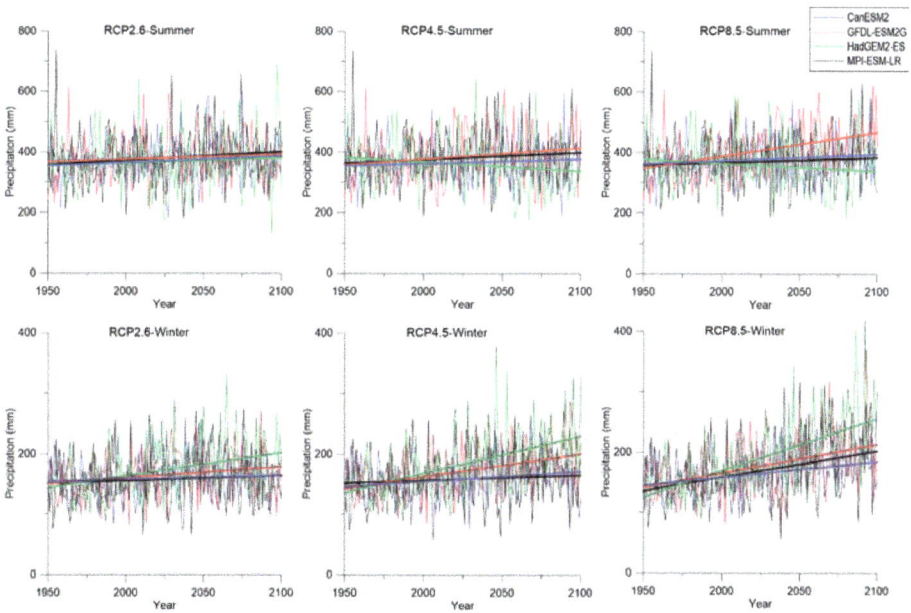

Figure 5. Multi-model ensembles for each representative concentration pathway (RCP) showing summer and winter precipitation with lines representing mean precipitation for each climate model. The spread between model simulations illustrates uncertainty.

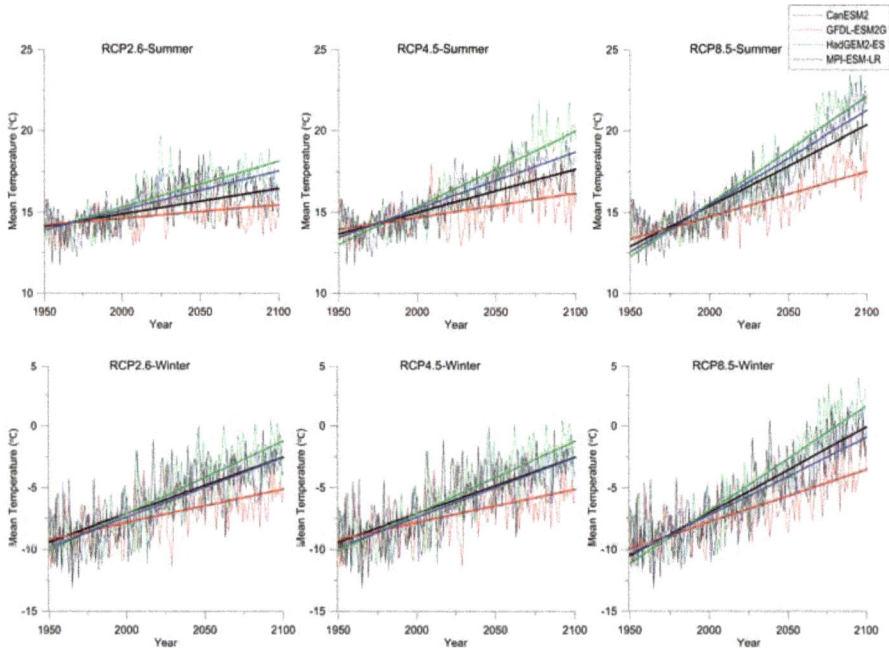

Figure 6. Multi-model ensembles for each representative concentration pathway (RCP) showing summer and winter temperature with lines representing mean temperature for each climate model. The spread between model simulations illustrates uncertainty.

2.2.1. Delta Method

As this study deals with the benefits of retention basins under future climate uncertainty, climate data were required to provide long term trends in future climate change that will impact the retention system's water volume. For this reason, a simple delta method application was chosen to simulate the two future climate periods. This method has been extensively used for studies that aim to provide a sensitivity analysis of future climate uncertainty [52,65–68]. The delta method applies changes in climate variables extracted from GCMs or Regional Climate Models (RCMs) for the study area to a baseline observed climatology [66,68]. Precipitation is adjusted multiplicatively and temperature is adjusted additively [62,66,68]. Spatial variability of the climate variables within the observed time series are preserved in any future climate simulations [62,66,68]. This assumption is a key limitation of the delta method [62,68]. However, for this study, it is more important to have accurate representation of future spring runoff volumes and precipitation volumes for the growing season [68].

Using the downscaled GCMs, climate outputs taken from PCIC future climate conditions were extracted for two ten-year time periods, 2050–2059 and 2090–2099. This allowed for representation of the middle (2050–2059) and end of the century (2090–2099). Multi-model ensemble means were used to calculate incremental changes in temperature and precipitation for the two future study periods based on the present-day time period (2005–2014). An ensemble mean value was used for this calculation as research has determined that ensemble means reduce bias within individual models, providing a more robust output than any individual GCM output [60]. Incremental changes in precipitation and temperature from the three RCP4.5 multi-model ensemble means were applied to 2005–2014 present day climate data within the hydrologic model (Table 4). Precipitation changes were applied to the baseline time period multiplicatively while temperature changes were applied additively. All other parameters within the modeling system remained the same as for the 2005–2014 time period.

Streamflow outputs from the hydrologic model provided the 2050s and 2090s hydrologic input for the modeling system.

Table 4. Mean season precipitation totals, mean season temperature increases and increases between study periods based on the multi-model ensemble means for representative concentration pathways (RCP)2.6, RCP4.5, and RCP8.5.

Decade	Mean Precipitation Total (mm)			Incremental Increase %			Mean Temperature (°C)			Temperature Increase (°C)		
	RCP			RCP			RCP			RCP		
	2.6	4.5	8.5	2.6	4.5	8.5	2.6	4.5	8.5	2.6	4.5	8.5
Summer (May–October)												
2005–2014	371.9	371.9	371.9	-	-	-	15.29	15.44	15.25	-	-	-
2050–2059	413.3	376.8	377.7	11.1	1.3	1.56	16.17	16.86	17.56	0.9	1.4	2.3
2090–2099	382.2	381.2	408.9	2.77	2.5	9.95	16.00	17.82	20.51	0.7	2.4	5.3
Winter (November–April)												
2005–2014	162.7	162.7	162.7	-	-	-	−6.867	−6.525	−6.857	-	-	-
2050–2059	174.5	176.9	199.3	7.25	8.8	22.5	−5.786	−4.489	−4.241	1.1	2.0	2.6
2090–2099	159.6	189.6	218.7	−1.91	16.6	34.4	−6.158	−3.178	−0.430	0.7	3.4	6.4

2.2.2. Uncertainty

It is important to note uncertainties associated with climate projections from GCMs as well as downscaled methods [69,70]. Future anthropogenic emission levels involve uncertainty. Models used for simulating future climate scenarios have uncertainties linked to imperfect representation of climate processes. Current understanding of climate conditions is imperfect leading to imperfect knowledge being fed into projections [69,70]. Finally, variability at the interannual and decadal level is difficult to represent accurately in long-term projections. However, this does not mean future climate projections are false as uncertainty can be quantified [69]. Each GCM and downscaling method has a unique set of parameters as initial conditions within the model. By using future climate scenario results for as many models as possible and producing a multi-model ensemble mean or median, a more probable future climate scenario can be determined. The spread in results between models illustrates the level of uncertainty in the obtained multi-model ensemble results [58,69,71].

3. Results

Annual gross margin was estimated from model simulations with and without irrigation for the middle of the century, 2050 to 2059, and the end of the century, 2090 to 2099 for three radiative forcing scenarios. The simulation results indicated that irrigated crops using water abstractions from the reservoir experienced a decrease in gross margin when compared to gross margins without irrigation and associated infrastructure for both study simulation periods. Results for the three radiative forcing scenarios, RCP2.6, RCP4.5, and RCP8.5 using incremental percentage increases to the 2005–2014 MESH climate data for the 2050s and 2090s are provided in Tables 5 and 6, respectively. The differences in gross margin when irrigation and associated infrastructure costs were considered are provided in the second column of Tables 5 and 6. The estimated yearly cost of the retention pond and irrigation infrastructure was $160.00/hectare. Any gross margin value above −$160.00 reflects that the increased crop yield enabled by irrigation water offset the yearly cost of the retention pond and irrigation infrastructure. A value above zero would indicate all retention pond and irrigation infrastructure costs are being offset by increased crop yield under irrigation application.

Table 5. Difference in crop gross margins without irrigation and crop gross margins with irrigation and associated variable and infrastructure costs, and increase in crop gross income under irrigation for the 2050s using three different radiative forcing scenarios.

Year	Difference in Crop Gross Margins without Irrigation and Crop Gross Margins with Irrigation and Associated Variable and Infrastructure Costs ($/hectare)			Increase in Crop Gross Margins under Irrigation ($/hectare)		
	RCP2.6	RCP4.5	RCP8.5	RCP2.6	RCP4.5	RCP8.5
2050	−174.00	−172.00	−173.00	−13.60	−11.80	−12.30
2051	−145.00	−151.00	−150.00	15.10	9.42	10.60
2052	−155.00	−153.00	−153.00	4.79	7.31	6.94
2053	−150.00	−146.00	−147.00	9.99	14.50	12.80
2054	−146.00	−143.00	−144.00	14.01	17.40	16.50
2055	−159.00	−159.00	−159.00	0.83	1.53	1.53
2056	−157.00	−146.00	−147.00	3.60	14.00	13.50
2057	−143.00	−136.00	−142.00	17.40	24.50	17.90
2058	−125.00	−127.00	−127.00	35.40	33.10	33.70
2059	−126.00	−126.00	−126.00	33.90	33.90	33.90
Average	−148.00	−146.00	−147.00	12.10	14.40	13.50

Note: Difference = Gross margin with retention pond used for irrigation - Gross margin without retention pond installation and associated irrigation.

Table 6. Difference in crop gross margins without irrigation and crop gross margins with irrigation and associated variable and infrastructure costs, and increase in crop gross income under irrigation for the 2090s using three different radiative forcing scenarios.

Year	Difference in Crop Gross Margins without Irrigation and Crop Gross Margins with Irrigation and Associated Variable and Infrastructure Costs ($/hectare)			Increase in Crop Gross Margins under Irrigation ($/hectare)		
	RCP2.6	RCP4.5	RCP8.5	RCP2.6	RCP4.5	RCP8.5
2090	−173.00	−173.00	−174.00	−13.10	−13.20	−13.50
2091	−151.00	−150.00	−142.00	9.42	10.10	18.00
2092	−154.00	−154.00	−155.00	6.39	6.35	4.86
2093	−146.00	−147.00	−153.00	14.30	12.80	7.15
2094	−143.00	−144.00	−146.00	17.00	16.30	14.00
2095	−159.00	−159.00	−159.00	1.53	1.53	0.82
2096	−147.00	−147.00	−157.00	13.30	13.00	3.60
2097	−135.00	−142.00	−144.00	25.20	17.90	16.30
2098	−125.00	−125.00	−125.00	35.30	35.30	35.40
2099	−126.00	−126.00	−129.00	34.10	33.90	30.90
Average	−146.00	−147.00	−148.00	14.30	13.40	11.80

Note: Difference = Gross margin with retention pond used for irrigation - Gross margin without retention pond installation and associated irrigation.

All years, with the exceptions of 2050 and 2090, under all radiative forcing scenarios, within the two study time periods would experience increased crop yields from irrigation water (Tables 5 and 6). The years 2050 and 2090 experienced the highest growing season precipitation amounts of the simulation periods (497–546 mm and 503–541 mm, respectively). A rainfall event on 14 July overwhelmed canola and wheat crops, reducing yields and subsequently triggering more irrigation water to be applied until the end of the growing season. This additional irrigation application after the crops had already received excess water was detrimental to crop yields. It is important to note that the model does not account for the dynamic decision making of farmers regarding irrigation application. In a high precipitation year, the farmer would recognize crop yield reductions were due to an excess of water and would abstain from irrigation application.

The average impact of irrigation application for the 2050s simulations was an increase in annual average gross income of $12.10 to $14.40/hectare, depending on the radiative forcing scenario (Tables 5 and 6). However, due to the cost of irrigation and reservoir installation this left the farmer with an average gross margin of $146.00 to $148.00/hectare, dependent on radiative forcing scenario, each year to cover the reservoir and irrigation infrastructure and operation costs. For the 2090s simulations, the average impact of irrigation application was an increase in annual average gross income of $11.80 to $14.40/hectare, decreasing as radiative forcing increased. This left the farmer with an average net cost of $146.00 to $148.00/hectare each year, increasing as radiative forcing increased, to cover the reservoir and irrigation infrastructure and operation costs. Therefore, although the availability of irrigation water did increase crop production, the increased gross income was insufficient to offset the costs of the irrigation water.

Yearly gross margins with and without irrigation, yearly precipitation amounts, and reservoir volumes are provided in Figures 7 and 8 for the 2050s and 2090s, respectively. There was one year under all radiative forcing scenarios in each simulation period, 2057 and 2097, which required the total reservoir storage volume for irrigation application. In addition, 2091 under RCP2.6 required the total reservoir storage volume for irrigation application. This year experienced the lowest precipitation levels of all years and radiative forcing simulations, which required substantial irrigation application to optimize crop yield. While initial reservoir volumes and irrigation withdrawal volumes varied between radiative forcing scenarios, during the course of the growing seasons the reservoir always filled to capacity, allowing for irrigation withdrawals. As the MESH simulations used climate data from 2005 to 2014, variability in precipitation and temperature reflected that time period.

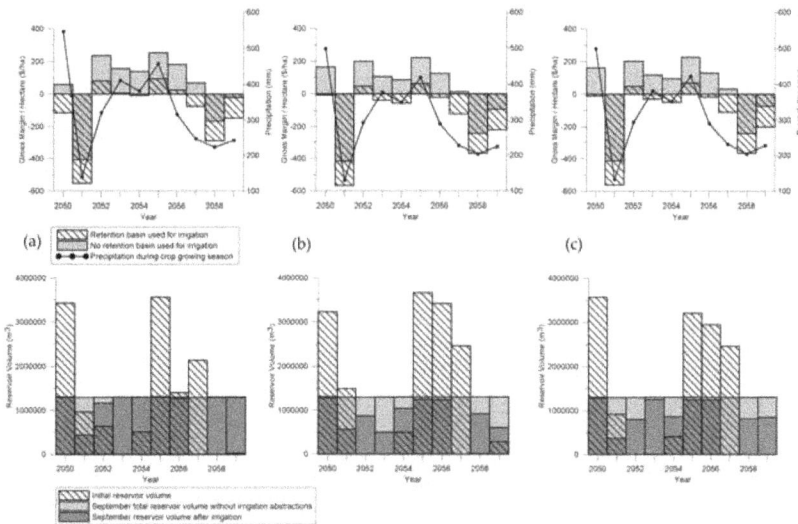

Figure 7. Yearly 2050–2059 crop gross margins with and without irrigation application and yearly water availability using incremental precipitation and temperature increases for: (**a**) Representative concentration pathway (RCP)2.6; (**b**) RCP4.5; and (**c**) RCP8.5. Reservoir levels for: (**a**) RCP2.6; (**b**) RCP4.5; and (**c**) RCP8.5 are also provided.

Results of this study can be explained by the precipitation and temperature increases experienced in the 2050s and 2090s. Incremental increases in temperature and precipitation during the 2090s increased from RCP2.6 to RCP8.5. These changes to temperature and precipitation caused crop gross income under irrigation to decrease from RCP2.6 to RCP8.5 and excess water overwhelmed the water capacity of the crops. Precipitation changes during the 2050s were less consistent with increases in radiative forcing. The 2050s experienced the largest increase in summer precipitation and smallest increase to winter precipitation under RCP2.6. Subsequently, the pattern of crop gross income increases does not align with increases in radiative forcing scenarios. Instead, RCP4.5 provided the highest gross income under irrigation and RCP2.6 provided the smallest increase to crop gross income under irrigation. Winter precipitation, which impacts initial reservoir volumes, increased for all scenarios except RCP2.6 for the 2090s. This however, did not provide enough water to fill the reservoir in all years. The increased summer precipitation levels did ensure the reservoir always filled to capacity by the end of the growing season. The variation in temperature and precipitation under different radiative forcing scenarios and time periods were not significant enough to have a significant impact on gross margins. Based on these simulation results, irrigation and reservoir installation at Pelly's Lake are not economically viable and do not generate positive crop gross margins in the middle or end of the century.

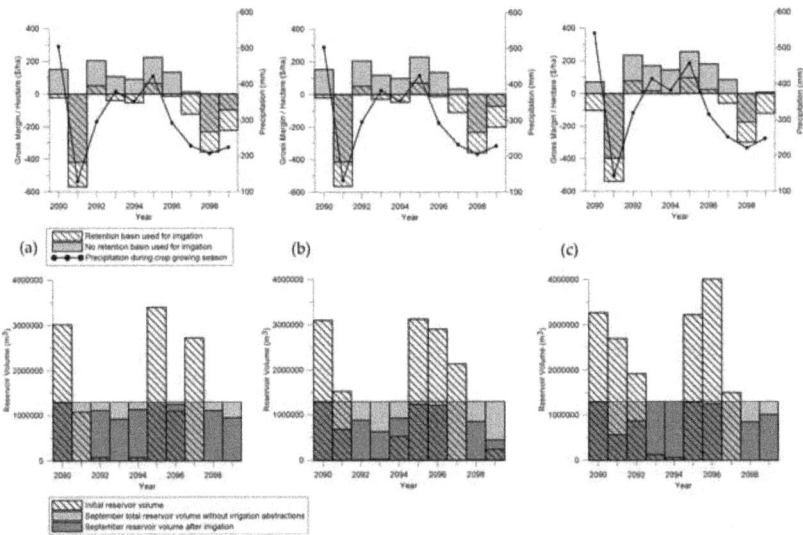

Figure 8. Yearly 2090–2099 crop gross margins with and without irrigation application and yearly water availability using incremental precipitation and temperature increases for: (a) Representative concentration pathway (RCP)2.6; (b) RCP4.5; and (c) RCP8.5. Reservoir levels for: (a) RCP2.6; (b) RCP4.5; and (c) RCP8.5 are also provided.

4. Discussion

The economic advantages of multi-purpose local farm retention pond systems and their use for irrigation under future climatic conditions were investigated using a dynamic modeling system. The middle of the century, 2050–2059, and the end of the century, 2090–2099, were simulated under three radiative forcing scenarios, RCP2.6, RCP4.5, and RCP8.5. The multi model ensemble future climate scenarios indicated precipitation increases will occur in the 2050s and 2090s under each RCP scenario, with the exception of winter precipitation under RCP2.6 for the 2090s, which decreased. Precipitation increases were higher for winter than summer. These results are supported by several studies which have predicted a general trend of increasing precipitation over Canada with higher projected increases to precipitation for winter months [1,3–5,72,73].

Using the delta method, incremental changes to temperature and precipitation from the present day period (2005–2014) were used to simulate future climate conditions. This method allowed for comparison of how changes in water volumes will impact irrigation water availability and subsequent crop gross margin. Due to the high costs of irrigation infrastructure and maintenance, irrigated cropping systems utilizing water abstractions from the reservoir experienced a decrease in gross margin when compared to gross margins without irrigation for both simulation periods under all RCP scenarios. To cover the costs of the irrigation and reservoir infrastructure, the farmer would have to earn incremental gross margins of $146.00 to $148.00/hectare of crop land/year in the 2050s and 2090s. Compared to present day simulations, the cost of irrigation and reservoir infrastructure to the farmer did not change under predicted future climate conditions [20]. However, increases in crop yield under irrigation were more consistent and increases in spring runoff and precipitation provided more stable reservoir water volumes which increased irrigation water availability.

The predicted increases in reservoir water volumes indicates that flooding may become more severe in the future, increasing multi-purpose surface water retention systems importance and value for flood reduction. However, this increase in reservoir water volumes also points to an opportunity for water storage for irrigation application. As future climate change is predicted to increase the severity and duration of floods and droughts [4,7,9,11], the ability to capture surface runoff in times of flood also provides water stores to draw on during times of drought. While irrigation installation remains costly, it is still an important adaptation strategy to reduce agricultural risk during times of drought. Policy providing irrigation subsidization may be implemented in the future to increase its adoption, in which case knowledge of strategies that provide sufficient water sources to utilize for irrigation application will be important.

Based on the findings of this research, multi-purpose retention basins are not economically justified when considering only the irrigation benefits provided to the participating farmers. However, a range of other benefits may be provided by the water retention system. As illustrated by Berry [20], Grosshans [17], and Dion and McCandless [35], multi-purpose retention systems provide significant economic benefits when you consider the ecosystem services they provide. Multi-purpose retention systems provide avoided flood damages downstream, sequester carbon, and reduce downstream nutrient and sediment runoff. There was substantial flow produced from spring melt that exceeded the capacity of the modelled reservoir under climate simulations for the middle and end of this century (Figure 7). A network of several multi-purpose retention systems, similar to the installed network in the South Tobacco Creek Watershed, may be required to deal with the future increases in spring runoff volumes. As part of the retention system network in the South Tobacco Creek Watershed in Manitoba, a multi-purpose dam reduced peak flow caused by spring snowmelt by an average of 72% per year, with a range of 38% to 100% peak flow reduction/year. Summer rainfall generated peak flow was reduced an average of 48% per year by the same multi-purpose dam. There is discussion regarding the construction of a second upstream reservoir at the Pelly's Lake retention site. This would increase the systems storage capacity by 1,600,000 m^3. As a result, the retention system would retain the predicted future volumes of spring melt runoff (Figure 9).

In additional to the economic benefits associated with avoided flood damages downstream, multi-purpose surface water retention systems can provide a biomass source. Harvesting cattail for biomass production and nutrient removal from the flooded area of multi-purpose retention systems is being commercialized in Manitoba and can provide monetary benefits from the production of carbon credits and bio products [17,28,34]. These ecosystem benefits of multi-purpose retention systems can provide additional economic revenue to compensate for the cost of the reservoir and irrigation infrastructure.

Using multi-purpose retention basins for avoided flood damages, nutrient retention, and biomass production also allows the public to benefit from these systems. The province of Manitoba has expressed interest in retention basins as a nutrient abatement option as part of their commitment to reducing downstream nutrient loading. The Manitoba Surface Water Management Strategy [14] states that water storage and associated release strategies should optimize production and harvest of biomass resources to remove phosphorus from the aquatic environment. As the South Tobacco Creek Watershed in Manitoba has illustrated, a series of retention systems on the Manitoba landscape has the potential to reduce downstream loading of phosphorus and nitrogen. Over a nine year period from 1999–2007, the retention system network decreased downstream nutrient loading above the Manitoba government's targets of 10% and 13% for phosphorus and nitrogen, respectively [21]. As the average phosphorus and nitrogen concentrations in the watershed were still in excess of recommended levels in the Canadian Prairies, Tiessen et al. [21] suggested using the reservoirs for local benefits, such as irrigation, to reduce downstream nutrient loading further. With the addition of cattail harvest, downstream loading of phosphorus and nitrogen would be further reduced [14,34,74,75]. Additionally, the removal of phosphorus during cattail harvest increases the wetlands ability to store more phosphorus, reducing downstream loading. This is essential for combating algal blooms and increasing water quality in aquatic environments such as Lake Winnipeg, Manitoba [34].

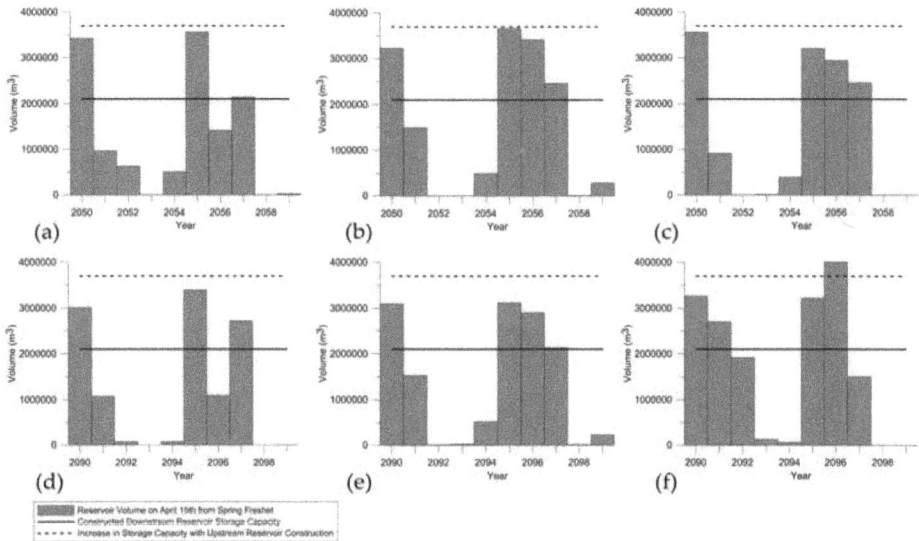

Figure 9. Reservoir capacities and initial reservoir volumes for future radiative forcing scenarios for the 2050s time period: (**a**) Representative concentration pathway (RCP)2.6; (**b**) RCP4.5; (**c**) RCP8.5; and the 2090s time period: (**d**) RCP2.6; (**e**) RCP4.5; (**f**) RCP8.5.

The current strategy for quickly removing water from the Manitoba landscape, via a series of ditches and drains, increases downstream flood peaks and decreases water quality. This method is only sustainable when there is adequate access to water and land use practices do not create nutrient pollution issues [3]. This quick drainage is already proving problematic for downstream nutrient loading into Lake Winnipeg in Manitoba. Future predictions of increased spring runoff volumes indicate increased issues with this strategy due to increased downstream flood peaks and increased nutrient loading. Moving forward, investing in multi-purpose local farm retention systems decreases flood peaks, increases water quality, while also providing water security during times of drought, as well as opportunities for biomass production and irrigation development. The reductions in phosphorus and nitrogen multi-purpose local farm retention systems can provide aid in Manitoba's goal of reducing nitrogen and phosphorus concentrations by 50% to Lake Winnipeg [14]. Rural municipalities and landowners benefit from the savings associated with avoided flooding damages while the province of Manitoba and its population benefit from the reduction to downstream nutrient loading and carbon storage providing climate regulation. Due to the economic and environmental gains multi-purpose retention systems provide to the province, subsidies could also be provided to incentivize widespread adoption.

Future Directions

The modelling system developed for this research could easily be adapted to include additional reservoirs within the catchment area, enabling an analysis of the regional impacts. As the current study is localized, it is difficult to state how well retention systems would work throughout the Red River Valley landscape. Regionalization of the study would also allow for the calculation of flow reductions over a larger area due to the installation of multiple retention systems. Comparisons could then be drawn between the effectiveness of water retention systems vs. current drainage systems on the Red River Valley landscape. The modelling system could also be easily expanded to include additional modules of interest to the researcher. Water samples are being collected for Pelly's Lake, upstream and downstream of the reservoir. Inclusion of a module on sediment and nutrient levels would allow for a more accurate economic assessment based on the amount of phosphorus and nitrogen loading the retention system at Pelly's Lake is reducing.

5. Conclusions

This paper aimed to estimate the monetary benefits of multi-purpose local farm retention systems under future climate scenarios. When there was insufficient precipitation to allow for maximum crop growth during the study time periods, irrigation using water stored from spring runoff and rainfall events in the water retention system provided increased annual crop gross income. However, a loss in gross margin was estimated due to the costs of developing the retention and irrigation systems. Retention basins' additional capacity for avoided flood damages, nutrient retention, biomass production, and carbon sequestration provide substantial economic and environmental gains. The addition of a second reservoir basin to the Pelly's Lake retention system would accommodate the predicted increases to future spring runoff, reducing downstream flood damage. The value multi-purpose retention basins can provide provincially supports the adoption of multi-purpose local farm surface water retention systems as an effective water management strategy. The recommended use of multi-purpose local farm retention systems is for avoided flood damages, nutrient retention, and biomass production to support farmers wanting to invest in irrigation. This use will subsequently provide economic and environmental benefits for the government of Manitoba.

Acknowledgments: This work was supported by Environment Canada's Lake Winnipeg Basin Stewardship Fund, Growing Forward 2, and SSHRC which is gratefully acknowledged.

Author Contributions: Pamela Berry, Ken Belcher, and Karl-Erich Lindenschmidt conceived and designed the study. Pamela Berry performed the experiment. Fuad Yassin performed the MESH hydrologic modelling. Pamela Berry wrote the paper.

Conflicts of Interest: The authors declare no conflict of interest.

References

1. Warren, F.J.; Lemmon, D.S. (Eds.) *Synthesis; in Canada in a Changing Climate: Sector Perspectives on Impacts and Adaptation;* Government of Canada: Ottawa, ON, Canada, 2014. Available online: http://www.nrcan.gc.ca/environment/resources/publications/impacts-adaptation/reports/assessments/2014/16309 (accessed on 3 January 2016).
2. Vincent, L.A.; Wang, X.L.; Milewska, E.J.; Wan, H.; Yang, F.; Swail, V. A second generation of homogenized Canadian monthly surface air temperature for climate trend analysis. *J. Geophys. Res. Atmos.* **2012**, *117*, 1–13. [CrossRef]
3. Venema, H.D.; Oborne, B.; Neudoerffer, C. *The Manitoba Challenge: Linking Water and Land Management for Climate Adaptation;* International Institute for Sustainable Development: Winnipeg, MB, Canada, 2010.
4. Bonsal, B.R.; Wheaton, E.E.; Chipanshi, A.C.; Lin, C.; Sauchyn, D.J.; Wen, L. Drought research in Canada: A review. *Atmos. Ocean* **2011**, *49*, 303–319. [CrossRef]
5. Intergovernmental Panel on Climate Change (IPCC). *Climate Change 2007: Synthesis Report. Contribution of Working Groups I, II, and III to the Fourth Assessment Report of the Intergovernmental Panel on Climate Change;* Core Writing Team, Pachauri, R., Reisinger, A., Eds.; IPCC: Geneva, Switzerland, 2007.

6. Hearne, R.R. Evolving water management institutions in the Red River Basin. *Environ. Manag.* **2007**, *40*, 842–852. [CrossRef] [PubMed]
7. Wheaton, E.; Kulshreshtha, S.; Wittrock, V.; Koshida, G. Dry times: Hard lessons from the Canadian drought of 2001 and 2002. *Can. Geogr.* **2008**, *52*, 241–262. [CrossRef]
8. Wall, E.; Smit, B. Climate Change Adaptation in Light of Sustainable Agriculture. *J. Sustain. Agric.* **2005**, *27*, 113–123. [CrossRef]
9. Pittman, J.; Wittrock, V.; Kulshreshtha, S.; Wheaton, E. Vulnerability to climate change in rural Saskatchewan: Case study of the Rural Municipality of Rudy No. 284. *J. Rural Stud.* **2011**, *27*, 83–94. [CrossRef]
10. Belcher, K. Agroecosystem Sustainability: An Integrated Modeling Approach. Ph.D. Thesis, University of Saskatchewan, Saskatoon, SK, Canada, 1999.
11. Samarawickrema, A.; Kulshreshtha, S. Value of irrigation water for drought proofing in the South Saskatchewan River Basin (Alberta). *Can. Water Resour. J.* **2008**, *33*, 273–282. [CrossRef]
12. Bower, S.S. Watersheds: Conceptualizing Manitoba's Drained Landscape, 1985–1950. *Environ. Hist. Durh. N. C.* **2007**, *12*, 796–819. [CrossRef]
13. La Salle Redboine Conservation District. La Salle River Watershed: State of the Watershed Report, 2007. pp. 1–296. Available online: https://www.gov.mb.ca/waterstewardship/iwmp/whitemud/documentation/summary_lasalle.pdf (accessed on 23 November 2015).
14. Government of Manitoba. *Surface Water Management Strategy*; Government of Manitoba: Winnipeg, MB, Canada, 2014. Available online: http://gov.mb.ca/waterstewardship/questionnaires/surface_water_management/pdf/surface_water_strategy_final.pdf (accessed on 1 October 2015).
15. Manitoba Government. Conservation and Water Stewardship. Available online: http://mli2.gov.mb.ca/ (accessed on 1 September 2014).
16. Pavelic, P.; Srisuk, K.; Saraphirom, P.; Nadee, S.; Pholkern, K.; Chusanathas, S.; Munyou, S.; Tangsutthinon, T.; Intarasut, T.; Smakhtin, V. Balancing-out floods and droughts: Opportunities to utilize floodwater harvesting and groundwater storage for agricultural development in Thailand. *J. Hydrol.* **2012**, *470–471*, 55–64. [CrossRef]
17. Grosshans, R.E.; Gass, P.; Dohan, R.; Roy, D.; Venema, H.D.; McCandless, M. *Cattail Harvesting for Carbon Offsets and Nutrient Capture: A Lake Friendly Greenhouse Gas Project*; International Institute for Sustaianble Development: Winnipeg, MB, Canada, 2012; Available online: http://www.iisd.org/library/cattails-harvesting-carbon-offsets-and-nutrient-capture-lake-friendly-greenhouse-gas-project (accessed on 3 November 2015).
18. Government of Canada. *Agriculture and Agri-Food Canada Positive Effects of Small Dams and Reservoirs: Water Quality and Quantity Findings from a Prairie Watershed*; WEBS Fact Sheet #7; Government of Canada: Ottawa, ON, Canada, 2012. Available online: http://www.agr.gc.ca/eng/?id=1351881784186 (accessed on 24 April 2016).
19. Arnold, J.G.; Stockle, C.O. Simulation of Supplemental Irrigation from On-Farm Ponds. *J. Irrig. Drain. Eng.* **1991**, *117*, 408–424. [CrossRef]
20. Berry, P. An economic Assessment of On-Farm Surface Water Retention Systems. Ph.D. Thesis, University of Saskatchewan, Saskatoon, SK, Canada, 2016.
21. Tiessen, K.H.D.; Elliott, J.A.; Stainton, M.; Yarotski, J.; Flaten, D.N.; Lobb, D.A. The effectiveness of small-scale headwater storage dams and reservoirs on stream water quality and quantity in the Canadian Prairies. *J. Soil Water Conserv.* **2011**, *66*, 158–171. [CrossRef]
22. Sharpley, A.; Smith, S.J.; Zollweg, J.A.; Coleman, G.A. Gully treatment and water quality in the Southern Plains. *J. Soil Water Conserv.* **1996**, *51*, 498–503.
23. Kovacic, D.A.; Twait, R.M.; Wallace, M.P.; Bowling, J.M. Use of created wetlands to improve water quality in the Midwest-Lake Bloomington case study. *Ecol. Eng.* **2006**, *28*, 258–270. [CrossRef]
24. Salvia-Castellvi, M.; Dohet, A.; Vander Borght, P.; Hoffmann, L. Control of the eutrophication of the reservoir of Esch-sur-Sûre (Luxembourg): Evaluation of the phosphorus removal by predams. *Hydrobiologia* **2001**, *459*, 61–71. [CrossRef]
25. Avilés, A.; Niell, F.X. The control of a small dam in nutrient inputs to a hypertrophic estuary in a Mediterranean climate. *Water Air Soil Pollut.* **2007**, *180*, 97–108. [CrossRef]
26. Uusi-Kämppä, J.; Braskerud, B.; Jansson, H.; Syversen, N.; Uusitalo, R. Buffer zones and constructed wetlands as filters for agricultural phosphorus. *J. Environ. Qual.* **2000**, *29*, 151–158. [CrossRef]

27. Chrétien, F.; Gagnon, P.; Thériault, G.; Guillou, M. Performance analysis of a wet-retention pond in a small agricultural catchment. *J. Environ. Eng.* **2016**, *142*, 4016005. [CrossRef]

28. Manitoba Liquor and Lotteries Corporation. Environmental Innovation. Available online: http://www.mbll.ca/content/environmental-innovation (accessed on 12 December 2016).

29. Brander, L.; Brouwer, R.; Wagtendonk, A. Economic valuation of regulating services provided by wetlands in agricultural landscapes: A meta-analysis. *Ecol. Eng.* **2013**, *56*, 89–96. [CrossRef]

30. Olewiler, N. *The Value of Natural Capital in Settled Areas of Canada*; Ducks Unlmited and the Nature Conservancy of Canada: Stonewall, MB, Canada, 2004; Available online: http://www.cmnbc.ca/sites/default/files/natural%2520capital_0.pdf (accessed on 5 May 2016).

31. Wilson, S. *Ontario's Wealth, Canada's Future: Appreciating the Value of the Greenbelt's Eco-Services*; David Suzuki Foundation: Vancouver, BC, Canada, 2008. Available online: http://www.davidsuzuki.org/publications/downloads/2008/DSF-Greenbelt-web.pdf (accessed on 5 May 2016).

32. Sohngen, B.; King, K.W.; Howard, G.; Newton, J.; Forster, D.L. Nutrient prices and concentrations in Midwestern agricultural watersheds. *Ecol. Econ.* **2015**, *112*, 141–149. [CrossRef]

33. Collins, A.R.; Gillies, N. Constructed Wetland Treatment of Nitrates: Removal Effectiveness and Cost Efficiency. *J. Am. Water Resour. Assoc.* **2014**, *50*, 898–908. [CrossRef]

34. Grosshans, R.; Grieger, L.; Ackerman, J.; Gauthier, S.; Swystun, K.; Gass, P.; Roy, D. *Cattail Biomass in a Watershed-Based Bioeconomy: Commercial-Scale Harvesting and Processing for Nutrient Capture, Biocarbon and High-Value Bioproducts*; International Institute for Sustainable Development: Winnipeg, MB, Canada, 2014. Available online: http://www.iisd.org/sites/default/files/publications/cattail-biomass-watershed-based-bioeconomy-commerical-scale-harvesting.pdf (accessed on 17 January 2015).

35. Dion, J.; McCandless, M. *Cost-Benefit Analysis of Three Proposed Distributed Water Storage Options for Manitoba*; International Institute for Sustainable Development: Winnipeg, MB, Canada, 2013. Available online: http://www.iisd.org/pdf/2014/mb_water_storage_options.pdf (accessed 23 December 2014).

36. McMinn, W.R.; Yang, Q.; Scholz, M. Classification and assessment of water bodies as adaptive structural measures for flood risk management planning. *J. Environ. Manag.* **2010**, *91*, 1855–1863. [CrossRef] [PubMed]

37. Government of Canada. *Government of Canada*; Statistics Canada; Government of Canada: Ottawa, ON, Canada, 2015; pp. 2–6. Available online: www.statcan.gc.ca/eng/start (accessed on 15 September 2015).

38. La Salle Redboine Conservation District. *Pelly's Lake Backflood Project*; La Salle Redboine Conservation District: Holland, MB, Canada, 2013; pp. 1–2. Available online: www.lasalleredboine.com/pellys-lake-backflood-project (accessed on 28 August 2015).

39. La Salle Redboine Conservation District. About. Available online: http://www.lasalleredboine.com/ (accessed on 24 November 2015).

40. Armstrong, R.N.; Pomeroy, J.W.; Martz, L.W. Estimating evaporation in a prairie landscape under drought conditions. *Can. Water Resour. J.* **2010**, *35*, 173–186. [CrossRef]

41. Pomeroy, J.; Fang, X.; Williams, B. *Modelling Snow Water Conservation on the Canadian Prairies*; University of Saskatchewan: Saskatoon, SK, Canada, 2011. Available online: http://www.usask.ca/hydrology/reports/CHRpt11_Prairie-Shelterbelt-Study_Apr11.pdf (accessed on 5 February 2015).

42. Costanza, R.; Duplisea, D.; Kautsky, U. Ecological Modelling on modelling ecological and economic systems with STELLA. *Ecol. Model.* **1998**, *110*, 1–4.

43. Pietroniro, A.; Fortin, V.; Kouwen, N.; Neal, C.; Turcotte, R.; Davison, B.; Verseghy, D.; Soulis, E.D.; Caldwell, R.; Evora, N.; et al. Development of the MESH modelling system for hydrological ensemble forecasting of the Laurentian Great Lakes at the regional scale. *Hydrol. Earth Syst. Sci.* **2007**, *11*, 1279–1294. [CrossRef]

44. Verseghy, D.L. CLASS—A Canadian Land Surface Scheme for GCMS. I. Soil Model. *Int. J. Climatol.* **1991**, *11*, 111–133. [CrossRef]

45. University of Saskatchewan. MEC—Surface and Hydrology (MESH). Available online: http://www.usask.ca/ip3/models1/mesh.htm (accessed on 15 October 2015).

46. Verseghy, D.L.; McFarlane, N.A.; Lazare, M. CLASS—A Canadian Land Surface Scheme for GCMS. II. Vegetation Model and Coupled Runs. *Int. J. Climatol.* **1993**, *13*, 347–370. [CrossRef]

47. Mahfouf, J.-F.; Brasnett, B.; Gagnon, S. A Canadian Precipitation Analysis (CaPA) project: Description and preliminary results. *Atmos. Ocean* **2007**, *45*, 1–17. [CrossRef]

48. Côté, J.; Desmarais, J.-G.; Gravel, S.; Methot, A.; Patoine, A.; Roch, M.; Staniforth, A. The operational CMC—MRB Global Environmental Multiscale (GEM) Model. Part II: Results. *Mon. Weather Rev.* **1998**, *126*, 1397–1418. [CrossRef]

49. Kouwen, N.; ASCE, M.; Soulis, E.D.; Pietroniro, A.; Donald, J.; Harrington, R.A. Grouped Response Units for Distributed Hydrologic Modeling. *J. Water Resour. Plan. Manag.* **1993**, *119*, 289–305. [CrossRef]

50. Government of Manitoba. *Guidelines for Estimating Crop Production Costs 2015*; Government of Manitoba: Winnipeg, MB, Canada, 2015. Available online: https://www.gov.mb.ca/agriculture/business-and-economics/financial-management/cost-of-production.html (accessed on 15 January 2015).

51. Statistics Canada. The Financial Picture of Farms in Canada. Available online: http://www.statcan.gc.ca/ca-ra2006/articles/finpicture-portrait-eng.htm (accessed on 25 January 2017).

52. Waelti, C.; Spuhler, D. Retention Basin. Available online: http://www.sswm.info/content/retention-basin (accessed on 14 August 2014).

53. Grinder, B. *Alberta's Irrigation Infrastructure*; Government of Alberta: Lethbridge, AB, Canada, 2000. Available online: http://www1.agric.gov.ab.ca/\protect\T1\textdollardepartment/deptdocs.nsf/all/irr7197 (accessed on 3 March 2015).

54. Government of Manitoba. Personal communication, 2015.

55. Government of Manitoba. *Manitoba Land Initiative: Core Maps—Data Warehouse*; Government of Manitoba: Winnipeg, MB, Canada. Available online: http://mli2.gov.mb.ca/mli_data/index.html (accessed on 23 September 2014).

56. Government of Canada. Historical Climate Data. Available online: http://climate.weather.gc.ca/ (accessed on 15 February 2015).

57. Belcher, K.W.; Boehm, M.M.; Fulton, M.E. Agroecosystem sustainability: A system simulation model approach. *Agric. Syst.* **2004**, *79*, 225–241. [CrossRef]

58. Werner, A.T. *BCSD Downscaled Transient Climate Projections for Eight Select GCMs over British Columbia, Canada*; Pacific Climate Impacts Consortium, University of Victoria: Victoria, BC, Canada, 2011. Available online: https://pdfs.semanticscholar.org/5dc0/a283b244d177e30c12e1d48d8a227f3720f9.pdf (accessed on 4 January 2016).

59. Pacific Climate Impacts Consortium Statistically Downscaled Climate Scenarios. Available online: https://www.pacificclimate.org/data/statistically-downscaled-climate-scenarios (accessed on 1 July 2016).

60. Randall, D.A.; Wood, R.A.; Bony, S.; Colman, R.; Fichefet, T.; Fyfe, J.; Kattsov, V.; Pitman, A.; Shukla, J.; Srinivasan, J.; et al. Climate Models and Their Evaluation. In *Climate Change 2007: The Physical Science Basis. Contribution of Working Group 1 to the Fourth Assessment Report of the Intergovernmental Panel on Climate Change*; Soloman, S., Qin, D., Manning, M., Chen, Z., Marquis, M., Averyt, K.B., Tignor, M., Miller, H.L., Eds.; Cambridge University Press: Cambridge, UK; New York, NY, USA, 2007; p. 996.

61. Schnorbus, M.; Werner, A.; Bennett, K. Impacts of climate change in three hydrologic regimes in British Columbia, Canada. *Hydrol. Process.* **2014**, *28*, 1170–1189. [CrossRef]

62. Hamlet, A.F.; Salathé, E.P.; Carrasco, P. *Statistical Downscaling Techniques for Global Climate Model Simulations of Temperature and Precipitation with Application to Water Resources Planning Studies*; Chapter 4 in Final Report for the Columbia Basin Climate Change Scenarios Project; Climate Impacts Group, Center for Science in the Earth System, Joint Institute for the Study of the Atmosphere and Ocean, University of Washington: WA, USA, 2010. Available online: https://cig.uw.edu/publications/statistical-downscaling-techniques-for-global-climate-model-simulations-of-temperature-and-precipitation-with-application-to-water-resources-planning-studies/ (accessed on 27 December 2015).

63. Werner, A.T.; Cannon, A.J. Hydrologic extremes—An intercomparison of multiple gridded statistical downscaling methods. *Hydrol. Earth Syst. Sci.* **2016**, *20*, 1483–1508. [CrossRef]

64. Shrestha, R.; Schnorbus, M.A.; Werner, A.T.; Zwiers, F.W. Evaluating Hydroclimatic Change Signals from Statistically and Dynamically Downscaled GCMs and Hydrologic Models. *J. Hydrometeorol.* **2014**, *15*, 844–860. [CrossRef]

65. Barrow, E. Climate Change Scenarios. In *The Availability, Characteristics and Use of Climate Change Scenarios*; Prairie Adaptation Research Collaborative: Regina, SK, Canada, 2001. Available online: http://www.parc.ca/pdf/conference_proceedings/jan_01_barrow1.pdf (accessed on 1 February 2017).

66. Diaz-Nieto, J.; Wilby, R.L. A comparison of statistical downscaling and climate change factor methods: Impacts on low flows in the River Thames, United Kingdom. *Clim. Chang.* **2005**, *69*, 245–268. [CrossRef]

Sustainability **2017**, *9*, 456

67. Graham, L.P.; Andreáasson, J.; Carlsson, B. Assessing climate change impacts on hydrology from an ensemble of regional climate models, model scales and linking methods—A case study on the Lule River basin. *Clim. Chang.* **2007**, *81*, 293–307. [CrossRef]

68. Chen, J.; Brissette, F.P.; Leconte, R. Uncertainty of downscaling method in quantifying the impact of climate change on hydrology. *J. Hydrol.* **2011**, *401*, 190–202. [CrossRef]

69. Trzaska, S.; Schnarr, E. *A Review of Downscaling Methods for Climate Change Projections*; Tetra Tech ARD: Burlington, VT, USA, 2014.

70. Intergovernmental Panel on Climate Change—Task Group on Data and Scenario Support for Impact and Climate Analysis (IPCC-TGICA). *General Guidelines on the Use of Scenario Data for Climate Impact and Adaptation Assessment*; Intergovernmental Panel on Climate Change: Geneva, Switzerland, 2007; Volume 2. Available online: http://www.ipcc-data.org/guidelines/TGICA_guidance_sdciaa_v2_final.pdf (accessed on 16 January 2016).

71. Charron, I. *A Guidebook on Climate Scenarios: Using Climate Information to Guide Adaptation Research and Decisions*; Ouranos: Montreal, QC, Canada, 2014. Available online: https://www.ouranos.ca/publication-scientifique/GuideCharron2014_EN.pdf (accessed on 5 January 2016).

72. Nyirfa, W.; Harron, B. *Assessment of Climate Change on the Agricultural Resources of the Canadian Prairies*; Prairie Adaptation Research Collaborative: Regina, SK, Canada, 2001. Available online: http://www.parc.ca/pdf/research_publications/agriculture4.pdf (accessed on 13 February 2016).

73. Sauchyn, D.J.; Barrow, E.M.; Hopkinson, R.F.; Leavitt, P.R. Aridity on the Canadian Plains. *Géogr. Phys. Quat.* **2002**, *56*, 247–259. [CrossRef]

74. Bourne, A.; Armstrong, N.; Jones, G. *A Preliminary Estimate of Total Nitrogen and Total Phosphorus Loading to Streams in Manitoba, Canada*; Government of Manitoba: Winnipeg, MB, Canada, 2002. Available online: http://www.gov.mb.ca/sd/eal/registries/4864wpgww/mc_nitrophosload.pdf (accessed 10 March 2016).

75. Lake Winnipeg Stewardship Board. *Reducing Nutrient Loading to Lake Winnipeg and Its Watershed: Our Collective Responsibility and Commitment to Action*; Government of Manitoba: Winnipeg, MB, Canada, 2006. Available online: https://www.gov.mb.ca/waterstewardship/water_quality/lake_winnipeg/lwsb2007-12_final_rpt.pdf (accessed on 5 April 2016).

sustainability

MDPI

Article

Financing High Performance Climate Adaptation in Agriculture: Climate Bonds for Multi-Functional Water Harvesting Infrastructure on the Canadian Prairies

Anita Lazurko [1] and Henry David Venema [2,*]

[1] Department of Environmental Sciences and Policy, Central European University, Budapest 1051, Hungary; anita.lazurko@mespom.eu

[2] Prairie Climate Centre, International Institute for Sustainable Development, Winnipeg, MB R3B 0T4, Canada

* Correspondence: hvenema@iisd.ca

Received: 3 May 2017; Accepted: 10 July 2017; Published: 14 July 2017

Abstract: International capital markets are responding to the global challenge of climate change, including through the use of labeled green and climate bonds earmarked for infrastructure projects associated with de-carbonization and to a lesser extent, projects that increase resilience to the impacts of climate change. The potential to apply emerging climate bond certification standards to agricultural water management projects in major food production regions is examined with respect to a specific example of multi-functional distributed water harvesting on the Canadian Prairies, where climate impacts are projected to be high. The diverse range of co-benefits is examined using an ecosystem service lens, and they contribute to the overall value proposition of the infrastructure bond. Certification of a distributed water harvesting infrastructure bond under the Climate Bond Standard water criteria is feasible given climate bond issue precedents. The use of ecosystem service co-benefits as additional investment criteria are recommended as relevant bond certification standards continue to evolve.

Keywords: climate change; agriculture; climate bonds; investment; distributed infrastructure; water harvesting; Canada

1. Introduction

The political success achieved by the 2015 Paris Climate Accord with respect to a broad political consensus to reduce greenhouse gas emissions and accelerate adaptation to climate change, was followed by further political commitments in 2016 to increase climate financing. The 2016 G20 Hangzhou Leader's summit communique stated, "We believe efforts could be made to … provide clear strategic policy signals and frameworks, promote voluntary principles for green finance, support the development of local green bond markets and promote international collaboration to facilitate cross-border investment in green bonds" [1].

The G20 leaders expressed support for a well-established trend—the rise of a new class of labeled infrastructure investment bond aligned with de-carbonization and climate de-risking objectives. Between 2011 and the 2015, the volume of "green" or "climate" labelled bonds issued increased from $3 billion to $95 billion, a large increase but still a small fraction of the estimated $93 trillion infrastructure investment requirements frequently cited as necessary to meet Paris accord objectives of limiting global warming to under 2 °C [2].

The large majority of labeled green and climate bonds have been designated for renewable energy, energy efficiency and low-carbon transport. In 2015 these sectors comprised 79% of the value of bond issues [3], whereas bonds specifically designated for climate adaptation had only a 4% market share—despite compelling evidence that investments in adaptation can provide very high rates of return [4]. The underlying issue is that although climate change is a global issue and its mitigation requires collective global action, climate change impacts are inherently localized and adaptation is necessarily a granular design process requiring highly localized climatic, socio-economic and ecosystem information—a challenge for harnessing the larger scale investment flows commensurate with the scale of the opportunity. In addition, bond financing requires that a large number of relatively small individual projects be aggregated to reach a sufficient scale. The scale at which local adaptation projects require financing is typically two to four orders of magnitude lower than the scale at which bonds are issued [3].

The Canadian Prairies are an interesting geographic context to analyse the logic for increasing market share for climate adaptation bonds and the associated challenges, by referencing the specific case of multi-functional water retention structures for agriculture. The Canadian Prairies comprise about 90% of Canada's agricultural land base, produce approximately 20% of internationally traded grains and oilseeds and thus are an important component of world food security. The Canadian Prairies also have a history of high vulnerability to climate shock for anthropogenic and climatological reasons, and a history of innovative ecosystem and water resources management based on distributed water harvesting (DWH) that could be revived in the context of climate adaptation [5]. Berry et al. [6] review a multi-purpose surface water retention system at Pelly's Lake, in the Canadian Prairie province of Manitoba that illustrates the economic case for water harvesting. Berry et al. conclude that when all economic benefits are evaluated; flood and drought risk reduction, irrigation and other ecosystem service benefits, the net value of retention storage (more than CAD \$25,000/hectare) far exceeded its land value as conventional agriculture. Nonetheless, the total investment requirement for this high performance, but highly local, climate adaptation project at under CAD \$1 million falls below the threshold for prioritization as conventional infrastructure spending. The urgency and logic for aggregating large numbers of such "precision infrastructure" projects for innovative climate financing through bond issues on the Canadian Prairies is, therefore, the focus of this paper.

This paper aims to explain and analyse the opportunity to finance high performance climate adaptation projects like multi-functional DWH infrastructure with certified climate bonds under the Water Criteria of the Climate Bond Standard, and to explore the concept of informing the project or bond value proposition with the economic value of ecosystem services and co-benefits. In addition, this paper aims to demonstrate the logic for aggregating a large number of relatively small projects to a scale appropriate for bond financing. This paper uniquely combines concepts and provides a new iteration upon leading solutions from seemingly disparate entities: engineers and scientists turning to distributed, localized, green infrastructure solutions, climate modelers increasingly understanding the importance of temporal variability and downscaling data to regional impacts, financers seeking to open new markets for green infrastructure and to find ways to aggregate localized projects into large-scale financing structures, and new entities like the Climate Bonds Initiative providing a new platform to set standards and increase visibility. The methodology of this paper includes articulating the direct benefits and enhanced ecosystem services of DWH solutions, presenting a general framework for a project and bond value proposition that aggregates those benefits using downscaled climate change data for assessing the value generated over future scenarios, and providing recommendations for the institutional, regulatory, and technical elements needed to finance this solution with government-issued bonds certified under the Water Criteria Climate Bond Standard Phase 1: Engineered Infrastructure [7]. This paper concludes with recommendations for implementation of DWH systems on the Canadian prairies and future development of CBS criteria for natural and semi-natural water infrastructure. The broad conclusions drawn in this report can be used to disseminate the DWH solution to other regions with similar climatic stressors and agricultural conditions.

2. Distributed Water Harvesting on the Canadian Prairies

2.1. Introduction to the Canadian Prairies

Climate change on the Canadian prairies manifests as temperature increases and changes to precipitation patterns that demand greater climate resilience in the agricultural sector. The size and shape of the continent of North America, its proximity to the Arctic Ocean, and other factors accelerate the climatic warming felt on the prairies. The Prairie Climate Centre has shown that Winnipeg may experience summer temperatures similar to the panhandle of Texas by the year 2080 [8]. The prairies are also vulnerable to precipitation changes, including an increase of spring precipitation and decrease of rainfall during the summer. Farmers will be forced to adapt their farming practices to stretch a variable hydrologic budget across a long, dry growing season. These rainfall challenges will be further exacerbated by the heightened temperatures through increased evapotranspiration rates [9]. In Saskatchewan and Manitoba, a large majority of agriculture is rain fed [10], and the patchwork of 150-acre quarter-sections of land separated by drainage ditches and culverts is designed to allow for limited groundwater percolation and rapid runoff into large reservoirs or natural water bodies. The use of fertilizer inputs in the region also results in accumulation of nutrients in runoff water and water bodies resulting in frequent eutrophication problems [11]. New precipitation patterns have already begun to strain the agriculture sector and government risk management practices, as seen during the Manitoba floods of 2011 [12]. Evidently, the current 'drainage culture' is in tension with the rainfall variability that will be introduced with the climatic pressures of the future, presenting the 21st century challenge of adaptation for farmers and governments.

2.2. The Engineered Solution

Multi-functional DWH infrastructure is a semi-natural climate adaptation solution that aims to overcome the climatic stresses that challenge the excessive drainage culture of agriculture in the region. It is a system of many small, controllable earthen dams that have been located and sequenced to enable control over current and future hydrologic cycles based on aggregated hydrologic and climate data. DWH mitigates floods in a similar manner to wetlands, but with a higher degree of control to overcome the risk of saturation and snow melt patterns that inhibit the ability of wetlands to buffer peak flows. By encouraging more groundwater percolation, maintaining a potentially higher groundwater table, and retaining standing water throughout the landscape, farmers will have the ability to access water during drought conditions. DWH is expected to have significantly less environmental disruption than hard infrastructure like dams and reservoirs, as well as a much lower infrastructure cost. Farmers upstream of the water harvesting system could have the option to drain their land more quickly to take advantage of early seeding dates, while farmers downstream of the system will be protected from seasonal flooding via controlled, intentional drainage patterns. Though innovative for the Canadian prairies, this solution is not new. India has met demand for seasonal water storage and lack of food security with similar technologies for millennia, though these systems were left abandoned or unmaintained in favor of groundwater irrigation in recent decades [13]. Sustainable development principles, cost-effectiveness, and environmental considerations are incenting a shift back toward such common-sense, localized solutions. Fortunately, the 21st century context of modern DWH systems presents new opportunities with this historic solution. For example, farmers may harvest biomass from "low spots" for energy generation, nutrient recovery, and profit, expanding the "bioeconomy" demonstrated in the Lake Winnipeg delta [14]. The multi-functional distributed water harvesting infrastructure as a climate adaptation solution inherently generates co-benefits and a business case at the intersection of the water–food–energy nexus.

2.3. Climate Change Adaptation and Enhanced Ecosystem Services

Climate change introduces new risks for governments, demanding innovative techniques for assessing and mitigating risk through adaptation. A higher frequency and severity of floods and droughts introduces significant challenges for governments, including infrastructure damage and loss of productivity in the agriculture sector. The 2011 floods in Manitoba caused CAD $1.2 billion of distributed infrastructure damage [12], triggering financial and stakeholder management challenges for the Province of Manitoba and the Government of Canada. Droughts may not directly cause property damage, but they have the potential to severely strain the agricultural sector and rural economies [15]. Assessing the impact of these climate change effects in terms of property and crop damage merely scratches the surface of the potential value of a well-managed flood mitigation and drought resilience program; assessing multiple dimensions of ecosystem services can highlight the full value of climate adaptation solutions. In addition, the economic valuation of such ecosystem services can inform a water pricing scheme that incorporates externalities and reflects full cost recovery [16], further incenting change toward water conservation and more appropriate water management. Robust assessments of risk and proposed value enable innovative solutions to emerge. These solutions demand resources, presenting the challenge of financing climate adaptation projects—a challenge insurance companies and the broader financial sector continue to grapple with. Balancing traditional institutional financing structures with the need to encourage granularity of high-performance adaptation projects informed by robust data and climate projections presents a unique design challenge for engineers, governments, and financers.

The main functional purpose of a multi-functional water harvesting system is to increase control over the hydrologic cycle to overcome climate change challenges to the agricultural sector. Climate change adaptation, a benefit derived from direct use of the infrastructure, is only part of the equation. An ecosystem services lens generates a more well-rounded picture of benefits and supporting services derived from DWH, generating a much stronger value proposition and informing better water management. Figure 1 depicts the network of potentially quantifiable climate change adaptation benefits and enhanced ecosystem services generated by a DWH system. The benefits in this figure could manifest similarly in different watersheds across the Canadian prairies, and so should be interpreted as a broad estimate of direct and co-benefits generated. In addition, this list of direct use and co-benefits could vary depending on the presence of agricultural irrigation or other climate adaptation measures in the region. The co-benefits in black typeface are significant and potentially quantifiable, while the co-benefits in grey typeface exist but are more difficult to quantify in economic terms in the value propositions described later in this paper. The following sections describe Figure 1 in more detail, which includes brief descriptions of the ecosystem services classified under the Millennium Ecosystem Assessment [17].

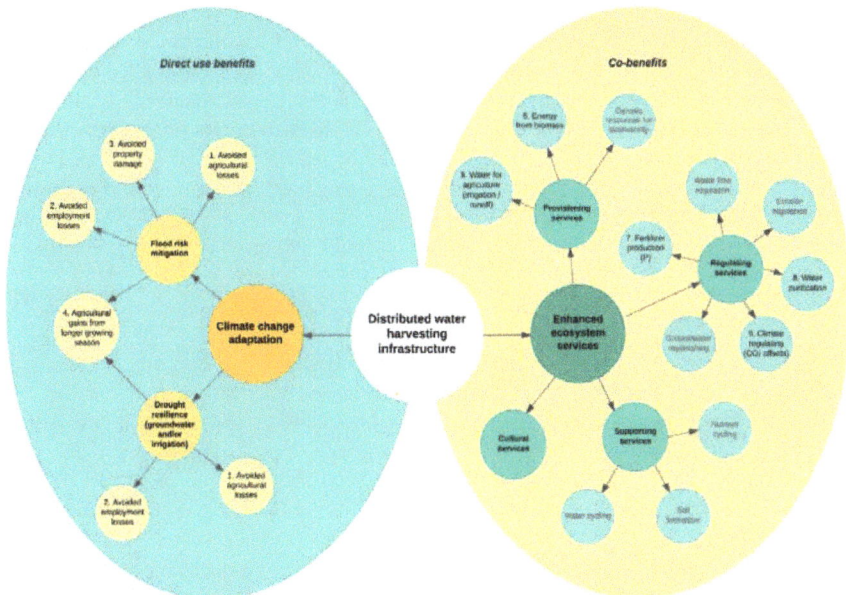

Figure 1. Distributed water harvesting infrastructure system as a network of direct use benefits and enhanced ecosystem services described in the following sections.

2.3.1. Flood Risk Mitigation and Drought Resilience

Climate change adaptation for flood risk mitigation and drought resilience can be easily connected to risk identification and management for governments and insurance entities. The need to consider climate change impacts, particularly property damages and crop loss but also ecosystem service benefits, will be increasingly important as governments begin to feel the monetary impacts. The flood risk mitigation benefit of the water harvesting system manages or avoids multiple hazards described in Figure 1, including agricultural losses due to loss of cultivable land or crop yield damages, property damages due to severe flood events or longer-term changes to the regional hydrology, and employment losses due to a decline in or local industry. The drought resiliency function of water harvesting systems manages similar hazards, including agricultural losses from lack of precipitation events that diminish crop yield and employment losses from reduced agricultural activity. DWH introduces the ability to control the hydrologic cycle with greater precision, presenting a valuable opportunity to increase crop yields with earlier seeding times and a longer growing season.

2.3.2. Provisioning Ecosystem Services

Provisioning ecosystem services are defined as 'the products obtained from ecosystems' [17]. These are the most relevant services provided in agriculture-based regions because of the direct economic benefit. Beyond agricultural crop yields, DWH may allow for provision of water for other uses such as irrigation or controlled runoff. The accumulation of biomass in low spots where water is retained by small earthen dams is an opportunity for farmers or private entities to harvest biomass seasonally for energy generation, similar to the bioeconomy of Lake Winnipeg [14]. This can lead to the secondary provisioning of phosphorus nutrients from the ash. Lastly, avoiding the environmental disruption of large dams and reservoirs may have a positive impact on the natural provision of biodiversity and genetic resources in the region, though this is difficult to quantify.

2.3.3. Regulating Ecosystem Services

Regulating ecosystem services are 'the benefits obtained from regulation of ecosystem processes' [17]. Water harvesting systems behave as a wetland during high water flow conditions, which can facilitate the natural purification of water and buffer peak water flows. Additional water purification functions are derived from biomass harvesting, by avoiding accumulation of phosphorus nutrients that are introduced to the landscape as chemical fertilizers in drainage basins. Water flow regulation is optimized by the higher degree of control over the hydrologic cycle facilitated by DWH systems. This flow regulation function may be a step away from the current, engineered drainage culture and closer to natural flow conditions, depending on the siting, sequencing, and control design of the system. Additional regulating services enhanced by the water harvesting infrastructure include erosion regulation from the more intentional drainage patterns and maintenance of the ground water table by encouraging more time for groundwater percolation.

2.3.4. Cultural Ecosystem Services

Cultural ecosystem services are 'the non-material benefits obtained from ecosystems', such as existence value, altruism, cultural benefits, educational value, and sense of place [17]. Because DWH is an engineering solution for a previously engineered landscape, it is very difficult to quantify the cultural services provided by this solution. However, opportunities may exist to derive cultural benefits, like educational value, if the systems are used intentionally by stakeholders in the social context.

2.3.5. Supporting Ecosystem Services

Supporting ecosystem services are 'the services necessary for the production of all other ecosystem services' [17]. For DWH, these supporting ecosystem services include the natural cycles enhanced by partially reversing or altering the current engineered drainage culture of the agricultural landscape on the Canadian prairies. This should improve the function of several supporting ecosystem services, including water cycling, nutrient cycling, and soil formation.

It is important to note that in addition to established monetary valuation techniques of many direct use and co-benefits, cultural ecosystem services are difficult to value in monetary terms. 'Willingness-to-pay' and related techniques have been used to justify monetary value of intangible assets. However, it cannot be assumed that an unwillingness to pay for an ecosystem service means that the service does not have value [18]. Several non-monetary valuation techniques exist, including Social Network Analysis, preference ranking, or the Q-methodology [18]. There is significant need for plural valuation that considers non-monetary value from such techniques alongside monetary values. However, until financing institutions are restructured to absorb such value into their more rigid frameworks, other important stakeholders may need to compromise and continue to use more easily quantified, less nuanced, monetary valuation techniques. The full list of ecosystem services depicted in Figures 1 and 2 is shown in Table 1.

Table 1. Key ecosystem services and monetization options from Figures 1 and 2.

Theme	Service	Examples of Service Monetization
Climate adaptation	Flood mitigation & drought resilience	Avoided agricultural losses *(estimated area loss x $ yield per unit area)* Avoided employment losses *(estimated job loss x employment insurance)* Avoided property damage *(estimated property damage as function of flood risk)* Crop yield increase from longer growing season *(Estimated yield increase x total affected area)*

Table 1. *Cont.*

Theme	Service	Examples of Service Monetization
Provisioning services	Irrigation water	Cost of equivalent agricultural irrigation *(Estimated irrigation costs for affected crop area)*
	Biomass harvesting	Cost of equivalent energy production *(Estimated energy from biomass x cost of alternative production)*
Regulating and supporting services	Nutrient cycling	Cost of purchasing chemical phosphorus fertilizers *(Estimated kg equivalent nutrient harvest from biomass x market price per kg)*
	Water purification	Cost of equivalent water treatment *(Estimated water quality improvement x cost of conventional water treatment methods)*
	CO_2 offsets	Cost of equivalent CO_2 offsets *(Estimated CO_2 offsets x price of carbon)*
Cultural services	Educational value, intrinsic natural value	Monetary valuation of cultural services *Willingness-to-pay* Non-monetary valuation of cultural services *Q-methodology, social network analysis, mental models,* etc.

2.4. The Design and Value Proposition of Climate Adaptation

Government risk management and strategic planning requires a balance of priorities. Robust quantification of the value proposition of climate change adaptation projects in economic terms, considering the direct benefits of flood risk mitigation and drought resilience, and the co-benefits of enhanced ecosystem services, can drive planning that reflects the multidimensional interests of society. This planning can feed into the project value proposition for DWH and better inform integrated water resource management via water pricing and other market-based mechanisms. The value proposition for DWH requires breaking down complexity and uncertainty with models informed by decades of detailed climate data that has been aggregated, downscaled to the appropriate region, and analyzed. The results of these models should quantify the difference between the impacts of future climate change scenarios with and without climate adaptation measures, such as a proposed distributed water harvesting system. The difference, in monetary terms, generates the measurable climate adaptation benefit over the long term with a relatively high degree of certainty.

Figure 2 below provides a broad framework to quantify the broad benefits derived from a DWH system. Internal rate of return (IRR) is the primary measure of the value or worth of an investment based on yield over the long term. Rather than quantifying the present worth or annual worth as separate entities, IRR calculates the break-even interest rate for which the project benefits are equal to the project costs [19]. In other words, IRR sets the sum of the Net Present Value (NPV) of all cash flows of a particular project equal to zero. The characterization of the NPV functions that make up the larger IRR function inherently takes into account the time-value of money, as the present value of each discrete Present Value function requires discounting the future value. This type of calculation is critical for DWH harvesting; without considering the up-front capital cost alongside the gradual increase of benefits over time, the true value of the project will not be revealed. The suggested formula for internal rate of return (IRR) as a function of {infrastructure cost, flood damage reduction, reservoir cost, drought resiliency benefit, employment benefits, crop yield benefits, ecosystem benefits from biomass, P, CO_2} offset on the diagram is thus an expansion of the more traditional IRR of flood mitigation infrastructure, with IRR as a function of {infrastructure cost, flood damage reduction}. In addition, the ability of governments to establish an institutional environment that supports innovation for biomass harvesting, energy production, and nutrient recovery significantly increases this project value proposition. There is uncertainty inherent in any IRR calculation given the use of NPV, which uses assumed interest rates. A robust assessment of uncertainty requires assessment of fluctuations of various categories of localized data, which can be assessed according to various interest rates. For example, Holopainen et al. [20] perform an uncertainty assessment for NPV calculations of forests. The study relates uncertainty to inventory data, growth models, and timber price fluctuation under assumed of 3, 4, and 5% interest rates. Similar studies must be performed to understand fluctuations of NPV, and ultimately IRR calculations, based on project valuation of DWH systems. For example, variability in hydrologic data or climate change projections will present uncertainty that must be addressed and understood to present a well-rounded assessment of present value and rate of return.

The mathematical expression for internal rate of return Figure 2 above is intentionally general, but further characterization of the mathematical expression may reflect the following, where r is the rate of return of the project, C_t is the net cash inflow during the period t, and C_0 is the net cash outflow during the same time period. As previously mentioned, the calculated IRR will be subject to uncertainty, which must be assessed on a case-by-case basis.

$$IRR = r \text{ when } \left[\sum_{t=1}^{T} \frac{C_t}{(1+r)^t} - C_0 \right]_{\text{employment losses}} + \left[\sum_{t=1}^{T} \frac{C_t}{(1+r)^t} - C_0 \right]_{\text{purchasing fertilizer}} + \left[\sum_{t=1}^{T} \frac{C_t}{(1+r)^t} - C_0 \right]_{CO_2 \text{ offsets}} + \ldots = 0 \tag{1}$$

The overall value derived from the methods described above inherently require a long-term view. This is particularly important considering the need for comparability between more conventional solutions for flood mitigation as governments choose between alternatives. High performance climate adaptation solutions require that the boundaries around the cost benefit analysis expand to include the co-benefits previously described, with an understanding of the full value proposition over several decades, hence the logic of a long-term view and bond finance. The threats of climate change manifest as significant costs for governments and individuals, but only if quantified over long time horizons informed by accurate data [8]. The IRR calculation described above helps capture this characteristic in monetary terms. Figure 3 below attempts to visualize the net increasing benefits over time, by separating the short term, medium term, and long term costs and benefits. The figure clearly shows that the peak monetary costs would likely occur within the first five years of the DWH project, while the maximum benefit may be realized on a much longer time horizon. The Red River Floodway in Manitoba, Canada, is a proven historical example of such benefits. The original floodway was built to protect the City of Winnipeg between 1962 and 1968 at a cost of CAD $63 million (in 2011 Canadian dollars) [21]. Premier Duff Roblin spearheaded project development, which required significant political persistence due to the massive project scale. Since 1969, "Duff's Ditch" has prevented over CAD $40 billion of flood damage in the City of Winnipeg [21]. The Red River Floodway is an excellent example of high up-front capital costs reaping long-term benefits, grounding the concept of Figure 3 in historical context.

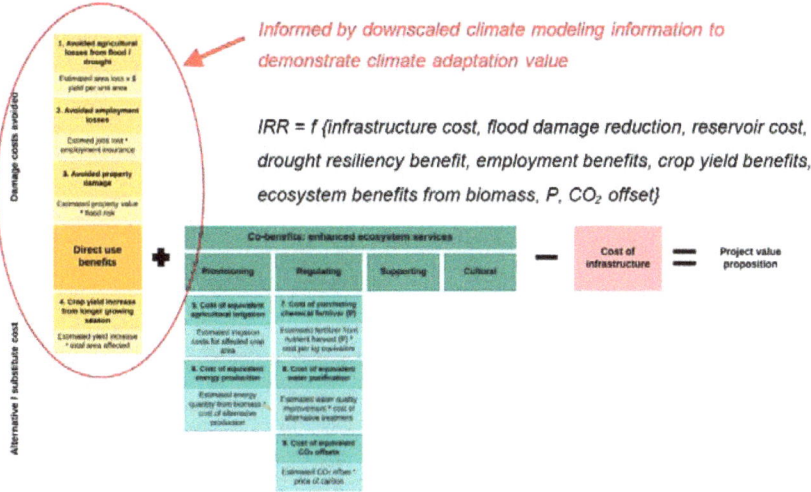

Figure 2. Framework for value proposition of distributed water harvesting infrastructure for consideration when quantifying project value in comparison to more traditional flood risk mitigation methods.

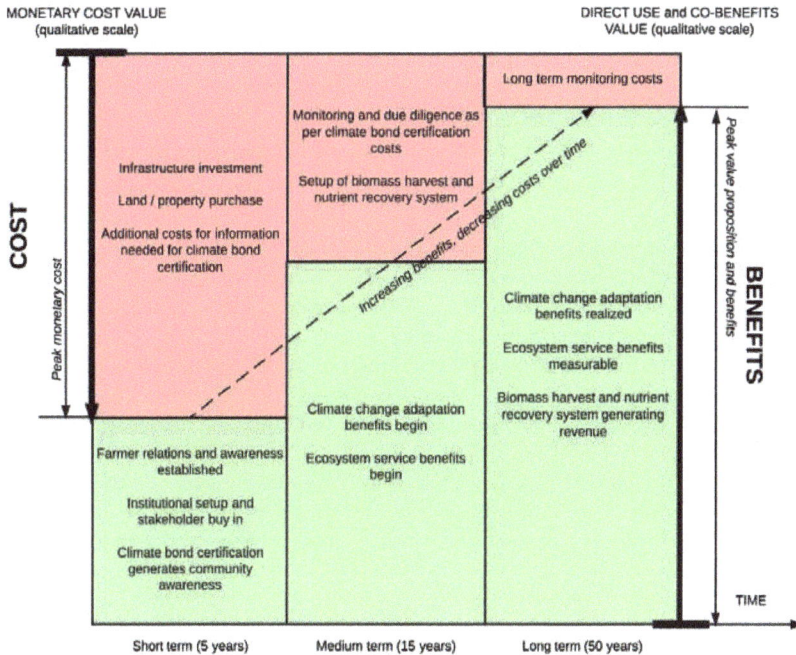

Figure 3. Temporal diagram depicting increasing benefits and decreasing costs over time, emphasizing the need to integrate the long term to understand the changing cost:benefit ratio of climate adaptation.

3. Climate Bonds for Financing Distributed Water Infrastructure

Multi-functional distributed water harvesting lies at the intersection of many challenges that are difficult for traditional debt instruments and government institutions to finance. Better climate adaptation solutions demand the sustainable development principle of subsidiarity, which in turn demands granularity in adaptation projects. Taking advantage of access to robust climate data and projections enables better engineering solutions, but it also places high demands on most aspects of financing including internal rate of return calculations, comparability to conventional projects, and the nuances of risk assessment. An emerging financing solution for climate-resilient and low carbon solutions is to use "climate-aligned bonds"—a twist on the traditional bond, a debt instrument when an investor loans money to a corporation or government for a predefined period of time on a fixed or variable interest rate [22]. These climate-aligned bonds are often unlabeled, but increasingly these bonds are certified as either "green" or "climate" bonds to provide a clear, reliable signal to investors.

3.1. Water Climate Bonds

The Climate Bonds Standard from the Climate Bonds Initiative ear-marks bonds that fund projects with very specific climate change adaptation and mitigation qualities [23]. The Canadian green bond market is growing, with Canadian labeled green bonds amounting to CAD \$2.9 billion and Canadian unlabeled climate-aligned bonds amounting to CAD \$30 billion [22]. The green bond label has been called into question recently, with some stakeholders questioning whether its criteria are restrictive enough to avoid "greenwashing" [22]. The Climate Bonds Initiative (CBI) uses its Climate Bond Standard (CBS), a rigorous certification and reporting process for climate adaptation and mitigation projects, to demonstrate the value of certification, incent a shift in public and investor perception, and provide a platform to highlight innovative climate-related projects. The Water Criteria under the Climate Bonds Standard were released in 2016, providing investors with "verifiable, sector-specific eligibility criteria to evaluated water-related bonds for low-carbon, climate resilient criteria" [23], with the first phase targeted toward engineered infrastructure. Adherence to the standard is determined after bond originators submit water-related issuances for certification of third party auditors [23]. Successful certification is a clear signal to investors that the project has rigorously considered its role in adapting to and mitigating climate change.

3.2. Government-Issued Bonds for Distributed Infrastructure

The water sector is beginning to embrace decentralized infrastructure as an emerging solution for modern water and climate challenges. For example, water utilities have found that distributed natural or semi-natural systems can help manage fluctuating demand and the strain on storm water and wastewater systems at a relatively low cost [24]. The decentralized nature of these systems, shared by many climate adaptation projects including DWH, is a major design challenge for financers. DWH systems are also distributed across many properties, some of which are privately owned, adding to the legal complexity. Statutory definitions that govern infrastructure projects and management of water systems have a long history, with some water governance regimes unable to accommodate for these project characteristics. As a result, many water utilities in the United States are forced to rely on cash financing of conservation and green infrastructure efforts and to save debt instruments for conventional infrastructure [24]. A report issued by Ceres identified four major themes that may enable legal authority for the issuance of bonds for distributed water infrastructure in the United States [24]. More legal analysis into public finance law and bond issuance requirements in various provinces in Canada is necessary to determine where uncertainties within the legal framework lie, but it can be assumed that the challenges are similar. The legal considerations for issuing bonds for distributed water harvesting infrastructure are outlined in Table 2 below. Financing distributed water infrastructure with bonds issued by public authorities presents some challenges, but to move forward with high performance climate adaptation and mitigation projects it is important to tap into these liquid markets.

Table 2. Legal considerations for issuance of bonds for distributed infrastructure [24].

Legal Consideration	Applicability to Distributed Water Harvesting
Bond issuer must have the legal authority to issue bonds for distributed infrastructure on private property.	Water harvesting requires financing to construct earthen dams on private property or to directly acquire the land.
The bond issuer or water utility must not be legally restrained from using enterprise revenue bonds to finance distributed infrastructure on private property, if applicable.	The provincial and federal government financing structure in Canada may limit acquisition of certain types of debt until existing debts are repaid.
	Constitutional clauses may prohibit the use of public credit for private benefit, though justifying based on the public benefit is possible (see Case Study Section 3.3.2)

Table 2. *Cont.*

Legal Consideration	Applicability to Distributed Water Harvesting
Bond issuer must structure the bond to maintain federal income tax exemptions.	Care must be taken to understand the role of farmers as private business, and to intentionally highlight and quantify public benefit.
Bond issuer must establish 'control' of the financed asset to conform to Generally Accepted Accounting Principles.	Conservation easements may act as intangible assets to ensure intended function of property and infrastructure.
	(Rebates have also been constituted as contracts with final customers in water efficiency programs.)

3.3. Case Studies

The available literature does not contain a precedent for funding DWH systems with bonds, in Canada or elsewhere. However, case studies from a variety of angles may inform the feasibility and methods for approaching the structure of a bond for this application.

3.3.1. Water Climate Bond Certified—San Francisco Public Utilities Commission [25]

The San Francisco Public Utilities Commission issued the first bond certified under the Water Criteria for the Climate Bond Standard in May of 2016. The USD $240 million will help fund projects under the Sewer System Improvement Program. The sewer and storm water systems in San Francisco are currently nearly 100 years old, and the aging infrastructure is expected to present increasingly significant risks to the region. In addition, San Francisco is located in a seismic zone and the aging structures are seismically vulnerable. By investing in large scale capital improvements now, the utilities commission hopes to avoid emergency repairs and regulatory fines, while creating broader public benefit from the improved system design. From a climate change perspective, San Francisco will experience increasing temperatures and greater intensity of downpours and storm systems that directly threaten the storm and waste water systems [26]. Certification of this project under the CBS Water Criteria is a positive signal for the possible certification of a bond financing DWH systems. Storm water and wastewater systems are distributed and decentralized by nature, involve many stakeholders, require long time horizons, and are informed by significant hydrologic complexity. These factors all exist as key institutional and technical considerations with DWH systems.

3.3.2. Bond Distributed on Private Property: Southern Nevada Water Authority [24]

The Southern Nevada Water Authority has financed its Water Smart Landscapes Program with government issued bonds. The water authority rebates customers USD $2 per square foot of grass removed and replaced with desert landscaping up to the first 5000 square feet converted per property per year. To satisfy the legal requirement to maintain control of the 'financed asset', a conservation easement is recorded against the property if the converted landscape is funded by bond funds. Again, this unique bond structure is a positive signal for the possibility to finance DWH with government bonds. The Southern Nevada Water Authority has justified the individual private benefit with the claim that public funds generate much greater public benefit. In addition, the use of conservation easements is a pertinent example of a legal structure that can overcome the legal requirement to maintain control of the asset being financed, which is also a pertinent consideration for DWH systems.

3.3.3. Canadian Green Bond: Province of Ontario [27]

The Province of Ontario Green Bond Program is leading the green bond market in Canada. The first bond issued as part of this program was a CAD $500 million bond to fund the Eglington Crosstown Light Rail Transit (LRT) project, which aims to generate public benefit and mitigate climate change impacts from multiple angles [28]. The new transit corridor will move people up to 60 percent faster than the current bus system. The LRT vehicles are electric and produce zero emissions, reducing the greenhouse gas footprint compared to the bus system. In addition, the shift of transport mode from auto to LRT is expected to further reduce the carbon footprint of the transport system. This project, and the successful issuance of a second CAD $750 million bond through the Province of Ontario Green Bond Program, demonstrates the potential liquidity of the market for financing rural projects certified under the international Certified Climate Bond Standard.

3.3.4. Asian and the Pacific Climate Bond: Asian Development Bank [29]

In early 2017 the Asian Development Bank (ADB) backed a climate bond for AP Renewables, Inc. of the Philippines. The local currency bond, equivalent to USD $225 million, is the first bond certified by the Climate Bonds Initiative to any country in Asia and the Pacific, and it is also the first ever single-project Climate Bond issued in an emerging market. The bond will finance AP Renewables' Tiwi-MakBan geothermal power generation facilities in the form of a guarantee of 75% of the principle and interest on the bond, in addition to a direct local currency ADB loan of USD $37.7 million equivalent. This landmark project demonstrates innovation in the financing realm from multiple dimensions—the opportunity for development institutions to assist developing and emerging economies in accessing new capital, the use of credit enhancement risk from the Credit Guarantee Investment Facility that has been established by ASEAN+3 governments and ADB to develop bond markets, and the proven importance of 'green' financing in emerging economies. The applicability of this financing mechanism, in addition to the DWH concept, is clearly transferable to economies all of the world with similar climatic and agricultural challenges, despite their different institutional structures and capacities.

4. Designing the System to Support Multi-Functional Distributed Water Harvesting Infrastructure and Climate Bond Certification

Implementation of a distributed water harvesting system is a complex design challenge with consideration of the engineering, property rights, environmental, institutional, and regulatory contexts. The following sections outline the starting point for implementing a DWH system on the Canadian prairies and ensuring that this setup increases the likelihood of successful bond certification under the Water Criteria of the Climate Bond Standard.

4.1. Engineering Considerations, Land and Property Ownership, and the Environment

There are several practical considerations when moving to implement water harvesting infrastructure. The list in Table 3 is not exhaustive but begins to frame the types of considerations to be made to successfully design and implement the technology solution, while incorporating the needs of various stakeholders and the technical requirements listed under the Climate Bond Standard.

Table 3. Considerations for technical/practical factors in implementing water harvesting infrastructure.

Theme	Relevant Factors to Consider
Engineering considerations	Hydrological modeling project boundaries must operate within provincial boundaries while considering river basin boundaries.
	Hydrological modeling must consider present and multiple climate change impact scenarios.

Table 3. *Cont.*

Theme	Relevant Factors to Consider
Engineering considerations	Hydrological modeling and engineering must take into account changes to water quality and water supply to all downstream.
	Siting and sequencing of location and scale of water harvesting dams and flow patterns should be optimized for physical context.
	Siting and sequencing of water harvesting dams and flow patterns should be adjusted based on external social or environmental factors if optimized physical considerations does not fit.
	Siting and sequencing of projects must meet regulated hydrological budgets based on current and future projections of water allocations.
Land and property ownership	Farmers or other property owners must be willing to sell land to municipal or provincial government.
	Farmers must be consulted on willingness to lease land back during periods when land is suitable for cultivation.
	Governments must be willing to consider easements or other mechanisms to incent farmers to allow for modifications to land and the landscape.
Environmental considerations	Siting and sequencing of projects must meet regulations on minimum environmental flows, water quality, etc.
	Water quality and flow monitoring must be in place to enable due diligence in project design and implementation.
Profit generating activities	System for harvest of biomass for local heating and/or sale for energy production must be set up for farmers to take advantage of the possible business case.

4.2. Institutional and Legal Structure

A multi-functional distributed water harvesting system requires the coordination of various stakeholders. The proper institutional and legal structure can ease project implementation and increase the likelihood of sustainable project outcomes. In addition to the institutional environment within Canada, it will be critical to consider the transboundary effects, given the shared water basins along the Canada–US border and the potential for changes to transboundary water allocation and environmental impacts. In addition to designing and implementing the technology solution, issuing bonds for the distributed, rural infrastructure and receiving certification for the bonds under the Climate Bond Standard requires an additional layer of stakeholder coordination. Table 4 identifies and explains key stakeholders involved and includes suggestions for possible stakeholders who may be well positioned to take on these roles and functions.

In addition to the key stakeholders in Table 3, the institutional environment for financing infrastructure includes several limitations and challenges. Provincial and federal governments may have limits to their debt, and bonds are only one of many avenues from which to obtain funding. If local governments are included in financing considerations, many municipal governments also face a patchwork of funding sources including provincial and federal grants. Perhaps most importantly, governments generally expect a 'net drain' on investments from infrastructure, unlike investments in other sectors such as electricity. This 'net drain' highlights the importance of implementing the biomass harvest and nutrient recovery system as soon as possible once the DWH system is operational [30]. A fundamental consideration for project design is the uncertainty of future system performance given future climate uncertainty, therefore, IRR estimates will necessarily have estimates of uncertainty that associate with the range of future climate projection, which investors should recognize and understand. The current state-of-the-art in hydraulic design is to use ensemble climate projections to analyze expected performance and variability [31,32]. A key hypothesis with respect to DWH design, and its bond value and risk management proposition is that the higher the degree of climate impact, the greater the system benefit as this class of infrastructure is designed specifically to modulate climate impacts.

Table 4. Stakeholders involved with institutional and legal structure of water harvesting infrastructure.

Role	Function	Possible Stakeholders
Project initiator	A government entity to initiate project under mandate to protect public and manage hydrology of a region.	Relevant municipal and provincial branches of governance, such as the Province of Manitoba, Province of Saskatchewan, or relevant municipalities.
Financing authority	A public lending institution that issues bonds on behalf of government entities.	Provincial lending institutions like Alberta Capital Financing Authority (ACFA), Ontario Financing Authority (OFA), or Infrastructure Ontario (IO).
Watershed management and environmental agencies and advisory committees	A broad role, this covers all agencies involved in watershed management, hydrological planning and monitoring of the region.	Canadian watershed-level entities such as Alberta Watershed Planning and Advisory Committees, Saskatchewan Watershed Advisory Committees, Manitoba Conservation Districts, Manitoba Water Council; Inter-province entities such as the Prairie Provinces Water Board; United States watershed-level entities such as North Dakota Water Resource Boards.
Regulator	Regulatory agencies that operate within and between jurisdictions with regulatory power.	A federal government agency such as Environment Canada; provincial government agencies such as Alberta Environment and Sustainable Resource Development and Department of Conservation and Water Stewardship in Manitoba; United States agency such as United States Environmental Protection Agency; transboundary agency such as International Joint Commission.
Property owners	Any individual or agency with private or public property involved with water harvesting project.	Individual property owners such as farmers; other property owners such as Ducks Unlimited.
Monitoring and verification	An agency that provides ongoing oversight into the operations, maintenance, and upgrades involved with water harvesting project.	An entity that already has monitoring responsibilities such as the Prairie Provinces Water Board, provincial water and environmental government bodies.

4.3. Climate Bond Standard Certification

Upon examination of the Climate Bond Standard Phase 1 Water Criteria [7] and the San Francisco Public Utilities Commission case, the DWH concept has the potential to be an eligible candidate for certification. Certainty requires a more in-depth analysis of the river basin in question and full scoring by the independent third party auditors commissioned by the CBI. In the case of the Canadian prairies, key stakeholders for certification include governmental stakeholders including the Government of Canada, the environmental departments of the provincial governments of Alberta, Saskatchewan, and Manitoba, inter-provincial or international (US-Canada) agencies of interest and all others listed in Table 3 above. If these stakeholders approach the project with the intention of bond financing and climate bond certification, several unique considerations emerge. For example, the CBS requires that the project boundaries for assessment only include the direct effect of the proceeds of the bond [33]. It is likely that the most suitable project boundary for a DWH system is a river basin, with additional consideration of provincial boundaries prompted by the CBS criteria. The project must also qualify under criteria for all certified bonds, criteria for sector-specific bonds, and broader human rights and environmental considerations for water management before being considered for CBS certification [33]. This requirement may also prompt more intentional engagement with community members and civil society.

The CBS Water Criteria are separated into two streams: projects primarily for climate adaptation and projects primarily for climate mitigation. Water harvesting clearly falls under the climate adaptation criteria. Evaluation for CBS certification is based on a Scorecard system, in which a range of criteria are evaluated for no points, half points, or full points. The evaluation starts with a Vulnerability Assessment, followed by an Adaptation Plan if deemed necessary by the Vulnerability Assessment. Rough consideration of the criteria and the integrated nature of DWH systems indicate that they would likely require the Adaptation Plan. The Vulnerability Assessment is split into three major categories described in Table 5 below.

In some cases, the water harvesting concept may exceed the criteria in the way they are currently written, while in other cases the criteria are limiting. In addition, DWH projects are inherently climate adaptation projects, and thus the requirement for an Adaptation Plan presents an opportunity to highlight this functional purpose. The following sections are based on the CBS Water Criteria for Phase 1: Engineered Infrastructure [7], and may inform upcoming iterations of the criteria for natural or semi-natural systems. The applicability of CBS water criteria to the water harvesting system is broken out in more detail in the following sections. This evaluation is partially informed by the 2015 Organization for Economic Cooperation and Development (OECD) report, *Water Resources Allocation: Sharing Risks and Opportunities* [34], which evaluates institutional gaps in water allocation policy in Alberta and Manitoba. The sections below focus on these two provinces.

Table 5. Vulnerability Assessment section themes (as per Climate Bonds Initiative (CBI) requirements).

Theme	Description
Allocation	Assesses how water is shared by users within a given basin or aquifer, concentrating on the potential impacts of bond proceeds on water allocation.
Governance	Assesses how or whether the proceeds of the bond take into account the ways in which water will be formally shared, negotiate, and governed.
	Assesses compliance with allocation mechanisms that protect water resources.
Diagnostic	Assesses how or whether the use of the proceeds takes into account changes to the hydrologic system over time.
Adaptation Plan	If Vulnerability Assessment reveals significant climate change impacts on the project, the Adaptation Plan must be created as a management response plan to the conclusions and findings of the Vulnerability Assessment, noting how identified climate risks will be addressed.

4.3.1. Meeting the Criteria

A strong institutional environment on the Canadian prairies already exists, increasing the likelihood for a DWH system on the Canadian prairies to be certified under the CBS criteria. Accountability mechanisms for management of water allocation at different institutional, spatial, and temporal scales are established by water management plans, water code statutes, and compliance mechanisms that are in place in the regions in question. For example, water monitoring is performed by the Prairie Provinces Water Board, Alberta Environment and Sustainable Resource Development (ESRD), and the Department of Water Conservation and Stewardship in Manitoba. Scientific hydrological services that inform monitoring of adherence to codes already exist in current institutions like Manitoba's Water Stewardship Division. Furthermore, some elements of water allocation policies are already designed as required by the CBS criteria. For example, Alberta and Manitoba have differentiated entitlements based on the level of security of supply or risk of water shortage [34]. Both provinces have sanctions for withdrawal over limits. New entitlements or the increase of existing entitlements requires assessment of third party impacts, an environmental impact assessment, and that existing users forgo use [34]. In Alberta, minimum environmental flows are considered, and monitoring and enforcement mechanisms are in place in both Manitoba and Alberta [34]. Manitoba's Water Use Licensing Section monitors compliance for agriculture, domestic, and industrial water use by metering [34]. Allocation is enforced through sanctions with fines, and conflicts are resolved through the normal application of principles of good governance [34]. Alberta ESRD monitors and enforces water allocation for agriculture, domestic use, energy production, and the environment through metering and drawing penalties for contravening the enforcement order. Part of the sanction actions may also include fines or imprisonment, and formal conflict resolution is included under Section 93 of the Alberta Water Act. These existing institutional frameworks are key components of climate bond certification.

4.3.2. Exceeding the Criteria

The nature of the multi-functional water harvesting solution for the Canadian prairies exceeds the CBS criteria in several ways, though these are not necessarily captured in the formal CBS Scorecard.

For example, the CBS criteria requires a connection between water resource management at the project and hydrologic scale. Because a DWH system is based entirely upon the hydrologic scale, the boundaries of the bond proceeds and the hydrologic scale are one and the same. The criteria also include requirements for specific data, flow criteria, modeling scenarios, and water users to be included in hydrologic modeling. The hydrologic models used to design the DWH systems on the Canadian prairies would easily integrate these requirements in a manner that complies with the CBS criteria. For example, a dynamic simulation model of a DWH climate adaptation system was recently conducted for a portion of a watershed downstream of Pelly's Lake, Manitoba, Canada. This simulation model integrates physical variables related to the landscape, energy balance, moisture fluxes, hydrologic cycle with operational climate forecasting tools to understand the multi-purpose benefits of the system and to estimate their economic value [6]. Furthermore, the use of downscaled climate data and quantification of future climate impact scenarios with and without the system increases certainty about the future success of the system, beyond the requirements of the CBS criteria. The quality and breadth of information put into these hydrological models, environmental impact assessments, and other assessment mechanisms that are part of the planning and design process benefit the climate bond certification process by informing a rigorous Adaptation Plan. More importantly, the use of downscaled climate change data with rigorous hydrologic modeling to design DWH systems demonstrates a fundamental shift towards greater certainty for context-specific system functionality as a climate adaptation solution under long range climate impacts.

4.3.3. Challenges with the Criteria

Some institutional gaps in water management on the Canadian prairies and the current structure of the CBS Water Criteria present some challenges for certification. Water allocation agreements must be dynamic to accommodate changes to flow scenarios with new water harvesting infrastructure, so adherence to the criteria may not be clear until the planning process is mature. Additionally, inconsistent provincial water allocation policies reveal weaknesses in water governance in some provinces. Manitoba does not define its environmental flows, and while freshwater biodiversity is considered on a project-by-project basis, terrestrial biodiversity is not considered. Return flow obligations are not specified, and the nature of water entitlements is based on the purpose of water allocation, maximum area irrigated, and the maximum volume removed, rather than as a proportion of total flow conditions. Alberta has a more rigorous policy framework, but its water allocation is currently classified as 'over-allocated'. These institutional gaps should not only be addressed to allow for climate bond certification, but also as part of an effort to establish best practices for water management.

5. Distributed Water Harvesting and Climate Bonds in the International Context

The Canadian Prairies are not the first or only agricultural region to be confronted with increasing pressure driven by climate change impacts—globally 80 percent of agricultural land is rainfed making up 65 to 70 percent of staple food crops [35]. Model output of mean climatic changes are far more robust than changes to climate variability, meaning that the full impacts of climate change are likely seriously underestimated [35]. However, just as with the Canadian example, the interactions of different climatic stresses on biological and food systems over time in different regions all over the world require investigation of localized changes over time. Variability in rainfall is demonstrated as the principle cause of inter-annual variability in crop yields at both aggregate and plot level [35]. Semi-arid and arid environments around the world are projected to face similar challenges that may be solved by DWH solutions or other distributed agricultural adaptation solutions financed by ear-marked climate bonds. For example, rainfall variability in the Middle East and the Mediterranean region is projected to result in an overall drier climate, with an impact on major river systems and food productivity [36]. Specific impacts are disparate across this region—rainfall is expected to decrease in southern Europe, Turkey, and the Levant, while rainfall in the Arabian Gulf may increase [36]. Still, in the former example, rainfall is expected to increase in the winter and decrease in the summer [36], affecting crop productivity differently in each growing season. In another locale, studies have also shown one of the highest agricultural productivity losses due to climate change scenarios is predicted in India [37]. Though temperatures are expected to rise and annual precipitation rates to remain stable, regional variability is expected to result in extreme changes to both surface and groundwater due to the changes in temporal rainfall variability [37]. Several countries in sub-Saharan Africa also rely heavily on rainfed agriculture, and expect a higher frequency of droughts and rainfall variability in the future [38]. DWH solutions, or some derivate of the technology, is likely to be necessary in regions with high dependence on rainfed agriculture and projected rainfall variations.

The existence of rainfed agriculture and current or projected climate change impacts is not enough to determine the suitability of DWH solutions or financing via the use of labeled or unlabeled climate bonds. An institutional environment conducive to such multi-stakeholder, rural-based solutions must exist or be managed to achieve the maximum return on project investment and ensure the system is used appropriately. Any institution or entity that is set up to issue a bond has the ability to issue a green bond, and if institutional capacity meets the requirements, may be certified under the Climate Bond Standard. Southern Europe and the Middle East may be well-served by such distributed engineering solutions, and may also be set up to access the pool of capital offered by green or climate bonds. In addition, developing countries face low visibility on low carbon projects because of the high cost of capital and higher interest rates, despite a significant need for climate-friendly infrastructure investment [39]. Development institutions, as demonstrated by the Asian Development Bank, are well positioned to facilitate and support such enabling environments. This paper has demonstrated the application of DWH harvesting and climate bond certification and financing in one locale, but several other contexts requires a similar approach, adapted to the local agricultural and climate system, institutional circumstance, and financing environment.

6. Conclusions and Recommendations

A multi-functional distributed water harvesting system on the Canadian prairies financed with government-issued bonds that are certified under the Water Criteria for the Climate Bond Standard presents a feasible, innovative climate adaptation solution for the increased temperatures and variable precipitation expected to strain agriculture in the region in the coming decades. Successfully implementing this solution requires stakeholder coordination, an institutional lens, and innovative engineering methods. In addition, lessons learned from the analysis contained in this paper can inform the establishment of CBS criteria for natural and semi-natural water infrastructure.

It is recommended that institutions involved with water management and public infrastructure on the Canadian prairies think creatively about their role in driving and supporting innovative climate adaptation projects. Taking advantage of the growing green bond market potential and learning from the success of the green bond initiatives in the Province of Ontario requires that more financial institutions recognize their value and build programs to support them. For example, the Liberal government's proposed Canadian Infrastructure Bank and other existing financers can consider green bonds as an opportunity to aggregate projects for risk reduction and public benefit and to access an otherwise exclusive pool of private capital. Assessing the true value of innovative solutions, particularly distributed climate adaptation projects, requires that governments consistently establish a long-term view that quantifies direct monetary ecosystem service benefits and co-benefits. This lens should not only be adopted to inform the full economic value for projects with direct environmental or climate adaptation benefits. A report from the Ministry of Environment in Sweden recommends the inverse view; that "...government should investigate different strategies to improve transparency regarding the dependence and impact of bond investments on the ecosystem services, including investments by the national pension funds" [40]. Taking care to involve existing stakeholders through all phases of visioning and implementation of a DWH system will take advantage of existing institutional capacity and help anticipate demands to fill institutional gaps. Stakeholder involvement should also include a comprehensive community benefits framework and active community engagement, as was established alongside the Eglington Crosstown LRT project under the Province of Ontario Green Bond program. Prairie Provinces may need to also consider tightening up water allocation policies to fill the identified gaps. Engineers, hydrologists, and environmental scientists must also consider their role in designing an effective system and using the requirements of the CBS Water Criteria to inform robust hydrological modeling and engineering practices. These stakeholders must also take care to build the business case and supply chain connections for farmers to harvest biomass, generate bioenergy, and recover nutrients, in order to capitalize on long-term project value and protect downstream water bodies from excess nutrient accumulation. All stakeholders that have a potential role in the design and implementation of a water harvesting system, financing the project under certified climate bonds, or creating an appropriate policy environment, must be aware of the complexity of the space and importance of demonstrating effective climate adaptation solutions.

A multi-functional distributed water harvesting system can enable agricultural productivity on the Canadian prairies in the face of climate change. Successfully implementing and financing a DWH project requires that stakeholders understand the value of the direct climate adaptation benefits and enhanced ecosystem services, actively pursue the business case generated alongside the public benefit, and generate buy-in and momentum through active institutional and community engagement. Financing a DWH project, and other distributed water infrastructure, with government bonds is possible if the bond is structured with consideration of the legal authority of the bond issuer. Seeking Climate Bond Standard certification creates an additional incentive for robust project design, takes advantage of an untapped pool of private capital, and demonstrates the full value that decades of climate data and refined hydrologic knowledge can bring to infrastructure solutions. Lastly, the Phase 1 Water Criteria for the CBS rewards water and wastewater projects that have shown adequate proof that climate adaptation and mitigation have been considered as design constraints. It is recommended that as the Climate Bonds Initiative develops water criteria for natural or semi-natural infrastructure, it might consider finding ways to explicitly reward projects that have a functional purpose of climate adaptation or mitigation rather than simply as a design consideration of a project with a different functional purpose. The analyses and recommendations contained in this paper are directed toward implementation of a DWH systems on a hypothetical river basin on the Canadian prairies, but it is evident that this solution is transferable to many regions with similar climate change effects and agricultural systems that will cause climate adaptation challenges in the future.

Acknowledgments: The manuscript was conceived through a partnership between the International Institute for Sustainable Development–Prairie Climate Centre and the Central European University Budapest under the direction of Laszlo Pinter at the Central European University. No grant funds were allocated to this project.

Author Contributions: Anita Lazurko and Henry David Venema both contributed to the analysis and wrote the manuscript together. All authors approved the final manuscript.

Conflicts of Interest: The authors declare no conflict of interest.

References

1. G20 Leaders' Communique. Hangzhou Summit. 4–5 September 2016. Available online: http://www. consilium.europa.eu/press-releases-pdf/2016/9/47244646950_en.pdf (accessed on 9 April 2017).
2. Granoff, I.; Hogarth, J.R.; Miller, A. Nested barriers to low-carbon infrastructure investment. *Nat. Clim. Chang.* **2016**, *6*, 1065–1071. [CrossRef]
3. OECD. *Mobilising Bond Markets for a Low-Carbon Transition*; OECD Publishing: Paris, France, 2017.
4. Neumann, J.E.; Strzepek, K. State of the literature on the economic impacts of climate change in the United States. *J. Benefit Cost Anal.* **2014**, *5*, 411–443.
5. Gray, J.H. *Men against the Desert*; Western Producer Prairie Book: Saskatoon, SK, Canada, 1967.
6. Berry, P.; Yassin, F.; Belcher, K.; Lindenschmidt, K. An economic assessment of local farm multi-purpose surface water retention systems under future climate uncertainty. *Sustainability* **2017**, *9*, 456. [CrossRef]

7. Climate Bonds Initiative (CBI). The Water Criteria of the Climate Bonds Standard: Phase 1: Engineered Infrastructure. Available online: https://www.climatebonds.net/files/files/Water_Criteria_of_the_ClimateBondsStandard_October2016.pdf (accessed on 15 January 2017).

8. Blair, D.; Mauro, I.; Smith, R.; Venema, H. Visualising Climate Change Projections for the Canadian Prairie Provinces. Available online: http://dannyblair.uwinnipeg.ca/presentations/blair-ottawa.pdf (accessed on 5 February 2017).

9. Betts, A.; Desjardins, R.; Worth, D.; Cerknowiak, D. Impact of land use change on diurnal cycle climate on the Canadian prairies. *J. Geophys. Res. Atmos.* **2013**, *118*, 11996–12011. [CrossRef]

10. Statistics Canada. Environment Accounts and Statistics Division. Agricultural Water Survey (Survey Number 5145). 2011. Available online: http://www.statcan.gc.ca/pub/16-402-x/2011001/ct002-eng.htm (accessed on 5 February 2017).

11. Environment and Climate Change Canada. Canadian Environmental Sustainability Indicators: Nutrients in Lake Winnipeg. Environment and Climate Change Canada, 2016. Available online: https://www.ec.gc.ca/indicateurs-indicators/55379785-2CDC-4D18-A3EC-98204C4C10C4/LakeWinnipeg_EN.pdf (accessed on 5 February 2017).

12. Manitoba 2011 Flood Review Task Force Report: Report to the Minister of Infrastructure and Transportation. Available online: https://www.gov.mb.ca/asset_library/en/2011flood/flood_review_task_force_report.pdf (accessed on 10 March 2017).

13. Van Meter, K.; Steiff, M.; McLaughlin, D.; Basu, N. The sociohydrology of rainwater harvesting in India: Understanding water storage and release dynamics across spatial scales. *Hydrol. Earth Syst. Sci.* **2016**, *20*, 2629–2647. [CrossRef]

14. Grosshans, R.; Grieger, L.; Ackerman, J.; Gauthier, S.; Swystun, K.; Gass, P.; Roy, D. Cattail Biomass in a Watershed-Based Bioeconomy: Commercial-Scale Harvesting and Processing for Nutrient Capture, Biocarbon, and High-Value Bioproducts. Available online: https://www.iisd.org/sites/default/files/publications/cattail-biomass-watershed-based-bioeconomy-commerical-scale-harvesting.pdf (accessed on 10 January 2017).

15. Quiring, S.; Papakryiakou, T. An evaluation of agricultural drought indices for the Canadian prairies. *Agric. For. Methodol.* **2003**, *118*, 49–62. [CrossRef]

16. Aylward, B.; Bandyopadhyay, J.; Belausteguigotia, J.; Borkey, P.; Cassar, A.; Meadors, L.; Saade, L.; Siebentritt, M.; Stein, R.; Sylvia, T.; et al. Chapter 7 Freshwater ecosystem services. In *Millennium Ecosystem Assessment*; Island Press: Washington, DC, USA, 2005.

17. Millenium Ecosystem Assessment (MEA). *Ecosystems and Human Well-Being Synthesis*; Island Press: Washington, DC, USA, 2005.

18. Gomez-Baggethun, E.; Martin-Lopez, B.; Barton, D.; Braat, L.; Kelemen, E.; Lorene, M.; Saarikoski, H.; van den Bergh, J. State-of-the-art report on integrated valuation of ecosystem services Deliverable D.4.1/WP4. 2014. Available online: http://www.openness-project.eu/sites/default/files/Deliverable%204%201_Integrated-Valuation-Of-Ecosystem-Services.pdf (accessed on 15 March 2017).

19. Newnan, D.; Eschenbach, T.; Lavelle, J. *Engineering Economic Analysis*, 9th ed.; Oxford University Press: New York, NY, USA, 2004.

20. Holopainen, M.; Mäkinen, A.; Rasinmäki, J.; Hyytiäinen, K.; Bayazidi, S.; Vastaranta, M.; Pietilä, I. Uncertainty in forest net present value estimations. *Forests* **2010**, *1*, 177–193. [CrossRef]

21. Province of Manitoba. Flood Information: Red River Floodway. 2011. Available online: http://www.gov.mb.ca/flooding/fighting/floodway.html (accessed on 10 February 2017).

22. CBI. Bonds and Climate Change: The State of the Market. Climate Bonds Initiative, 2016. Available online: https://www.climatebonds.net/files/files/reports/cbi-hsbc-state-of-the-market-2016.pdf (accessed on 10 January 2017).

23. CBI. Climate Bonds Standard, Version 2.1. 2016. Available online: https://www.climatebonds.net/standards/standard_download (accessed on 10 January 2017).

24. Leurig, S.; Brown, J. Bond Financing Distributed Water Systems: How to Make Better Use of Our Most Liquid Market for Financing Water Infrastructure. Report for Ceres. 2014. Available online: https://www.ceres.org/resources/reports/bond-financing-distributed-water-systems-how-to-make-better-use-of-our-most-liquid-market-for-financing-water-infrastructure (accessed on 10 January 2017).

25. San Fransisco Public Utilities Commission, Climate Bond Certified. Announcement from CBI. Available online: https://www.climatebonds.net/standards/certification/SFPUC (accessed on 10 January 2017).

26. Ekstrom, J.A.; Susanne, C.M. Climate Change Impacts, Vulnerabilities, and Adaptation in the San Francisco Bay Area: A Synthesis of PIER Program Reports and Other Relevant Research. California Energy Commission; Publication Number: CEC-500-2012-071; 2012. Available online: http://www.energy.ca.gov/2012publications/CEC-500-2012-071/CEC-500-2012-071.pdf (accessed on 15 March 2017).

27. Ontario Financing Authority (OFA). Green Bond Presentation: Province of Ontario. Available online: https://www.ofina.on.ca/pdf/ontario_greenbonds_presentation_jan2016_en.pdf (accessed on 5 February 2017).

28. Stear Davies Gleave. *Eglington Crosstown Rapid Transit Benefits Case Final Report*; Prepared for Metrolinx; Stear Davies Gleave: Toronto, ON, Canada, 2009; Available online: http://www.metrolinx.com/en/regionalplanning/projectevaluation/benefitscases/Benefits_Case-Eglinton_Crosstown_2009.pdf (accessed on 15 March 2017).

29. Asian Development Bank (ADB). ADB Backs First Climate Bond in Asia in Landmark \$225 Million Philippines Deal. 2016. Available online: https://www.adb.org/news/adb-backs-first-climate-bond-asia-landmark-225-million-philippines-deal (accessed on 15 March 2017).

30. Siemiatycki, M. Creating an Effective Canadian Infrastructure Bank. Independent Research Study prepared for Residential and Civil Construction Alliance of Ontario. 2016. Available online: http://www.rccao.com/research/files/02_17_RCCAO_Federal-Infrastructure-Bank2016WEB.pdf (accessed on 20 February 2017).

31. Tabari, H.; Troch, R.D.; Giot, O.; Hamdi, R.; Termonia, P.; Saeed, S.; Brisson, E.; Lipzig, N.V.; Willems, P. Local impact analysis of climate change on precipitation extremes: Are high-resolution climate models needed for realistic simulations? *Hydrol. Earth Syst. Sci.* **2016**, *20*, 3843–3857. [CrossRef]

32. Kavvas, M.L.; Ishida, K.; Trinh, T.; Ercan, A.; Darama, Y.; Carr, K.J. Current issues in and an emerging method for flood frequency analysis under changing climate. *Hydrol. Res. Lett.* **2017**, *11*, 1–5. [CrossRef]

33. Matthews, J.; Timboe, I. Guidance Note for Issuers and Verifiers: Phase 1: Engineered Infrastructure. Supplementary note to the Water Criteria. Climate Bonds Initiative, 2016. Available online: https://www.climatebonds.net/files/files/Water_Criteria_Guidance_Note_to_Issuers%26Verifiers_October_2016(1).pdf (accessed on 10 January 2017).

34. OECD. *Water Resources Allocation: Sharing Risks and Opportunities*; OECD Studies on Water; OECD Publishing: Paris, France, 2015; Available online: http://www.oecd.org/fr/publications/water-resources-allocation-9789264229631-en.htm (accessed on 5 February 2017).

35. Thornton, P.; Ericksen, P.; Herrero, M.; Challinor, A. Climate variability and vulnerability to climate change: A review. *Glob. Chang. Biol.* **2014**, *20*, 3313–3328. [CrossRef] [PubMed]

36. Lelieveld, J.; Hadjinicolaou, P.; Kostopoulou, E.; Chenoweth, J.; El Maayar, M.; Giannakopoulos, C.; Hannides, C.; Lange, M.A.; Tanarhte, M.; Tyrlis, E.; et al. Climate change and impacts in the Eastern Mediterranean and the Middle East. *Clim. Chang.* **2012**, *114*, 667–687. [CrossRef] [PubMed]

37. Asha latha, K.V.; Gopinath, M.; Bhat, A.R.S. Impact of climate change on rainfed agriculture in India: A case study of Dharwad. *Int. J. Environ. Sci. Dev.* **2012**, *3*, 368–371.

38. Cooper, P.J.M.; Dimes, J.; Rao, K.; Shapiro, B.; Shiferaw, B.; Twomlow, S. Coping better with current climatic variability in the rain-fed farming systems of sub-Saharn Africa: An essential first step in adapting to future climate change? *Agric. Ecosyst. Environ.* **2008**, *126*, 24–35. [CrossRef]

39. Nelson, D.; Shrimali, G. Finance Mechanisms for Lowering the Cost of Renewable Energy in Rapidly Development Countries. Report from Climate Policy Initiative Series 2014. Available online: https://climatepolicyinitiative.org/wp-content/uploads/2014/01/Finance-Mechanisms-for-Lowering-the-Cost-of-Clean-Energy-in-Rapidly-Developing-Countries.pdf (accessed on 10 January 2017).

40. Schultz, M. *Making the Value of Ecosystem Services Visible*; Summary of the Report of the Inquiry M 2013:01 Ministry of the Environment; Swedish Government Inquiries: Stockholm, Sweden, 2013. Available online: https://www.cbd.int/financial/hlp/doc/literature/sammanfattning_engelska_1301105.pdf (accessed on 5 February 2017).

sustainability

MDPI

Article

Evaluation of the Agronomic Impacts on Yield-Scaled N₂O Emission from Wheat and Maize Fields in China

Wenling Gao and Xinmin Bian *

College of Resources and Environmental Sciences, Nanjing Agricultural University, 1st Weigang Road, Xuanwu District, Nanjing 210095, China; 2007203003@njau.edu.cn
* Correspondence: bxm@njau.edu.cn; Tel.: +86-025-5867-1383

Received: 22 March 2017; Accepted: 29 June 2017; Published: 7 July 2017

Abstract: Contemporary crop production faces dual challenges of increasing crop yield while simultaneously reducing greenhouse gas emission. An integrated evaluation of the mitigation potential of yield-scaled nitrous oxide (N_2O) emission by adjusting cropping practices can benefit the innovation of climate smart cropping. This study conducted a meta-analysis to assess the impact of cropping systems and soil management practices on area- and yield-scaled N_2O emissions during wheat and maize growing seasons in China. Results showed that the yield-scaled N_2O emissions of winter wheat-upland crops rotation and single spring maize systems were respectively 64.6% and 40.2% lower than that of winter wheat-rice and summer maize-upland crops rotation systems. Compared to conventional N fertilizer, application of nitrification inhibitors and controlled-release fertilizers significantly decreased yield-scaled N_2O emission by 41.7% and 22.0%, respectively. Crop straw returning showed no significant impacts on area- and yield-scaled N_2O emissions. The effect of manure on yield-scaled N_2O emission highly depended on its application mode. No tillage significantly increased the yield-scaled N_2O emission as compared to conventional tillage. The above findings demonstrate that there is great potential to increase wheat and maize yields with lower N_2O emissions through innovative cropping technique in China.

Keywords: climate change; food security; cropping system; soil management; greenhouse gas emission

1. Introduction

Nitrous oxide (N_2O) is a long-lasting greenhouse gas that significantly contributes to stratospheric ozone depletion and global warming. It is estimated that about 60% of total anthropogenic N_2O is emitted from agricultural soil, which is mainly produced by nitrification and denitrification processes of reactive nitrogen (N) in soil [1,2]. Reducing the N_2O emission from soil is urgent in contemporary crop production for the mitigation of global warming. However, global crop production is also facing a great challenge of growing by 70~100% by 2050 to meet an expected 34% increase in world population [3,4]. Meeting this goal will result in increased pressure to use more N fertilizer, thereby potentially increasing N_2O emission [5–7]. Maize (*Zea mays* L.) and wheat (*Triticum aestivum* L.) account for the largest and second largest global consumption of all fertilizer N in major cereal crops [8]. Therefore, it is necessary and urgent to study how to increase maize and wheat yields with lower N_2O emissions in the future.

Agronomic practices such as cropping systems and soil management options are the primary factors regulating N_2O emission from cropland soil. Improving these practices (e.g., reducing inorganic N fertilizer, use of enhanced-efficiency fertilizers and no tillage) has the potential to reduce N_2O emission from soil [9]. However, changes to agronomical practices often simultaneously affect crop yield. It is still early to decide which option is optimal for the balance of mitigating N_2O emission and increasing crop yield. A particular practice beneficial to reducing N_2O emissions may or may not favor crop yield enhancement. For example, replacing N fertilizer with manure can mitigate N_2O emission but could decrease crop yields compared to inorganic N fertilizer application only [10]. Application of enhanced-efficiency N fertilizers can reduce N_2O emissions but can either increase [11,12] or decrease crop yields [13,14]. Therefore, integrating assessment on both N_2O emissions and crop yield is essential in optimizing cropping practices. Although many studies have evaluated the impact and mitigation potential of cropping practices on N_2O emission [15–22], few studies have been linked to crop yield [23–25]. Recent studies suggest that comprehensive assessments of cropping practices per unit yield (yield-scaled) rather than land area (area-scaled) could benefit sustainable intensification of cropping practices and policy selection with a trade-off of N_2O emission mitigation and food security [23,24,26].

China takes the first and the second positions, respectively, in global wheat and maize production. Wheat and maize production in China was 121.7 and 217.8 million tons in 2013, approximately 17.1% and 21.4% of global output, respectively [27]. Meanwhile, N_2O emissions from croplands in China occur mostly during wheat and maize growing seasons in China [28]. As a result, mitigating yield-scaled N_2O emissions during these growing seasons in China plays an important role in the sustainable development of global cereal crop production. Using meta-analysis, this study integrated the results of field measurements to assess the mitigation potential of major agronomic practices on yield-scaled N_2O emissions from croplands during wheat and maize growing seasons in China.

2. Materials and Methods

2.1. Data Selection

A literature review of English and Chinese language peer-reviewed studies on N_2O emissions from Chinese wheat and maize fields prior to January 2017 was conducted using Thomson Reuters' ISI-Web of Science research database (http://thomsonreuters.com/thomson-reuters-web-of-science) and the China Knowledge Resource Integrated Database (www.cnki.net), the largest Chinese academic journal database. The 52 studies including 186 wheat and 167 maize measurements were selected based on the following criteria: (1) measurements were conducted under field conditions; (2) N_2O flux rates were measured during an entire crop growth period using the static chamber method; (3) N_2O emission and grain yield were determined simultaneously. (See Supplementary Materials Table S1 for details).

2.2. Data Analysis

For every study, the value of N_2O emission was converted to global warming potential (GWP) using a 100-year radiative forcing potential coefficient of 298 [29]. Area- and yield-scaled N_2O emissions were calculated in GWP of N_2O emission per unit of cropland and yield, respectively. In some studies, measurements were taken during more than one year; the mean value of the results measured in different years was calculated as a single observation.

Based on the field experiments conducted in selected studies, two kinds of major agronomic practices (cropping system and soil management practices) were assessed in the current study. Cropping systems were divided into four groups: winter wheat-upland crops rotation (W-U), winter wheat-rice rotation (W-R), single spring maize (M) and summer maize-upland crops rotation (M-U). W-U is mostly practiced in the semi-arid regions of northern China such as Shandong, Henan and Hebei provinces, where about 60% of China's wheat supply is produced. The winter wheat in W-U is usually planted between late September and early October and harvested between late May and early June. W-R is practiced in the humid regions along the Yangtze River of southern China; it accounts for about 28% of China's wheat production. The winter wheat in W-R is usually planted between late October and mid-November and harvested from late May to early June. M is mostly practiced in the northeast and northwest regions of China and accounts for about 44% of China's maize production.

It is usually planted between late April and early May and harvested from late September to early October. M-U is mostly practiced in the semi-arid or arid regions of eastern China and accounts for about 33% of China's maize production. It is summer maize and is usually planted in late June after the harvest of previous crops; it is harvested during similar periods to spring maize.

Weighted mean values of area-scaled N_2O emission, crop yield and yield-scaled N_2O emission were used as effect size indexes in current study to compare the difference between the four cropping systems. The equations used were as follows [24,30]:

$$Mean = \sum (y_i \times wt_i) / \sum wt_i \qquad (1)$$

$$wt_i = n \times f/o \qquad (2)$$

The details of these formulas can be found in Feng et al. [30]. Briefly, Equation (1) was used to calculate the weighted mean values of cropping systems. *Mean* is the mean value of area-scaled N_2O emission, crop yield and yield-scaled N_2O emission. Whereas y_i is the observation of area-scaled N_2O emission, crop yield and yield-scaled N_2O emission at the ith site, respectively. wt_i is the weight of the observations from the ith site and was calculated using Equation (2), in which, n is the number of replicates in the field experiment. f is the number of N_2O flux measurements per month and o is the total number of observations from the ith site. This weighting approach assigned more weight to the field measurements that were well replicated and in which more precise fluxes were estimated. The approach adjusted the weights according to total number of observations from one site to avoid dominating the dataset with studies with many observations from one site.

Four types of soil management practices were assessed in the study. These included inorganic N fertilizer application, enhanced-efficiency N fertilizers application, organic amendments and soil tillage. Their impact on area-scaled N_2O emission, crop yield, and yield-scaled N_2O emission were evaluated by the response ratio (Rr) [31]. Only studies including side-by-side comparisons were selected in the analysis of soil management practices. The rates of inorganic N fertilizer were empirically divided into six levels (N < 100, $100 \leq N < 150$, $150 \leq N < 200$, $200 \leq N < 250$, $250 \leq N < 300$ and N > 300 kg N ha^{-1} per season). The enhanced-efficiency fertilizers were categorized into two groups: nitrification inhibitors (NI) and controlled-release fertilizers (CRF). The organic amendments were classified as crop straw retention and three modes of manure application: (1) equal inorganic N fertilizer as the control with additional manure application (Equal IN + manure), (2) reduced inorganic N fertilizer with additional manure application (Reduced IN + manure), and (3) manure only application with N amount equal to the control (Manure alone). The mean retention amount of crop straw was 5768 kg ha^{-1} in selected studies. As in the three modes of manure application, the mean input rates of inorganic N fertilizer and manure were 175 kg and 194 kg N ha^{-1} for Equal IN + manure, 95 and 74 kg N ha^{-1} for Reduced IN + manure, and 0 and 154 kg N ha^{-1} for Manure alone, respectively, in selected studies. Finally, two groups of soil tillage practices (no tillage and reduced tillage) were analyzed.

The response ratio (Rr) of each management practice was calculated using Equation (3):

$$\ln Rr = \ln(x_t/x_c) \qquad (3)$$

where, x_t and x_c are the measurements for treatments and controls, respectively. The controls were non-fertilization, conventional N fertilizer, non-organic amendments and conventional tillage, respectively, which corresponded to inorganic N fertilizer application, enhanced-efficiency N fertilizers application, organic amendments and conservational tillage.

In addition, the mean of the response ratios was calculated from lnRr of individual studies using Equation (4):

$$Mr = EXP\left(\sum [\ln r(i) \times wr(i)] / \sum wr(i)\right) \qquad (4)$$

In Equation (4), $w(i)$ is the weighting factor and is estimated by Equation (5):

$$wr(i) = n \times f \qquad (5)$$

where, n is the number of experiment replicates and f is the number of N_2O flux measurements per month.

Additionally, we further analyzed the effects of cropping systems and soil management practices under different aridity regions. Aridity is an integrated indicator of rainfall and potential evapotranspiration. Following the generalized climate classification scheme for Global-Aridity values, study sites with an aridity index < 0.65 were classified as "arid"; whereas study sites with a higher index (>0.65) were classified as "humid" [25].

The meta-analysis was performed using MetaWin 2.1 (Sinauer Associates Inc., Sunderland, UK) [32]. Mean effect sizes were estimated using the random-effects model. The 95% confidence intervals (CIs) of the mean effect sizes were calculated using the bootstrapping with 4999 iterations [24,32].

3. Results and Discussion

3.1. Mitigation Potential of Cropping Systems

As shown in Figure 1, there were significant differences in area-scaled N_2O emission, crop yield and yield-scaled N_2O emission between the cropping systems. Area-scaled N_2O emission during the wheat season of W-R was significantly higher (256%) than that of W-U (Figure 1a), although average N application amounts were similar (W-R, 171.4 kg N ha^{-1}; W-U, 161.7 kg N ha^{-1}). There are two possible reasons that might explain this. Firstly, continuous flooding during the rice season of W-R could have provided more substrate and favorable soil conditions for N_2O production in the following wheat season [33]. As a result, W-R stimulated more N_2O emission during the following wheat season compared to W-U. Secondly, W-R and W-U were respectively located in the humid subtropical and semi-arid temperate regions of China. The mean annual temperature and precipitation were higher for W-R (16–24 °C, 1000–2000 mm) compared to W-U (9–15 °C, 520–980 mm) [34], A relatively higher temperature and precipitation might have increased the N_2O emission during the wheat season of W-R [35].

However, wheat yields did not significantly differ between W-U and W-R (Figure 1b). The yield-scaled N_2O emission during the wheat season of W-U was 107.8 kg CO_2 eq Mg^{-1}, which was close to the estimation of N_2O emission of global wheat production [24]. The yield-scaled N_2O emission of W-R was 304.7 kg CO_2 eq Mg^{-1}, which was significantly higher than that of W-U. Thus, increasing the planting area of W-U and reducing W-R could reduce the yield-scaled N_2O emission by 64.6% (196.9 kg CO_2 eq Mg^{-1}) without wheat yield loss.

There was no significant difference in area-scaled N_2O emissions between M and M-U during the maize season (Figure 1d). However, the maize yield of M was significantly higher than that of M-U by 25.7% (Figure 1e). In China, spring maize is usually planted between late April and early May and harvested from late September to early October [36], while summer maize is usually planted in late June after harvesting previous crops and harvested at the same time as spring maize [10]. As a result, the longer growth period of spring maize contributed to the relatively higher yield.

The yield-scaled N_2O emission of M-U was 144.1 kg CO_2 eq Mg^{-1} (Figure 1f), which was also close to the estimation of N_2O emission of global maize production [24]. But the yield-scaled N_2O emission of M (86.1 kg CO_2 eq Mg^{-1}) was significantly lower than that of M-U. Although increasing M did not reduce the N_2O emission per unit of cropland, the N_2O emission per unit of maize yield could be mitigated by 40.2% (58.0 kg CO_2 eq Mg^{-1}) due to the relatively higher yield.

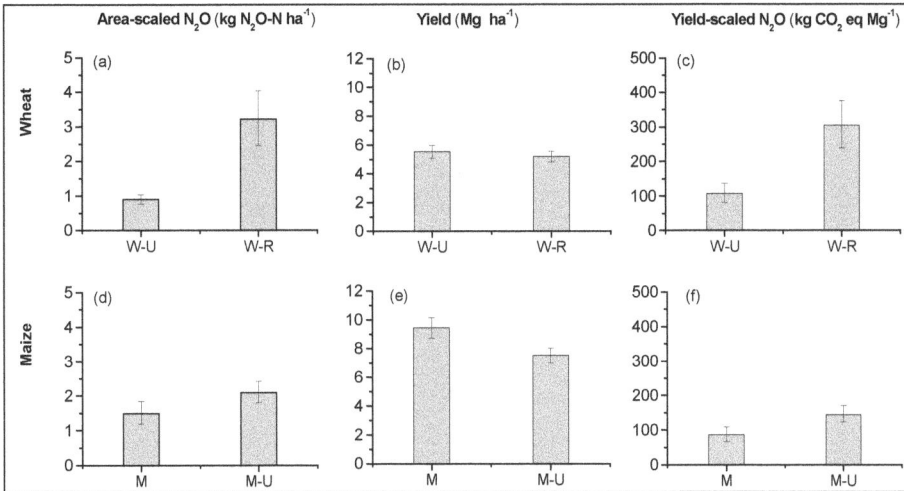

Figure 1. Impacts of cropping systems on area-scaled N_2O emission, crop yield, and yield-scaled N_2O emission during wheat and maize growing seasons ((a): area-scaled N_2O of wheat; (b) yield of wheat; (c): yield-scaled N_2O of wheat; (d): area-scaled N_2O of maize; (e) yield of maize; (f) yield-scaled N_2O of maize). The observations for winter wheat-upland crops rotation system (W-U), winter wheat-rice rotation system (W-R), single spring maize (M), and summer maize-upland crops rotation system (M-U) were 104, 76, 45, and 124, respectively. The error bars represent 95% confidence intervals.

We further analyzed the effects of aridity on the performance of the cropping system on the area-scaled N_2O emission, crop yield and yield-scaled N_2O emission. In China, W-U, M and M-U were located in both arid and humid regions, while W-R was mainly located in humid regions. So, we analyzed the differences of W-U, M and M-U in arid and humid regions (Figure 2). Though the mean N rate for M in arid area (201 kg N ha^{-1}) was higher than that in humid region (182 kg N ha^{-1}); the mean area- and yield-scaled N_2O emissions for M was significantly lower in arid than humid regions. As for W-U and M-U, the mean N rates were also higher in arid (172 kg and 175 kg N ha^{-1} for W-U and M-U,) than humid regions (119 kg and 126 kg N ha^{-1} for W-U and M-U); however, the higher N rate raised both the area-scaled N_2O emission and crop yield in arid than humid regions, resulting in no significant difference in yield-scaled N_2O emission between arid and humid regions. These results indicated that an arid climate was favorable for wheat and maize to control the yield-scaled N_2O emissions. This was possible because that low soil moisture inhibited N_2O production [25].

These results suggest that adjusting cropping systems had great potential in the mitigation of yield-scaled N_2O emission. Replacing W-R and M-U with W-U and M was the recommend strategy to mitigate yield-scaled N_2O emissions in national wheat and maize productions, especially in arid regions. During the past 20 years, the planting areas of W-R and W-U have been respectively reduced by 17.3% and 8.1%, which has been effective in mitigating yield-scaled N_2O emission. These changes are mostly affected by the comparative profits and consumption of wheat and maize in different agro-eco regions [37]. However, there has been almost no attention placed on the mitigation of N_2O emissions. Therefore, a national-scale plan is needed to balance the N_2O emission mitigation and food security by adjusting cropping systems in future wheat and maize production.

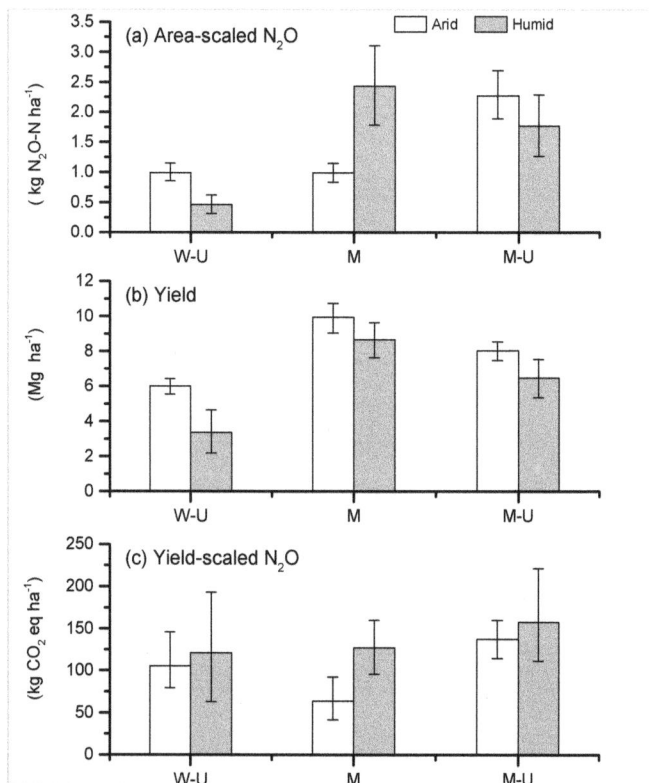

Figure 2. The impacts of aridity on the area-scaled N_2O emission (**a**), yield (**b**) and yield-scaled N_2O emission (**c**) of three cropping system.

3.2. Mitigation Potential of Inorganic N Fertilizer

The application of inorganic N fertilizer is essential for high crop production; however, it also directly provides the substrate for N_2O production. Comparing its contribution to crop yield and N_2O emission is essential for deciding the optimal N rate to mitigating yield-scaled N_2O emission. Results showed that the response ratios of N_2O emission to N addition were higher than that of crop yield at all N levels (Figure 3a), indicating that the application of inorganic N fertilizer could stimulate more N_2O emission than crop yield compared to no N fertilizer. In addition, the differences in response ratios between N_2O emission and crop yield increased with N input rates. This result was inconsistent with that of paddy fields, which showed that N fertilizer application raised more rice yield than total GWP of CH_4 and N_2O emissions [30]. In paddy fields, CH_4 emission contributed more than 80% of total GWP. When the N application rate was above 140 kg ha^{-1}, the inorganic N fertilizer began to inhibit CH_4 emission [38].

Therefore, it was difficult to obtain an optimal N rate that increased more crop yield than N_2O emission. Reducing the N application rate is the most promising option for mitigating N_2O emissions; however, it can affect crop yield. Based on this, the level that can achieve maximum economical returns or N uptake efficiency is usually suggested as the optimal N rate for the balance of crop yield enhancement and N_2O emission mitigation, because the addition of N beyond this level only slightly increases crop yield but produces far more N_2O emissions [23,39]. In our results, when the N fertilizer application rate was below 211 kg N ha^{-1}, the response ratios of N_2O emission and crop yield showed insignificant differences. However, when the N addition rate increased to 282 kg N ha^{-1}, the response ratio of N_2O emission became significantly higher (233%) than that of crop yield and the response ratio of yield-scaled N_2O emission increased significantly (Figure 3b). Thus, the suggested N rate for the balance of N_2O emission and crop yield was below 211 kg N ha^{-1}.

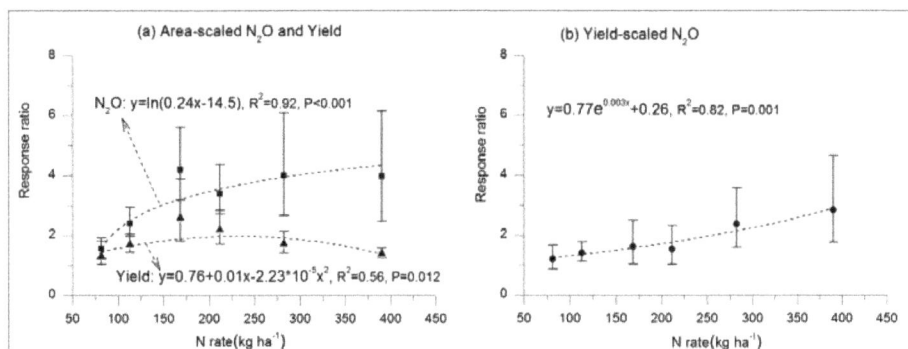

Figure 3. The relationship between N application rates and response ratios of area-scaled N_2O emission (**a**), crop yield (**a**), and yield-scaled N_2O emission (**b**). The data is expressed as mean response ratios of six N levels (N < 100, $100 \leq N < 150$, $150 \leq N < 200$, $200 \leq N < 250$, $250 \leq N < 300$ and N > 300 kg N ha^{-1} per season) with 95% confidence intervals. The observations for six N levels are 5, 12, 21, 14, 13 and 6, respectively. (Note: Only the subgroups of $100 \leq N < 150$ and $150 \leq N < 200$ have enough observations to differentiate the effects of N application under arid or humid areas; the study sites of other subgroups were all located in arid or humid regions. The results of the subgroups of $100 \leq N < 150$ and $150 \leq N < 200$ under different aridity regions were listed in the Supplementary Materials (Figure S1)).

In order to improve use efficiency of inorganic N fertilizer and reduce environmental impact, a more precise inorganic N application scheme had been recommended in major cereal crops planting regions in China since 2013 based on a national project of soil testing and fertilizer recommendation [40]. The suggested inorganic N application rates were 103–127 kg N ha^{-1}, 144–209 kg N ha^{-1} and 236–258 kg N ha^{-1} for low-yield (<6 Mg ha^{-1}), medium-yield (6–9 Mg ha^{-1}) and high-yield (>9 Mg ha^{-1}) croplands of wheat production respectively, and 105–167 kg N ha^{-1}, 136–206 kg N ha^{-1}, and 190–235 kg N ha^{-1} for low-yield (<7.5 Mg ha^{-1}), medium-yield (7.5–10.5 Mg ha^{-1}) and high-yield (>10.5 Mg ha^{-1}) croplands of maize production respectively. Only the N rates for high-yield croplands of wheat and maize production exceeded 211 kg N ha^{-1}. Thus, reducing N application rates in the high-yield croplands is essential for the mitigation of N_2O emission. However, reducing the N rate in high-yield croplands could decrease crop production and farmer's profits, because the output of high-yield croplands makes up a large part of farmers' profits. Therefore, financial incentives might be required to compensate farmers for reducing N application rates. Additionally, more work is needed to optimize the application options of inorganic N fertilizer (such as N source, placement and application time) that allow for N-rate reductions to better match crop growth demand and mitigate N_2O emissions without yield loss in high-yield croplands [41]. Improving these options could lessen the need for financial compensation [42].

3.3. Mitigation Potential of Enhanced-Efficiency N Fertilizers

Enhanced-efficiency N fertilizers have been developed to increase crop N use efficiency and decrease N loss to the environment. Our results (Figure 4) showed that, compared to conventional N fertilizer, NI significantly reduced N_2O emissions by 34.2%, which was similar to the report by [18]. NI can delay the bacterial oxidation of ammonium to nitrite and subsequently reduce the denitrification, which is an important process of N_2O production in upland soil [43]. So the application of NI can mitigate N_2O emission from soil. Additionally, the delay of nitrification also provides a better opportunity for the crop to uptake N fertilizer. Our results showed that wheat and maize yield increased 12.9% due to NI application compared to conventional N fertilizer, thereby resulting in a significant reduction in yield-scaled N_2O emission by 41.7% (Figure 4). Aridity did not affect the performance of NI. The effect sizes of NI on area-scaled N_2O emission, crop yield and yield-scaled N_2O did not show significant difference.

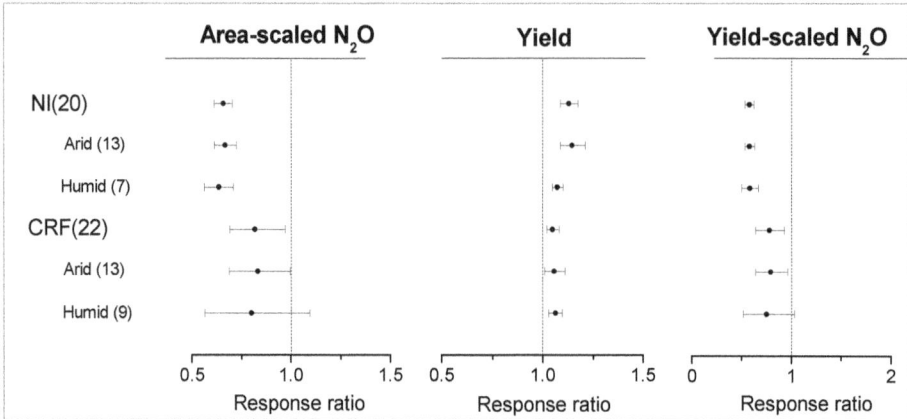

Figure 4. Impacts of enhanced-efficiency N fertilizers on area-scaled N_2O emission, crop yield and yield-scaled N_2O emission. The data is expressed as mean response ratio with 95% confidence intervals. The numbers of observations are indicated in the parentheses.

CRF also showed significant effect size on N_2O emission and crop yield (Figure 4). Compared to conventional inorganic fertilizer, N_2O emission was reduced 18.2% by CRF, which was lower than the report by [18]. In addition, crop yield increased by 4.9%. Yield-scaled N_2O emission was significantly mitigated by 22.0% due to CRF. The effect size of CRF was affected by aridity. CRF performed better in arid than humid regions. Though CRF significantly enhanced the crop yield in humid regions, its effects on area- and yield-scaled N_2O emissions were not significant. In this analysis, the CRF in selected studies was polymer-coated urea; this coating can slow down the release of N and subsequently reduce the loss of N_2O emission [44]. High soil moisture in humid regions may weaken the effect of CRF on controlling N release, and increase the N release from CRF [45], which may raise the N_2O production.

CRF did not perform as well as NI. The mitigation effect of CRF on area- and yield-scaled N_2O emissions was weaker than NI, and showed a greater 95% CI. A possible reason for this was that the release of nitrogen from CRF was easily affected by environmental factors such as soil moisture and temperature [13,46]. If N released from the CRF did not synchronize with crop N demands, the redundant N in favorable environmental conditions could raise the amount of N_2O emissions from denitrification [14,18]. The effect of CRF on N_2O emission might depend on field condition, climate aridity and crop growth.

Both NI and CRF showed a significant ability in mitigating yield-scaled N_2O emissions, and could be recommended to mitigate N_2O emissions without yield loss in wheat and maize production in China. However, NI and CRF are not widely used by farmers in cereal crop production in China since the additional costs of NI and CRF only increase limited crop yields. As a result, additional studies are needed to optimize the management options such as application time and irrigation approaches to improve the effectiveness of enhanced-efficiency fertilizers, especially CRF, on crop productivity [47] and to encourage farmers to use enhanced-efficiency fertilizers in maize and wheat production.

3.4. Mitigation Potential of Organic Amendments

No significant effect of crop straw retention was found on area-scaled N_2O emission, crop yield and yield-scaled N_2O emission (Figure 5). Existing evidence showed that the effect of straw retention can be either positive or negative on N_2O emissions [48,49]. On one hand, straw retention can increase soil temperature and/or moisture, which can stimulate the microbial process of nitrification and denitrification, and thereby raise N_2O emissions [49]. On the other hand, straw with a high C/N ratio can immobilize soil mineral N and decrease soil N availability, consequently leading to a reduction in the substrate N for N_2O production [48,50]. Additionally, the allelochemicals produced from the decomposition of crop straw can reduce the activity of nitrifiers and inhibit N_2O production [51]. The integrated impact of these effects was mostly determined by basal inorganic N application rate, retention timing and straw type [52,53]. As shown in Figure 6, the response ratio of N_2O emissions increased with the application rates of basal inorganic N fertilizer. Field experiments also reported that incorporation of straw with low N application rate could reduce N_2O emission compared to no straw retention [48]. Additionally, the performance of crop straw was affected by climate aridity. Crop straw returning significantly increased the N_2O emission in arid region, but did not affect N_2O emission in humid regions. In arid regions, the positive effect of crop straw, such as increased soil moisture and substrate C, may raise the N_2O emission. As in humid regions, the decomposition of straw possibly intensified the O_2 limitation due to the rapid microbial decomposition, and active the further reduction from N_2O to N_2 [20].

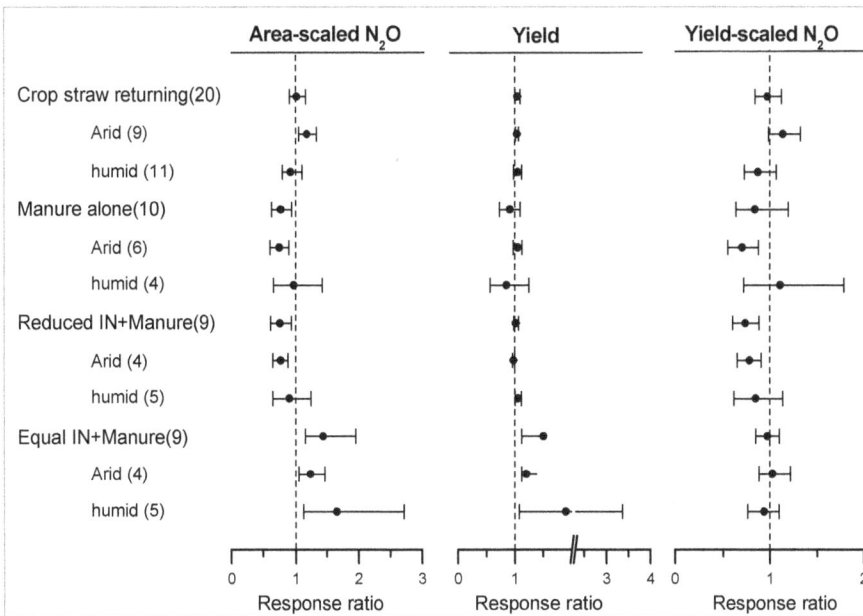

Figure 5. Impacts of organic amendments on area-scaled N_2O emission, crop yield and yield-scaled N_2O emission.

Contradictory effects (either an increase or a reduction) of manure application on N_2O emissions have been demonstrated in previous field experiments [10,54]. Our results showed that the mode of manure application was an important factor influencing the impacts on N_2O emissions (Figure 5). Manure application without inorganic N fertilizer (manure alone) significantly reduced N_2O emissions by 22.8% compared to inorganic N fertilizer. Generally, incorporation of manure in agricultural soil can provide abundant easily decomposable C and cause N_2O to be completely denitrified to N_2 [55]. Therefore, although the total N amount in manure was the same as that in the inorganic N fertilizer control; the N_2O emission was significantly lower under manure alone. However, manure alone did not reduce yield-scaled N_2O emission due to decreases in wheat and maize yields (Figure 5). Aridity affected the effect of manure alone treatment. In arid regions, manure alone significantly mitigated the N_2O emission by 29.2%, which may be primarily due to the enhancement of the crop yield. Soil water was an important factor affecting the crop yield in arid regions. The application of manure could increase the rainfall use efficiency of crop plants by improving soil penetration [56], which provided a benefit to the enhancement of crop yield.

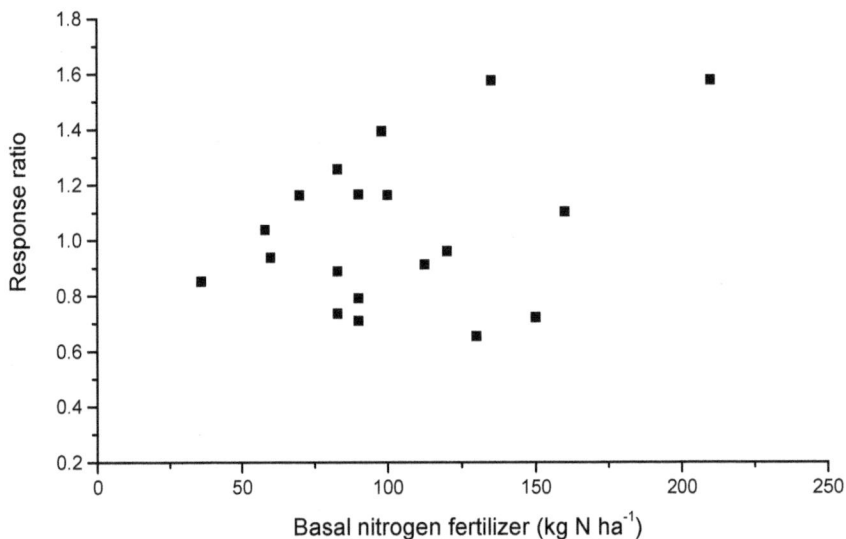

Figure 6. The relationship between inorganic N application rates of basal fertilizer and response ratios of crop straw returning on area-scaled N_2O emission.

Partial substitution of inorganic N with manure in basal fertilizer (mean: 43.6%, range: 22% to 50% in selected studies) (Reduced IN + manure) significantly reduced N_2O emission by 24.0% but had no significant effects on wheat and maize yields. Consequently, the yield-scaled N_2O emission was significantly reduced by 25.8%. However, additional manure application with an equal inorganic N fertilizer amount to the control (Equal IN + manure) significantly increased N_2O emission by 44.6% (Figure 5). Under the same chemical N conditions, manure application can provide additional N and available C for the microbial processes of nitrification and denitrification [57], and thereby significantly stimulate N_2O emission. Although crop yield increased by 50.5% under Equal IN + manure, yield-scaled N_2O emission showed no significant difference between Equal IN + manure and control. Therefore, partial substitution of inorganic N with manure can be suggested as a climate smart practice for balancing crop yield increase and N_2O emission mitigation. Recently, a long-term field experiment in North China also demonstrated that replacing 50% of inorganic N with manure significantly reduced N_2O emission by 41.7% without significant decreases in wheat and maize yields [10].

3.5. Mitigation Potential of Soil Tillage

As shown in Figure 7, no tillage significantly increased N_2O emission (26.6%) compared to conventional tillage. This was due to the fact that no tillage tended to increase soil moisture and bulk density and maintained the N fertilizers on the soil surface [13], consequently resulting in a significant stimulation in N_2O emissions. Wheat and maize yields were lower under no tillage than conventional tillage (Figure 7), which was consistent with a previous report [58]. Therefore, yield-scaled N_2O emission increased significantly by 42.6% as a result of no tillage compared to conventional tillage (Figure 7). As to reduced tillage, no significant effects were found on area-scaled N_2O emission, crop yield and yield-scaled N_2O emission compared to conventional tillage (Figure 7).

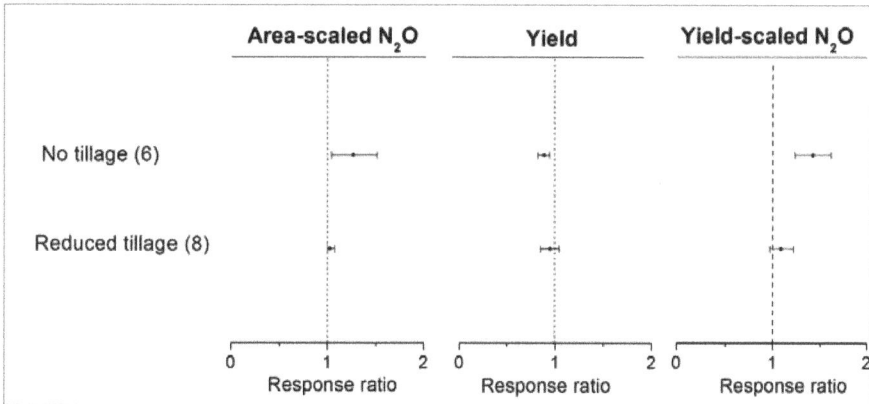

Figure 7. Impacts of reduced and no tillage on area-scaled N_2O emission, crop yield and yield-scaled N_2O emission. (Note: only one observation located in arid region for NT and two observations located in humid area for RT. So, the impact of climate aridity was not analyzed.).

The effect of soil tillage on N_2O emission was affected by N placement, duration and environmental factors [25]. The different effects on N_2O emission between reduced tillage and no tillage could be attributed to the placement depth of N fertilizer. In the select studies of this analysis, the N fertilizer was generally placed on soil surface under no-tillage and incorporated into soil layers (5–10 cm) under reduced tillage. A previous study had reported that tillage interacted with N fertilizer placement depth to regulate N_2O emission; no tillage with surface N placement tends to stimulate N_2O emission compared to reduced tillage [59]. Thus, deep N placement is suggested in no tillage to reduce N_2O emission. In addition, some studies have suggested that tillage duration was an important factor influencing the impact on N_2O emission [16]. Based on a meta-analysis, for example, Kessel et al. [25] reported that reduced and no tillage significantly mitigated the N_2O emission by 14% when experiment durations lasted > 10 years, especially in a dry climate. Recently, a 10-year tillage experiment in the North China Plain reported that reduced tillage (rotary tillage and subsoiling) mitigated N_2O emissions and improved crop productivity in wheat-maize rotation system [60]. However, the field experiments on the effects of tillage on N_2O emissions and crop yields in China were still limited; experiment durations were less than five years in the selected experiments of our meta-analysis. Therefore, additional field experiments are needed to investigate the long-term effects of tillage on area and yield-scaled N_2O emissions, and both short- and long-term effects should be considered in the evaluation of tillage impacts.

4. Conclusions

Agronomic practices affect both crop yield and N_2O emission. Ecological intensification of agronomic practices plays an important role in the sustainable development of future crop production. Our study comprehensively evaluated the impacts of main agronomic practices on area- and yield-scaled N_2O emissions, and analyzed the mitigation potential of N_2O emission by optimizing cropping practices during wheat and maize seasons in China. Results demonstrated that adjusting cropping systems, NI, CRF and reduced IN+ manure were recommend for the mitigation of yield-scaled N_2O emission during wheat and maize growing seasons. Policy options are essential to encourage the application of these strategies. For example, a projected macroscopic plan is needed to adjust cropping systems in national wheat and maize production for the mitigation of N_2O emission. Policies that provided sufficient financial compensation for farmers are required to change the agricultural practices.

Due to limited data, this study did not analyze N_2O emission and crop yield in non-wheat and non-maize growing seasons of four cropping systems. More studies should be conducted to investigate year-round N_2O emissions and crop yield during complete durations of these cropping systems. Additionally, this study only evaluated direct N_2O emissions from soil during wheat and maize growing seasons. In the future, indirect N_2O emissions and carbon cost should be considered in the assessment of the mitigation potential of cropping practices. For example, although no tillage increased direct N_2O emissions from soil, it reduced machine and diesel oil input. As a result, a life-cycle assessment of cropping practices could provide more precise references for the recommendation of management practices.

Supplementary Materials: The following are available online at www.mdpi.com/2071-1050/9/7/1201/s1.

Acknowledgments: This work was supported by the National Key Technology Support Program of China (2011BAD16B14).

Author Contributions: Wenling Gao collected the data, conducted the data analysis and drafted the paper. Xinmin Bian designed the data analysis and revised the paper.

Conflicts of Interest: The authors declare no conflicts of interest.

References

1. Wrage, N.; Velthof, G.L.; Van Beusichem, M.L.; Oenema, O. Role of nitrifier denitrification in the production of nitrous oxide. *Soil Biol. Biochem.* **2001**, *33*, 1723–1732. [CrossRef]
2. Intergovernmental Panel on Climate Change. Summary for Policymakers. In *Climate Change 2013: The Physical Science Basis*; Contribution of Working Group I to the Fifth Assessment Report of the Intergovernmental Panel on Climate Change; Stocker, T.F., Qin, D., Plattner, G.-K., Tignor, M., Allen, S.K., Boschung, J., Nauels, A., Xia, Y., Bex, V., Midgley, P.M., Eds.; Cambridge University Press: Cambridge, UK; New York, NY, USA, 2013.
3. Alexandratos, N. How to feed the world in 2050. In Proceedings of the Technical Meeting of Experts, Rome, Italy, 24–26 June 2009; FAO: Rome, Italy, 2009.
4. Tilman, D.; Balzer, C.; Hill, J.; Befort, B.L. Global food demand and the sustainable intensification of agriculture. *Proc. Natl. Acad. Sci. USA* **2011**, *108*, 20260–20264. [CrossRef] [PubMed]
5. Cai, X.; Zhang, X.; Wang, D. Land availability for biofuel production. *Environ. Sci. Technol.* **2010**, *45*, 334–339. [CrossRef] [PubMed]
6. Popp, A.; Lotze-Campen, H.; Bodirsky, B. Food consumption, diet shifts and associated non-CO_2 greenhouse gases from agricultural production. *Glob. Environ. Chang.* **2010**, *20*, 451–462. [CrossRef]
7. Van Beek, C.L.; Meerburg, B.G.; Schils, R.L.M.; Verhagen, J.; Kuikman, P.J. Feeding the world's increasing population while limiting climate change impacts: Linking N_2O and CH_4 emissions from agriculture to population growth. *Environ. Sci. Policy* **2010**, *13*, 89–96. [CrossRef]
8. Heffer, P. *Assessment of Fertilizer Use by Crop at the Global Level*; International Fertilizer Industry Association (IFA): Paris, France, 2013; Available online: www.fertilizer.org//En/Statistics/Agriculture_Committee_Databases.aspx (accessed on 21 May 2014).
9. Smith, P.; Martino, D.; Cai, Z.; Gwary, D.; Janzen, H.; Kumar, P.; McCarl, B.; Ogle, S.; O'Mara, F.; Rice, C.; et al. Agriculture. In *Climate Change 2007: Mitigation*; Contribution of Working Group III to the Fourth Assessment Report of the Intergovernmental Panel on Climate Change; Metz, B., Davidson, O.R., Bosch, P.R., Dave, R., Meyer, L.A., Eds.; Cambridge University Press: Cambridge, UK; New York, NY, USA, 2007.
10. Cai, Y.; Ding, W.; Luo, J. Nitrous oxide emissions from Chinese maize-wheat rotation systems: A 3-year field measurement. *Atmos. Environ.* **2013**, *65*, 112–122. [CrossRef]

11. Parkin, T.; Hatfield, J. Influence of nitrapyrin on N_2O losses from soil receiving fall-applied anhydrous ammonia. *Agric. Ecosyst. Environ.* **2010**, *136*, 81–86. [CrossRef]

12. Ma, Y.; Sun, L.; Zhang, X.; Yang, B.; Wang, J.; Yin, B.; Yan, X.; Xiong, Z. Mitigation of nitrous oxide emissions from paddy soil under conventional and no-till practices using nitrification inhibitors during the winter wheat-growing season. *Biol. Fert. Soils* **2013**, *49*, 627–635. [CrossRef]

13. Venterea, R.T.; Bijesh, M.; Dolan, M.S. Fertilizer source and tillage effects on yield-scaled nitrous oxide emissions in a corn cropping system. *J. Environ. Qual.* **2011**, *40*, 1521–1531. [CrossRef] [PubMed]

14. Hu, X.; Su, F.; Ju, X.; Gao, B.; Oenema, O.; Christie, P.; Huang, B.; Jiang, R.; Zhang, F. Greenhouse gas emissions from a wheat-maize double cropping system with different nitrogen fertilization regimes. *Environ. Pollut.* **2013**, *176*, 198–207. [CrossRef] [PubMed]

15. Cole, C.V.; Duxbury, J.; Freney, J.; Heinemeyer, O.; Minami, K.; Mosier, A.; Paustian, K.; Rosenberg, N.; Sampson, N.; Sauerbeck, D.; et al. Global estimates of potential mitigation of greenhouse gas emissions by agriculture. *Nutr. Cycl. Agroecosyst.* **1997**, *49*, 221–228. [CrossRef]

16. Six, J.; Ogle, S.M.; Conant, R.T.; Mosier, A.R.; Paustian, K. The potential to mitigate global warming with no-tillage management is only realized when practised in the long term. *Glob. Chang. Biol.* **2004**, *10*, 155–160. [CrossRef]

17. Rochette, P.; Worth, D.E.; Lemke, R.L.; McConkey, B.G.; Pennock, D.J.; Wagner-Riddle, C.; Desjardins, R. Estimation of N_2O emissions from agricultural soils in Canada. I. Development of a country-specific methodology. *Can. J. Soil Sci.* **2008**, *88*, 641–654. [CrossRef]

18. Akiyama, H.; Yan, X.; Yagi, K. Evaluation of effectiveness of enhanced-efficiency fertilizers as mitigation options for N_2O and NO emissions from agricultural soils: Meta-analysis. *Glob. Chang. Biol.* **2010**, *16*, 1837–1846. [CrossRef]

19. Kim, D.G.; Hernandez-Ramirez, G.; Giltrap, D. Linear and nonlinear dependency of direct nitrous oxide emissions on fertilizer nitrogen input: A meta-analysis. *Agric. Ecosyst. Environ.* **2013**, *168*, 53–65. [CrossRef]

20. Chen, H.; Li, X.; Hu, F.; Shi, W. Soil nitrous oxide emissions following crop residue addition: A meta-analysis. *Glob. Chang. Biol.* **2013**, *19*, 2956–2964. [CrossRef] [PubMed]

21. Zhao, X.; Liu, S.; Pu, C.; Zhang, X.; Xue, J.; Zhang, R.; Wang, Y.; Lal, R.; Zhang, H.; Chen, F. Methane and nitrous oxide emissions under no-till farming in China: A meta-analysis. *Glob. Chang. Biol.* **2016**, *22*, 1372–1384. [CrossRef] [PubMed]

22. Xia, L.; Lam, S.K.; Chen, D.; Wang, J.; Tang, Q.; Yan, X. Can knowledge-based N management produce more staple grain with lower greenhouse gas emission and reactive nitrogen pollution? A meta-analysis. *Glob. Chang. Biol.* **2017**, *23*, 1917–1925. [CrossRef] [PubMed]

23. Van Groenigen, J.W.; Velthof, G.L.; Oenema, O.; Van Groenigen, K.J.; Van Kessel, C. Towards an agronomic assessment of N_2O emissions: A case study for arable crops. *Eur. J. Soil Sci.* **2010**, *61*, 903–913. [CrossRef]

24. Linquist, B.; Groenigen, K.J.; Adviento-Borbe, M.A.; Pittelkow, C.; Kessel, C. An agronomic assessment of greenhouse gas emissions from major cereal crops. *Glob. Chang. Biol.* **2012**, *18*, 194–209. [CrossRef]

25. Van Kessel, C.; Venterea, R.; Six, J.; Adviento-Borbe, M.A.; Linquist, B.; Van Groenigen, K.J. Climate, duration, and N placement determine N_2O emissions in reduced tillage systems: A meta-analysis. *Glob. Chang. Biol.* **2013**, *19*, 33–44. [CrossRef] [PubMed]

26. Intergovernmental Panel on Climate Change. Agriculture, Forestry and Other Land Use (AFOLU). In *Climate Change 2014, Mitigation of Climate Change*; Contribution of Working Group III to the Fifth Assessment Report of the Intergovernmental Panel on Climate Change; Edenhofer, O., Pichs-Madruga, R., Sokona, Y., Farahani, E., Kadner, S., Seyboth, K., Adler, A., Baum, I., Brunner, S., Eickemeier, P., Eds.; Cambridge University Press: Cambridge, UK; New York, NY, USA, 2014; Chapter 11.

27. FAOSTAT. Available online: http://faostat.fao.org (accessed on 12 October 2014).

28. Xing, G. N_2O emission from cropland in China. *Nutr. Cycl. Agroecosyst.* **1998**, *52*, 249–254. [CrossRef]

29. Intergovernmental Panel on Climate Change. *Climate Change 2007: The Physical Science Basis*; Solomon, S., Qin, D., Manning, M., Chen, Z., Marquis, M., Averyt, K.B., Tignor, M., Miller, H.L., Eds.; Cambridge University Press: Cambridge, UK; New York, NY, USA, 2007.

30. Feng, J.; Chen, C.; Zhang, Y.; Song, Z.; Deng, A.; Zheng, C.; Zhang, W. Impacts of cropping practices on yield-scaled greenhouse gas emissions from rice fields in China: A meta-analysis. *Agric. Ecosyst. Environ.* **2013**, *164*, 220–228. [CrossRef]

31. Hedges, L.V.; Gurevitch, J.; Curtis, P.S. The Meta-analysis of response ratios in experimental ecology. *Ecology* **1999**, *80*, 1150–1156. [CrossRef]

32. Rosenberg, M.S.; Adams, D.C.; Gurevitch, J. *MetaWin-Statistical Software for Meta-Analysis*; Sinauer Associates Inc.: Sunderland, UK, 2000.

33. Peng, S.; Hou, H.; Xu, J.; Yang, S.; Mao, Z. Lasting effects of controlled irrigation during rice-growing season on nitrous oxide emissions from winter wheat croplands in Southeast China. *Paddy Water Environ.* **2013**, *11*, 583–591. [CrossRef]

34. Zhao, G. Study on Chinese wheat planting regionalization (I). *J. Triticeae Crop.* **2010**, *30*, 886–895. (In Chinese with English abstract).

35. Smith, K.A.; Thomson, P.E.; Clayton, H.; McTaggart, I.P.; Conen, F. Effects of temperature, water content and nitrogen fertilisation on emissions of nitrous oxide by soils. *Atmos. Environ.* **1998**, *32*, 3301–3309. [CrossRef]

36. Hou, P.; Gao, Q.; Xie, R.; Li, S.; Meng, Q.; Kirkby, E.A.; Römheld, V.; Müller, T.; Zhang, F.; Cui, Z.; et al. Grain yields in relation to N requirement: Optimizing nitrogen management for spring maize grown in China. *Field Crop. Res.* **2012**, *129*, 1–6. [CrossRef]

37. Deng, Z.; Feng, Y.; Zhang, J.; Wang, J. Regional pattern change and its influencing factors of cereals crops production in China. *Macroeconomics* **2014**, *3*, 94–100. (In Chinese with English abstract).

38. Banger, K.; Tian, H.; Lu, C. Do nitrogen fertilizers stimulate or inhibit methane emissions from rice fields? *Glob. Chang. Biol.* **2012**, *18*, 3259–3267. [CrossRef]

39. Hoben, J.; Gehl, R.; Millar, N.; Grace, P.; Robertson, G. Nonlinear nitrous oxide (N_2O) response to nitrogen fertilizer in on-farm corn crops of the US Midwest. *Glob. Chang. Biol.* **2011**, *17*, 1140–1152. [CrossRef]

40. Ministry of Agriculture of the People's Republic of China. Fertilizer Recommendation for Rice, Wheat and Maize Production in Major Growing Regions. Available online: http://www.moa.gov.cn/govpublic/ZZYGLS/201307/t20130729_3541508.htm (accessed on 29 July 2013). (In Chinese)

41. Decock, C. Mitigating nitrous oxide emissions from corn cropping systems in the Midwestern U.S.: Potential and data gaps. *Environ. Sci. Technol.* **2014**, *48*, 4247–4256. [CrossRef] [PubMed]

42. Venterea, R.T.; Halvorson, A.D.; Kitchen, N.; Liebig, M.A.; Cavigelli, M.A.; Grosso, S.J.D.; Motavalli, P.P.; Nelson, K.A.; Spokas, K.A.; Singh, B.P.; et al. Challenges and opportunities for mitigating nitrous oxide emissions from fertilized cropping systems. *Front. Ecol. Environ.* **2012**, *10*, 562–570. [CrossRef]

43. Liu, C.; Wang, K.; Zheng, X. Effects of nitrification inhibitors (DCD and DMPP) on nitrous oxide emission, crop yield and nitrogen uptake in a wheat-maize cropping system. *Biogeosciences* **2013**, *10*, 2427–2437. [CrossRef]

44. Ji, Y.; Liu, G.; Ma, J.; Xu, H.; Yagi, K. Effect of controlled-release fertilizer on nitrous oxide emission from a winter wheat field. *Nutr. Cycl. Agroecosyst.* **2012**, *94*, 111–122. [CrossRef]

45. Feng, J.; Li, F.; Deng, A.; Feng, X.; Fang, F.; Zhang, W. Integrated assessment of the impact of enhanced-efficiency nitrogen fertilizer on N_2O emission and crop yield. *Agric. Ecosyst. Environ.* **2016**, *231*, 218–228. [CrossRef]

46. Jiang, J.; Hu, Z.; Sun, W.; Huang, Y. Nitrous oxide emissions from Chinese cropland fertilized with a range of slow-release nitrogen compounds. *Agric. Ecosyst. Environ.* **2010**, *135*, 216–225. [CrossRef]

47. Abalos, D.; Jeffery, S.; Sanz-Cobena, A.; Guardia, G.; Vallejo, A. Meta-analysis of the effect of urease and nitrification inhibitors on crop productivity and nitrogen use efficiency. *Agric. Ecosyst. Environ.* **2014**, *189*, 136–144. [CrossRef]

48. Ma, E.; Zhang, G.; Ma, J.; Xu, H.; Cai, Z.; Yagi, K. Effects of rice straw returning methods on N_2O emission during wheat-growing season. *Nutr. Cycl. Agroecosyst.* **2010**, *88*, 463–469. [CrossRef]

49. Liu, C.; Wang, K.; Meng, S.; Zheng, X.; Zhou, Z.; Han, S.; Chen, D.; Yang, Z. Effects of irrigation, fertilization and crop straw management on nitrous oxide and nitric oxide emissions from a wheat-maize rotation field in northern China. *Agric. Ecosyst. Environ.* **2011**, *140*, 226–233. [CrossRef]

50. McKenney, D.; Wang, S.; Drury, C.; Findlay, W. Dentrification and mineralization in soil amended with legume, grass, and corn residues. *Soil Sci. Soc. Am. J.* **1993**, *57*, 1013–1020. [CrossRef]

51. Huang, Y.; Zhang, F.; Liu, S.; Cao, Q. Effect of allelochemicals on N_2O emission from soil. *Acta Sci. Circumst.* **1999**, *19*, 478–482. (In Chinese with English abstract).

52. Hao, X.; Chang, C.; Carefoot, J.M.; Janzen, H.H.; Ellert, B.H. Nitrous oxide emissions from an irrigated soil as affected by fertilizer and straw management. *Nutr. Cycl. Agroecosyst.* **2001**, *60*, 1–8. [CrossRef]

53. Baggs, E.M.; Stevenson, M.; Pihlatie, M.; Regar, A.; Cook, H.; Cadisch, G. Nitrous oxide emissions following application of residues and fertiliser under zero and conventional tillage. *Plant Soil* **2003**, *254*, 361–370. [CrossRef]

54. Adviento-Borbe, M.; Kaye, J.; Bruns, M.; McDaniel, M.; McCoy, M.; Harkcom, S. Soil greenhouse gas and ammonia emissions in long-term maize-based cropping systems. *Soil Sci. Soc. Am. J.* **2010**, *74*, 1623–1634. [CrossRef]

55. Sánchez-Martín, L.; Vallejo, A.; Dick, J.; Skiba, U. The influence of soluble carbon and fertilizer nitrogen on nitric oxide and nitrous oxide emissions from two contrasting agricultural soils. *Soil Biol. Biochem.* **2008**, *40*, 142–151. [CrossRef]

56. Wang, X.; Jia, Z.; Liang, L.; Yang, B.; Ding, R.; Nie, J.; Wang, J. Impacts of manure application on soil environment, rainfall use efficiency and crop biomass under dryland farming. *Sci. Rep.* **2016**, *6*, 20994. [CrossRef] [PubMed]

57. Van Groenigen, J.W.; Kasper, G.J.; Velthof, G.L.; Van den Pol-van Dasselaar, A.; Kuikman, P.J. Nitrous oxide emissions from silage maize fields under different mineral nitrogen fertilizer and slurry applications. *Plant Soil* **2004**, *263*, 101–111. [CrossRef]

58. Van den Putte, A.; Govers, G.; Diels, J.; Gillijns, K.; Demuzere, M. Assessing the effect of soil tillage on crop growth: A meta-regression analysis on European crop yields under conservation agriculture. *Eur. J. Agron.* **2010**, *33*, 231–241. [CrossRef]

59. Venterea, R.T.; Stanenas, A.J. Profile analysis and modeling of reduced tillage effects on soil nitrous oxide flux. *J. Environ. Qual.* **2008**, *37*, 1360–1367. [CrossRef] [PubMed]

60. Tian, S.; Wang, Y.; Ning, T.; Zhao, H.; Wang, B.; Li, N.; Li, Z.; Chi, S. Greenhouse gas flux and crop productivity after 10 years of reduced and no tillage in a wheat-maize cropping system. *PLoS ONE* **2013**, *8*, e73450. [CrossRef] [PubMed]

sustainability

MDPI

Article

Social Vulnerability Assessment by Mapping Population Density and Pressure on Cropland in Shandong Province in China during the 17th–20th Century

Yu Ye [1,2,*], Xueqiong Wei [1], Xiuqi Fang [1] and Yikai Li [1]

[1] School of Geography, Faculty of Geographical Science, Beijing Normal University, Beijing 100875, China; weixueqiong1988@126.com (X.W.); xfang@bnu.edu.cn (X.F.); liyikai2016@foxmail.com (Y.L.)
[2] Key Laboratory of Environment Change and Natural Disaster, Ministry of Education, Beijing Normal University, Beijing 100875, China
* Correspondence: yeyuleaffish@bnu.edu.cn; Tel.: +86-132-4188-4630

Received: 31 May 2017; Accepted: 30 June 2017; Published: 5 July 2017

Abstract: Cropland area per capita and pressure index on cropland are important parameters for measuring the social vulnerability and sustainability from the perspective of food security in a certain region in China during the historical periods. This study reconstructed the change in spatial distribution of cropland area per labor/household and pressure index on cropland during the 17th–20th century by using historical documents, regression analysis, pressure index model, and GIS (geographic information system). Following this, we analyzed the impacting process of climate change and sustainability of cropland use during the different periods. The conclusions of this study are as follows: (i) there was an obvious spatial difference of labor/household density, as there was higher density in three agricultural areas, which had the same pattern as cropland distribution during the same periods; (ii) Cropland area per capita was relatively higher during the 17th–18th century, which were above 0.4 ha/person in the majority of counties and were distributed homogenously. Until the 19th century and the beginning of 20th century, cropland area per capita in a considerable proportion of regions decreased below 0.2 ha/person, which embodies the increase in social vulnerability and unsustainability at that time; (iii) The pressure index on cropland also showed a spatial pattern similar to cropland area per capita, which presented as having a lower threshold than nowadays. During the 17th–18th century, there was no pressure on cropland. In comparison, in the 19th century and at the beginning of 20th century, two high-value centers of pressure index on cropland appeared in the Middle Shandong and the Jiaodong region. As a result, pressure on cropland use increased and a food crisis was likely to have been created; (iv) A higher extent of sustainable cropland use corresponded to the cold period, while a lower extent of sustainable cropland use corresponded to the warm period in Shandong over the past 300 years. The turning point of the 1680s from dry to wet was not distinctively attributed to the decrease in the extent of sustainable cropland use in Shandong. Since the beginning of the 20th century, the increasing pressure on the sustainability of cropland use finally intensified the social conflict and increased the probability of social revolts.

Keywords: pressure on cropland; labor/household density; Shandong Province in China; 17th–20th century

1. Introduction

Human interference has occurred with the climate system, with climate change posing a threat to natural systems and human sustainable development. The core concept of the fifth assessment report of Work Group II of the Intergovernmental Panel on Climate Change (IPCC WGII AR5) is the theme of impact, adaptation, and vulnerability related to climate change. It illustrates that climate-related risk results from the interaction of natural hazards (including hazardous events and trends) with the vulnerability and exposure of human and natural systems [1,2]. In the traditional agricultural society of historical China, climate change first impacts the level of food production, which hinders the improvement in living standards and social development by a transmission of forcing–responding chain. The forcing–responding chain means that the impact of climate change passes on from climate change to agriculture harvest to food supply, and finally to famine and social stability [3]. Overall, the extent of the impact on society by climate change depends on the social vulnerability and human adaptation actions. Socially sustainable development depends on whether the contradiction between human and land is resolved.

In the majority of current studies on the impact of historical climate change and human adaptation, climate change and societal stability have been well studied. For example, some typical researchers analyzed the impact of climate change on violent conflict in Europe over the last millennium [4]; the relationship between climate and the collapse of Maya civilization [5]; North Atlantic seasonality and implications for Norse colonies [6]; the relationship between sun, climate, hunger, and mass migration [7]; linkage of climate with Chinese dynastic change [8], and so on. However, some intermediate factors in this influencing process (e.g., population, agricultural production, and policy adjustment) were not fully considered. In addition, social vulnerability has not been stressed in these similar international studies.

In China, some researchers discussed the Chinese population and cropland area mainly from the perspective of historical geography or agricultural history. For example, Ge [9] studied the history of Chinese demographic composition, population change, and distribution; Li [10] analyzed the impact of climate change on several instances of Chinese historical population fluctuation; He [11] firstly evaluated the ancient land data in China; and other researchers [12–16] evaluated Chinese historical land data. They have produced methods for the data estimation of Chinese population and cropland area. Recently, scientists working in the field of global change have reconstructed the spatial distribution of historical cropland cover in China [17,18] or regions in China [19–22]. However, to understand the dynamics of climate-related risks, it would be better to combine agricultural production with the impacts of historical climate change. Pressure index on cropland was first put forward by Cai et al. [23], and has been extensively used to evaluate food security in certain regions [24,25]. In addition, the indexes (e.g., population density, cropland area per capita, pressure index on cropland, and so on) are important parameters that represent societal vulnerability from the perspective of food security. They are available to be used for research on the impact of historical climate change and social sustainability development.

North China is located in the northern temperate monsoon belt. The variability of temperature and precipitation is significant here. It has both higher sensitivity and certain adaptation ability to the impact of climate change. In addition, it was the administrative center of the traditional agricultural area in China during the Qing Dynasty. The impacting and responding processes of climate change in this region directly relate to the social stability, which is often preferentially considered by the central government. There have been many studies on the impact of climatic disasters and its response in this region. They include the analysis of relationship between revolt and drought–flood in Shandong Province during middle and late Qing Dynasty [26]; a case study on the impact of extreme climate events on migration and land reclamation in the early Qing Dynasty [27]; various types of responses in Northeast China to climatic disasters in North China over the past 300 years [28]; revolts frequency in the North China Plain during 1644–1911 and its relationship with climate [29]; social responses in Eastern Inner Mongolia to flood/drought-induced refugees from the North China Plain during 1644–1911 [30], and so on. In these similar national studies, the impacting and responding processes of historical climate change were mostly based on the method of time series comparison, with less attention paid to social vulnerability from the perspective of food security.

This article explores the spatial difference of factors such as the labor/household density, cropland area per capita, and pressure index on cropland in Shandong Province during the 17th–20th century. It would be used to estimate change of social vulnerability and sustainability from the perspective of food security in this region during the historical periods. It also provides fundamental data for research on historical climate change impact and adaptation.

2. Research Area

This paper takes the modern Shandong Province in China as the research area. It is located in the mid-latitude area of the northern hemisphere, within the range of 34°22′52″ N 114°19′53″ E–38°15′02″ N 122°42′18″ E, including 110 cities or counties. In the Qing dynasty, Shandong Province had 10 districts (named as Fu), 3 states directly under the central government (named as Zhili states), 8 scattered states under Fu (named as San states), and 96 counties [31]. The administrative boundaries of some counties in Shandong Province have changed, although this was mainly attributed to the split and combination of counties. Therefore, for ease of comparison with modern results, we converted historical data into the following indexes based on modern county boundaries (the base map comes from the 1:400 base data of China including administrative boundaries, rivers, roads, cities, etc.).

Shandong Province is located on eastern coast area of China, which is the lower reach of the Yellow River (the middle and northern part of Beijing). The Hangzhou Great Channel runs through Shandong Province. Shandong mainly consists of plain and hilly area, which occupies 55% and 28.7% of the total land area, respectively. This is located in the Middle and South Shandong. The northwest area includes the Northwest Plain of Shandong, an alluvial plain formed by the Yellow River. The eastern peninsula is mostly gently fluctuating hilly areas (Figure 1). Shandong has a semi-humid monsoon climate in the warm temperate zone. The climate is mild, and four seasons are discernible. The mean annual temperature is 11–14 °C. The mean annual precipitation is 550–950 mm. Shandong Province is an important agricultural production area in China. Cropland is distributed extensively, being found mainly on the Northwest Plain, Southwest Plain, and Jiaolai Plain. Forest and grassland are mainly distributed on the mountain, and hilly areas in Middle and South Shandong, Jiaodong Peninsula, and the Yellow River Delta. Wetland occurs mainly along the coast (Figure 2).

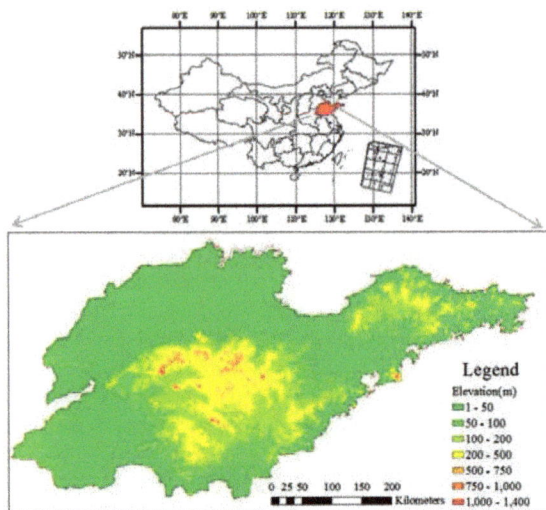

Figure 1. Location of Shandong Province.

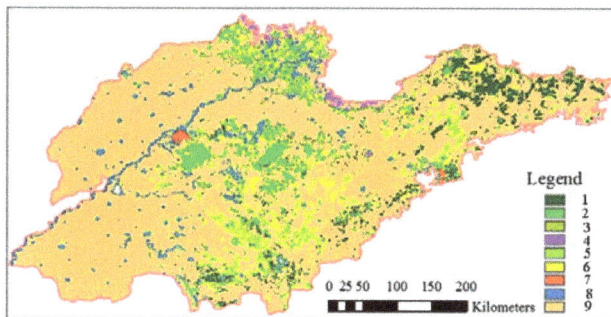

Figure 2. Land use/cover in Shandong Province. 1 = Evergreen coniferous forest; 2 = Deciduous broadleaved forest; 3 = Shrub; 4 = Coastal wetland; 5 = Grassland; 6 = Meadow; 7 = Urban land; 8 = River and lake; and 9 = Cropland.

3. Data Sources and Methods

3.1. Sources of Historical Climatic Data

The temperature data were sourced from the decadal mean temperature change series of North China from 1380s to 1980s [32] and that of Eastern China over the past 1000 years [33]. Precipitation data were sourced from precipitation (drought/flood) change series of North China over the past 2000 years [33]. These climatic data were all based on historical documents.

3.2. Sources and Processing of Population Data

Population data were sourced from gazetteers of counties in Shandong Province during the Qing Dynasty (1644–1911) and the period of Republic of China (1912–1949). There was a total of 244 volumes, which cover the 110 cities or counties in the research area. For one county, there are sometimes 2–4 versions of gazetteers in different periods. The records on the amount of labor, households, and population in different versions of gazetteers were validated with each other. The units of labor and household amount are "Ding" and "Hu", respectively. Ding and Hu are the population tax units in China during the historical periods.

First, we estimated the calculation ratio of labor, household, and population. "Ding" is mostly defined as an adult male aged from 16–60, who must pay the labor tax. One household has 2–3 laborers and includes 5–6 persons in Shandong Province generally during the 17th–20th century. The ratios of labor, household, and population were calculated from county data recording these numbers in 244 volumes of gazetteers, which is relatively reasonable at that time.

Secondly, we reconstructed the numbers of laborer of the 17th century and 18th century as well as the number of households of the 19th century and 20th century by interpolation. The number of records for labor, households, and population during the past four centuries used in this paper are listed in Table 1. The correlation analysis results used for interpolation are as follows:

(1) $Y = 1.444x$, $R^2 = 0.843$; (x = laborer in the 17th century; Y = laborer in the 18th century)

(2) $Y = 0.461x$, $R^2 = 0.869$; (x = laborer in the 18th century; Y = households in the 19th century)

(3) $Y = 1.14x$, $R^2 = 0.802$; (x = household in the 19th century, Y = household in the 20th century).

Table 1. The number of records for labor, household, and population during the past 400 years.

	17th Century	18th Century	19th Century	20th Century
Laborer (Ding)	53	52		
Laborer/Household (Ding/Hu)			40	46
Population (Person)			26	44

3.3. Spatial Distribution Change of Labor/Household Density and Cropland Areas per Capita

According to the above data for labor, household, and cropland area [34] in each county in Shandong during the 17th–20th century, by the equations of 1 Hu = 2–3 Ding, 1 Hu = 5–6 person the labor/household density and cropland area per capita in the four time-sections were calculated and spatially analyzed using the inputs of 1 Hu = 2–3 Ding and 1 Hu = 5–6 people (Figures 3 and 4). To represent the social vulnerability from the perspective of food security, we used 0.05 ha as the basic unit of division referring to the warning line of cropland area per capita (0.053 ha) put forward by Food and Agriculture Organization of United Nations (FAO) [35]. The legend of cropland area per capita is expressed as the segmentations separated by 1, 2, 4, 8, 16, 32 times 0.05 ha.

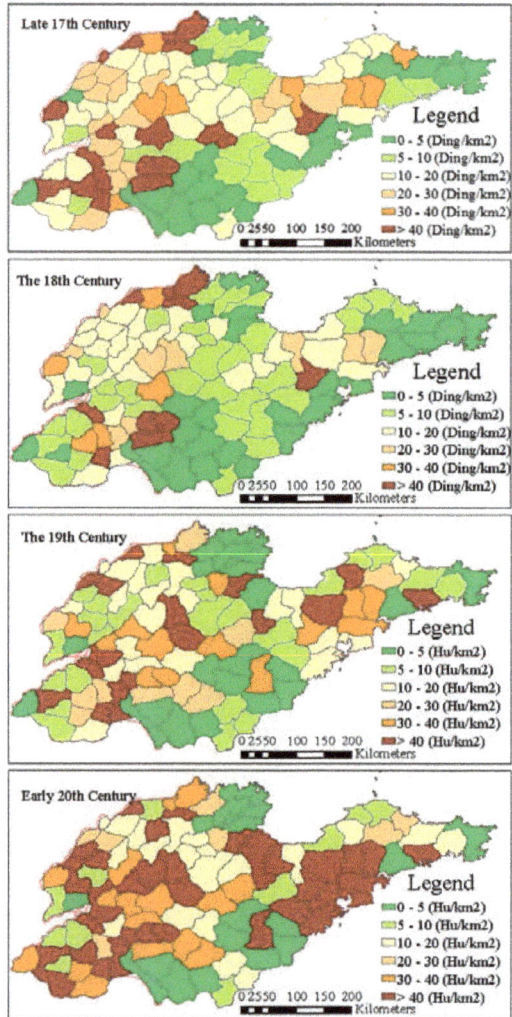

Figure 3. Spatial distribution change of labor/household density in Shandong from the late 17th century to the beginning of 20th century.

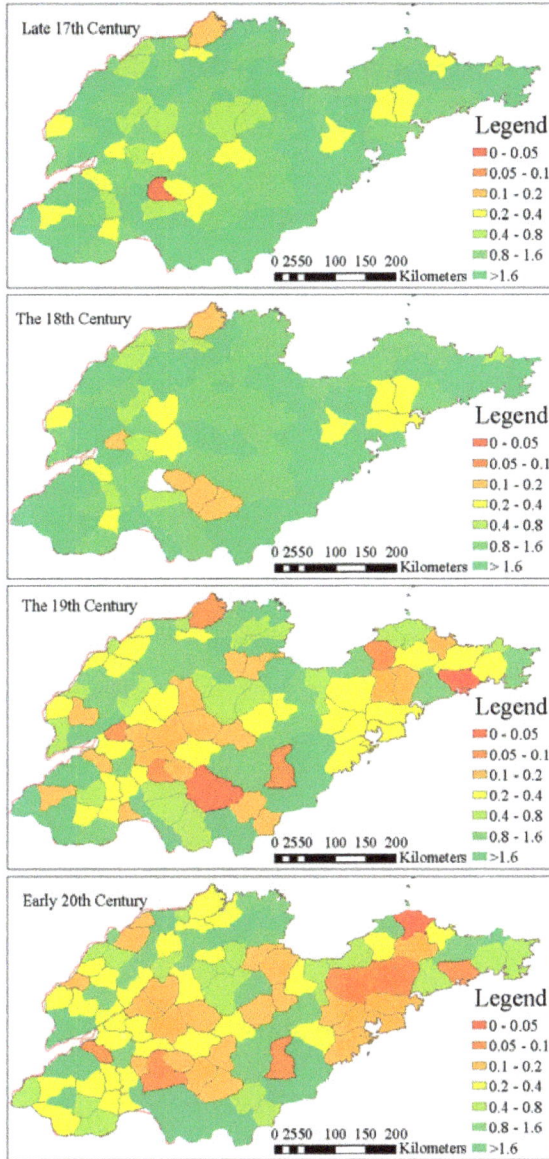

Figure 4. Spatial distribution change of cropland area per capita in Shandong from the late 17th century to the beginning of 20th century (Unit: ha).

3.4. Spatial Distribution of Pressure Index on Cropland

Pressure index on cropland measures the degree of shortage of cropland resources in a certain region. It also reflects the pressure on cropland and the social vulnerability from the perspective of food security during the historical periods. The minimum cropland area per capita represents the necessary cropland area to satisfy a person's basic food consumption at the normal living level in a certain region. Since the Qing dynasty, the "warning line" of cropland area per capita in Shandong has changed, as crop yield had been improved from about 1500 kg/ha to 6000 kg/ha [36,37]. Dietary structure based on grain has most likely remained the same. To make historical research easier, the model put forward by Cai et al. (2002) [23] is simplified by assuming the minimum cropland area per capita in four time-sections to be four times greater than those in modern times, while minimum cropland area per capita in modern times applies the warning line of cropland area per capita (0.053 ha) put forward by FAO (Wang, 2001) [35]. Pressure index on cropland (K) is the ratio of minimum cropland area per capita (S_{min}) and actual cropland area per capita (S_a). Its formulation is:

$$K = S_{min}/S_a \qquad (1)$$

Assume cropland area per labor and cropland area per household are Sd and Sh, respectively. According to the ratios of the numbers of laborer, household, and population discussed above (1 Hu = 3 Dings, 1 Hu = 6 people), it can be obtained that:

$$K = (0.053 \times 4)/S_a = (0.053 \times 4)/(S_d \times 6/3) = 0.424/S_d \text{ or}$$
$$K = (0.053 \times 4)/S_a = (0.053 \times 4)/6S_h = 1.272/S_h$$

The pressure index on cropland in each county in four time-sections was calculated. Following this, we produced the spatial distribution map of K-values by the ArcGIS software (ArcGIS is a word-leading application platform which can be used for collecting, organizing, managing, analyzing, communicating, and releasing geographic information) and analyzed its change (Figure 5). Finally, K value is interpolated by the inverse distance weighted method to identify the vulnerable center in the 19th century and at the beginning of 20th century, which is shown as the brown area in Figure 6. We divided K into 1–6 grades, which are 0–0.8, 0.8–1.6, 1.6–3.2, 3.2–4, and >4 by considering the actual discrete distribution of K-value and lower historical agricultural production level. The higher index grade represents a heavier pressure on sustainable cropland use and a larger possibility of an impending food crisis.

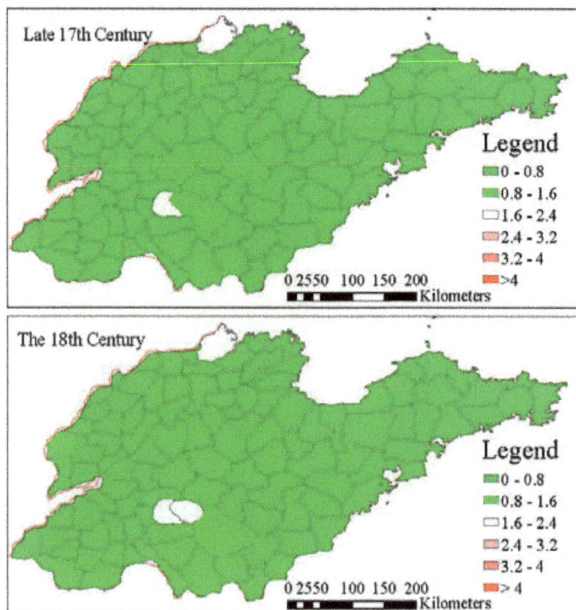

Figure 5. *Cont.*

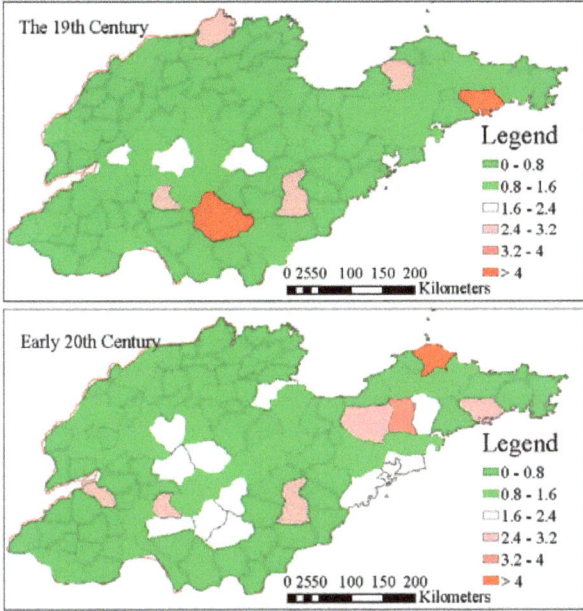

Figure 5. Spatial distribution change of pressure index on cropland in Shandong from the late 17th century to the beginning of the 20th century.

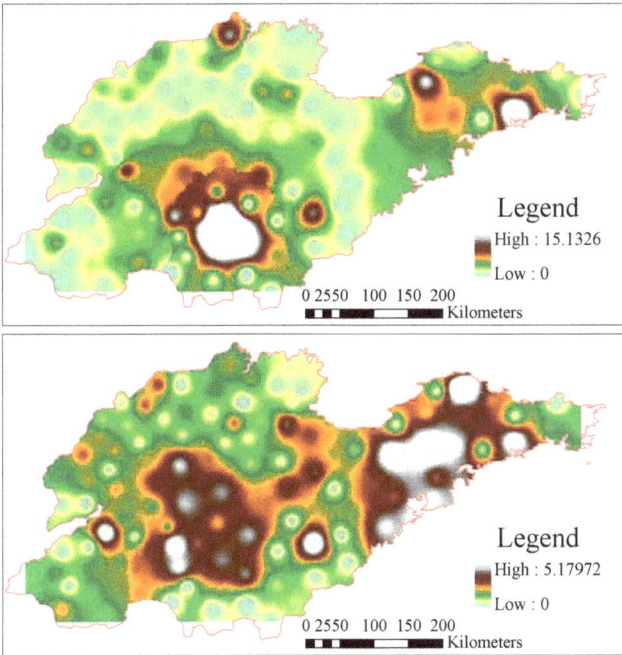

Figure 6. Spatial pattern of pressure index on cropland in Shandong during the 19th century (**up**) and at the beginning of the 20th century (**down**).

3.5. Analysis of Climate Change and Sustainable Cropland Use

The sustainability of cropland use can be measured by the above indexes, including population density, cropland area per capita, and pressure index on cropland. We compared climate change phases with the extent of sustainable cropland use in Shandong during different periods over the past 300 years. By the linkage of some intermediate elements, such as population, agricultural production, policy adjustment, and so on, society vulnerability and food sustainability during different periods were analyzed and discussed.

4. Results of Analysis

4.1. Climate Change in North China over the Past 300 Years

It shows that five regions in Eastern China (including North China) all had two distinctive cold periods (1620s–1710s and 1800s–1860s). The warmest period occurred in the 20th century during the past 500 years. The annual average temperature in the coldest hundred years (1800s–1900s) was lower than that of 20th century by 1.0 °C, with the coldest 30 years having happened in 1650s–1680s (Ge et al., 2012 [33]. The series of average temperatures in North China since the 1380s show that two cold periods occured (1550s–1690s and 1800s–1860s) (Wang et al., 1991 [32]).

The climate in Eastern China during the Qing Dynasty (1644s–1911s) was generally humid, although decadal variation was very distinctive. There were continuous droughts in 1720, 1785, 1810, and 1877. Climate in the 20th century tended to be dry but fluctuating, with the middle of 1940s being wetter than the middle of the 1960s. After this, the weather tended to be dry since the 1980s. It also shows that three sub-regions in Eastern China had a high consistency in the dry/wet change since the 1680s (all humid relatively), although the change in the dry/wet trend of North China (dry relatively) during 1520s–1680s was opposite to that of Jianghuai and Jiangnan regions (wet relatively) (Ge et al., 2012 [33]).

4.2. Spatial Distribution Change of Labor/Household Density in Shandong over the Past 300 Years

From the spatial distribution map of labor/household density (Figure 3), it was found that there existed an obvious spatial difference in labor/household density, which showed a similar pattern to cropland area in the corresponding periods. In essence, there was a relatively greater proportion of the population in the agricultural area appropriate for cultivation, while only a minority of the population settled in the regions not appropriate for cultivation. This embodies the impact of land suitability for cultivation on population distribution. In agricultural areas such as Northwest and Southwest Shandong as well as the Jiaolai Plain, the density of labor/household in the majority of cities/counties was above 10 Ding/km² during the 17th–18th century. At the beginning of the 20th century, most of cities/counties reached above 10 Hu/km². In comparison, there was a smaller distribution of the population in the hilly areas of middle and south Shandong, the Jiaodong Peninsula, and coastal swamp area. The density of labor/household in many cities/counties in these regions was below 5 Ding/km² or 5 Hu/km² from the 17th century to the beginning of the 20th century.

The population density of Shandong Province over the past 300 years has been increasing, especially in the three agricultural areas. Furthermore, the spatial difference of population density decreased from the 19th century to the beginning of the 20th century. In the Northwest and Southwest agricultural areas, the labor densities were above 10 Ding/km² during the 17th–18th century. Until the 19th century and the beginning of 20th century, the population densities in the whole research area were increasing, and reached 20 Hu/km² in many regions of the three agricultural areas.

4.3. Spatial Distribution Change of Cropland Area per Capita in Shandong over the Past 300 Years

From the spatial distribution map of cropland area per capita in Shandong from the late 17th century to the beginning of the 20th century (Figure 5), it was found that the cropland area per capita was distributed relatively uniformly and the spatial difference was less obvious than the population density.

The values of cropland area per capita during the 17th–18th century were higher, with those of the majority of counties being above 0.4 ha. This is eight times higher than the modern warning line of cropland area per capita put forward by the FAO. However, cropland area per capita in a few counties in Binzhou, Linyi, and Jining was below four times higher than the modern warning line, which means that food security in these regions might be at risk with stronger social vulnerability and would most likely be affected by climatic disasters.

The value of cropland area per capita from the 19th century to the beginning of 20th century generally decreased. The strength of social vulnerability and the possibility of social turbulence resulting from threatened food security increased. During the 19th century, the cropland area per capita in the majority of counties in the middle Shandong and Jiaodong Peninsula was below 0.1 ha, which was two times higher than the modern warning line of cropland area per capita. Food security in Binzhou, Linyi, Jining, Laiwu, Tai'an, Yantai, and Weihai were under threat. At the beginning of the 20th century, the cropland area per capita in the whole research area decreased universally, especially in counties in Jiaolai Plain and middle Shandong, as the area in these places decreased to below 0.2 ha. In Northwest and Southwest of Shandong, it appeared that the cropland area per capita of many counties were lower than 0.4 ha. The numbers of vulnerable regions increased. New areas under threat (e.g., Jinan, Dezhou, and Qingzhou) appeared at the beginning of 20th century with previous areas under threat to food security in the 19th century still having this risk (Figure 5).

4.4. Spatial Distribution Change of Pressure on Cropland in Shandong over the Past 300 Years

From the spatial distribution map of the pressure index on cropland in Shandong from the late 17th century to the beginning of the 20th century (Figure 6), it was found that the pressure index on cropland showed a similar spatial distribution to cropland area per capita, and its threshold was lower than that in modern times. The pressure index on cropland distributed homogeneously. It was relatively lower in the 17th–18th century and increased during the 19th–20th century.

Pressure index on cropland in the majority of counties during the 17th–18th century was below 0.8, which means that there was no pressure on cropland. The exceptions included minority regions in Jinan, Linyi, and Binzhou, which probably had threat to food security with the pressure index being above 0.8 (Figure 6). Until the 19th century and the beginning of the 20th century, two high-value centers of pressure on cropland appeared in middle Shandong and the Jiaodong Peninsula (Figure 7), including some cities or counties in Linyi, Tai'an, Jinan, Laiwu, Yantai, Qingzhou, and Weihai, with their pressure index on cropland being above 1.6. There is the possibility of a food crisis in these areas.

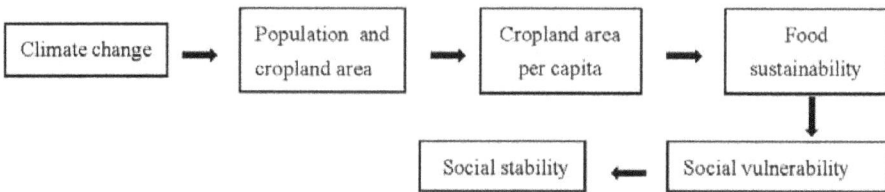

Figure 7. The impacting process of climate change.

4.5. Comparison between Climate Change Phases and the Extent of Sustainable Cropland Use in Shandong

First, it seems that a higher extent of sustainable cropland use occurred in the cold period, while a lower extent of sustainable cropland use occurred in the warm period in Shandong over the past 300 years. In the two cold periods (1620s–1710s and 1800s–1860s), the population density in Shandong was relatively lower. At this time, the labor/household densities of the majority of cities or counties during the 17th–18th century were below 10 Ding/km^2, which reached above 10 Hu/km^2

at the beginning of the 20th century during the warm period. In particular, the labor densities of the Northwest and Southwest agricultural areas increased from above 10 Ding/km^2 during the 17th–18th century to reach 20 Hu/km^2 in many regions until the beginning of 20th century. The cropland area per capita during the 17th–18th century was higher above 0.4 ha, which was eight times higher than the modern warning line of cropland area per capita put forward by the FAO. From the 19th century to the beginning of the 20th century, the cropland area per capita in the majority of counties in middle Shandong and the Jiaodong Peninsula decreased to below 0.1 ha, which is two times higher than the modern warning line. This meant that the strength of social vulnerability and the possibility of social turbulence resulting from a threat to food security increased. Similarly, the pressure index on cropland in the majority of counties during the 17th–18th century was below 0.8, which means that there was no pressure on cropland. Until the 19th century and the beginning of the 20th century, two high-value centers of pressure on cropland appeared in middle Shandong and the Jiaodong Peninsula. In these places, the pressure index on cropland in some cities or counties reached above 1.6, meaning that there was the possibility of a food crisis in these areas.

Second, the turning point of the 1680s from dry to relatively wet in North China seems to be attributed to the decrease in the extent of sustainable cropland use in Shandong, although this was not very distinctive. During the 17th–18th century, the population density in Shandong was relatively lower, the cropland area per capita was higher (above 0.4 ha), and the pressure index on cropland in majority of counties was below 0.8. This means that there was no pressure on cropland during the period around the 1680s. It is likely that the effect of dry/wet change on the sustainability of cropland use was not as obvious as that of temperature change.

5. Discussion

5.1. Impacting Process of Climate Change and the Sustainability of Cropland Use

Many researchers have analyzed the relationships between climate change or extreme climatic events with refugees and social stability in North China (Ye, et al., 2004; Xiao, 2011; Fang et al., 2007; Ye and Fang, 2013; Xiao et al., 2013). These case studies all showed that the impact of historical climate change on social stability in this area was often influenced by a failure in food production. In this present study, the intermediate elements of population, cropland area per capita, and pressure on sustainable cropland use were emphasized as a means of obtaining a better understanding of the impacting and responding processes of climate change.

It appears that a warm climate was beneficial in driving an increase in population and agricultural development, which finally resulted in pressure on the sustainability of cropland use. In the cold periods, the population density in Shandong was relatively lower and the cropland area per capita was higher. During these periods, there was no pressure on sustainable cropland use. In the warm periods of the 20th century, the population density in Shandong increased more quickly, while cropland area increased slowly. This led to a decrease in cropland area per capita, which was followed by a decrease in the sustainability of cropland use. This would intensify the social conflict and increase the probability of social revolts. Therefore, the impacting process of climate change can be depicted as shown in Figure 7.

In addition, the turning point of the 1680s from dry to wet in North China seemed to contribute to the decrease in the extent of sustainable cropland use in Shandong, although this was not very distinctive.

5.2. Special View of this Research and Its Scientific Value

Although some deviation still exists in the reconstructed results of population and cropland area per capita, this paper provides a perspective of food security and social vulnerability in the research area. To a certain extent, it makes up for the deficiency of paying more attention to the comparison of a series of climate change and social results while neglecting some intermediate links in the impact process and social vulnerability in this research area. By combining social vulnerability and human adaptation actions with climate-related hazards and physical exposure, it aims to accurately evaluate the impact of historical climate change or risk on human society. It also provides references to modern sustainable cropland use.

6. Conclusions

By analyzing historical documents, regression analysis, model of pressure index on cropland, and geographic information system (GIS), this paper reconstructed spatial patterns of labor/household density, cropland area per capita, and pressure index on cropland at the county level in Shandong Province during the 17th–20th century. Following this, we analyzed the impacting process of climate change and the sustainability of cropland use during the different periods. The conclusions of this study are as follows:

There was a distinct spatial difference in labor/household density which showed the effect of land suitability for reclamation on the population distribution. There was a greater proportion of the population in the agricultural area of Northwest and Southwest Shandong as well as the Jiaolai Plain. The population density of Shandong Province over the past 300 years has been increasing, especially in the three agricultural areas.

The spatial distribution of cropland area per capita in Shandong over the past 300 years has been relatively uniform. From the 19th century to the beginning of the 20th century, cropland area per capita decreased extensively and social vulnerability was strengthened. There was likely a threat to food security in Binzhou, Linyi, Jining, Laiwu, Taian, Yantai, Weihai, Jinan, Dezhou, Qingzhou.

The pressure index on cropland also showed a similar spatial distribution to cropland area per capita, but its threshold was lower than that in modern times. During the 19th century and the beginning of the 20th century, two high-value centers of pressure on cropland appeared in Middle Shandong and the Jiaodong Peninsula.

A warm climate was beneficial to driving an increase in population and agricultural development, which finally resulted in increasing pressure on the sustainability of cropland use. This increase in pressure on cropland was also related to the growth of the population as well as the spatial differences in land quality between plain and hill areas. The impacting process of climate change was sketched as following the flowchart shown in Figure 7: climate change—population and cropland area—cropland area per capita—food sustainability and society vulnerability—social stability.

Acknowledgments: This work was supported by the National Natural Science Foundation of China (Grant No. 41471156), the National Key Research and Development Program of China (Grant No. 2016YFA0602500), and the Strategic Priority Research Program from Chinese Academy of Sciences (XDA05080102).

Author Contributions: Yu Ye wrote this text and produced the figures. Xueqiong Wei, Xiuqi Fang and Yikai Li listed the references and help to edit the text.

Conflicts of Interest: The authors declare no conflict of interest.

References

1. McCarthy, J.J.; Canziani, O.F.; Leary, N.A.; Dokken, D.J.; White, K.S. *Climate Change 2001: Impacts, Adaptation and Vulnerability, Contribution Of Working Group II to the Third Assessment Report of the Intergovernmental Panel on Climate change*; Cambridge University Press: Cambridge, UK, 2001.
2. Field, C.B.; Barros, V.R.; Dokken, D.J.; Mach, K.J.; Mastrandrea, M.D.; Bilir, T.E.; Chatterjee, M.; Ebi, K.L.; Estrada, Y.O.; Genova, R.C.; et al. *Climate Change 2014: Impacts, Adaptation and Vulnerability, Contribution of Working Group II to the Fifth Assessment Report of the International Panel on Climate change*; Cambridge University Press: Cambridge, UK, 2014.
3. Fang, X.Q.; Zheng, J.Y.; Ge, Q.S. Historical climate change impact-response processes under the framework of food security in China. *Sci. Geogr. Sin.* **2014**, *34*, 1291–1298.
4. Richard, S.J.; Wagner, T.S. Climate change and violent conflict in Europe over the last millennium. *Clim. Chang.* **2010**, *99*, 65–79.

5. Haug, G.H.; Gunther, D.; Peterson, L.C.; Sigman, D.M.; Hughen, K.A.; Aeschlimann, A. Climate and the collapse of Maya civilization. *Science* **2003**, *299*, 1731–1735. [CrossRef] [PubMed]
6. Patterson, W.P.; Dietrich, K.A.; Holmden, C. Two millennia of North Atlantic seasonality and implications for Norse colonies. *Proc. Natl. Acad. Sci. USA* **2010**, *107*, 5306–5310. [CrossRef] [PubMed]
7. Hsu, K.J. Sun, climate, hunger and mass migration. *Sci. China Ser. D Earth Sci.* **1998**, *41*, 449–472. [CrossRef]
8. Zhang, D.E.; Li, H.C.; Ku, T.L.; Lu, L.H. On linking climate to Chinese dynastic change: Spatial and temporal variations of monsoonal rain. *Chin. Sci. Bull.* **2010**, *55*, 77–83. [CrossRef]
9. Ge, J.X. *The History of China's Population*; Fudan University Press: Shanghai, China, 2002.
10. Li, B.Z. Climate change and several times of Chinese historical population fluctuation. *Popul. Res.* **1999**, *23*, 15–19.
11. He, B.D. Verification and Evaluation of Ancient and Today's Land Data in China. unpublished work.
12. Bian, L. Folk measure method and essence of farmland area in South China during Ming dynasty and Qing dynasty. *China Agric. Hist.* **1995**, *14*, 49–56.
13. Wan, H. Historical comparison between cropland numbers in China during Ming dynasty and early Qing dynasty. *China Agric. Hist.* **2000**, *19*, 34–40.
14. Shi, Z.X. Analysis on change of population and land in the area of Gan, Ning and Qing during the late Qing dynasty. *China Agric. Hist.* **2000**, *19*, 72–79.
15. Geng, Z.J. Tentative analysis of question of men's and farmland's discount in Shanxi during Qing dynasty. *China Agric. Hist.* **2000**, *19*, 67–71.
16. Zhou, R. Integrated review and new calculation of cropland area during the early Qing dynasty. *Jianghan Tribune* **2001**, *9*, 57–61.
17. Ge, Q.S.; Dai, J.H.; He, F.N.; Zheng, J.Y. Change of the amount of cropland resource and analysis of driving forces in partial provinces in China during the past 300 years. *Prog. Natl. Sci.* **2003**, *13*, 825–832.
18. He, F.N.; Li, S.C.; Zhang, X.Z. The reconstruction of cropland area and its spatial distribution pattern in the mid-northern Song Dynasty. *Acta Geogr. Sin.* **2011**, *66*, 1531–1539.
19. Xie, Y.W.; Wang, X.Q.; Wang, G.S.; Yu, L. Cultivated land distribution simulation based on grid in middle reaches of Heihe River Basin in the historical periods. *Adv. Earth Sci.* **2013**, *28*, 71–78.
20. Tian, Y.C.; Li, J.; Ren, Z.Y. Analysis of cropland change and spatial-temporal pattern inLoess Plateau over the recent 300 years. *J. Arid Land Resour. Environ.* **2012**, *26*, 94–101.
21. Ye, Y.; Fang, X.Q. Expansion of cropland area and formation of the eastern farming-pastoral ecotone in northern China during the twentieth century. *Reg. Environ. Chang.* **2012**, *12*, 923–934. [CrossRef]
22. Ye, Y.; Fang, X.Q.; Ren, Y.Y.; Zhang, X.Z.; Chen, L. Cropland cover change in Northeast China during the past 300 years. *Sci. China Ser. D Earth Sci.* **2009**, *52*, 1172–1182. [CrossRef]
23. Cai, Y.L.; Fu, Z.Q.; Dai, E.F. The minimum area per capita of cultivated land and its implication for the optimization of land resource allocation. *Acta Geogr. Sin.* **2002**, *57*, 127–134.
24. Ren, G.Z.; Zhao, X.G.; Chao, S.J.; Dong, L.L.; Zhao, Y.M. Temporal-spatial analysis of cultivated land pressure in China based on the ecological tension indexes of cultivated land. *J. Arid Land Resour. Environ.* **2008**, *22*, 37–41.
25. Liu, X.T.; Cai, Y.L. Grain security of basic cropland pressure index in Shandong Province. *Popul. Resour. Environ.* **2010**, *20*, 334–337.
26. Ye, Y.; Fang, X.Q.; Ge, Q.S.; Zheng, J.Y. Response and adaptation to climate change indicated by the relationship between revolt and drought-flood in Shandong Province during middle and late Qing Dynasty. *Sci. Geogr. Sin.* **2004**, *24*, 680–686.
27. Fang, X.Q.; Ye, Y.; Zeng, Z.Z. Extreme climate events, migration for cultivation and policies: A case study in the early Qing Dynasty of China. *Sci. China Ser. D Earth Sci.* **2007**, *50*, 411–421. [CrossRef]
28. Ye, Y.; Fang, X.Q.; Khan, M. Migration and reclamation in Northeast China in response to climatic disasters in North China during the past 300 years. *Reg. Environ. Chang.* **2012**, *12*, 193–206. [CrossRef]
29. Xiao, L.B.; Ye, Y.; Wei, B.Y. Revolts frequency during 1644–1911 in North China Plain and its relationship with climate. *Adv. Clim. Res.* **2011**, *2*, 218–224. [CrossRef]
30. Xiao, L.B.; Fang, X.Q.; Ye, Y. Reclamation and revolt: social responses in Eastern Inner Mongolia to flood/drought-induced refugees from the North China Plain 1644–1911. *J. Arid Environ.* **2013**, *88*, 9–16. [CrossRef]

Sustainability **2017**, *9*, 1171

31. Niu, P.H. *Table of Administrative Division Evolution during the Qing Dynasty*; China Cartographic Publishing House: Beijing, China, 1990.

32. Wang, S.W.; Wang, R.S. Reconstruction of temperature series of North China from 1380s to 1980s. *Sci. China Ser. B Chem.* **1991**, *34*, 751–759.

33. Ge, Q.S.; Zheng, J.Y.; Hao, Z.X.; Liu, H.L. General characteristics of climate changes during the past 2000 years in China. *Sci. China Earth Sci.* **2012**, *42*, 934–942. [CrossRef]

34. Ye, Y.; Wei, X.Q.; Li, F.; Fang, X.Q. Reconstruction of cropland cover changes in Shandong Province over the past 300 years. *Sci. Rep.* **2015**, *5*. [CrossRef] [PubMed]

35. Wang, W.M. The perspective of 0.8 Mu warning line of cropland per capita. *Chin. Land* **2001**, *10*, 32–33.

36. Guo, S.Y. Food production in rainfed agricultural region in North China in the Qing Dynasty. *Res. Chin. Econ. Hist.* **1995**, *1*, 22–44.

37. Agriculture Department of the Shandong Province. Agriculture Information Website of Shandong Province. 2014. Available online: www.sdny.gov.cn (accessed on 4 July 2017).

sustainability

MDPI

Article

Effect of Planting Date on Accumulated Temperature and Maize Growth under Mulched Drip Irrigation in a Middle-Latitude Area with Frequent Chilling Injury

Dan Wang [1], Guangyong Li [1,*], Yan Mo [1,2], Mingkun Cai [1] and Xinyang Bian [3]

[1] College of Water Resources & Civil Engineering, China Agriculture University, Beijing 100083, China; wangdan_9090@cau.edu.cn (D.W.); cmk1993@cau.edu.cn (M.C.)
[2] Department of Irrigation and Drainage, China Institute of Water Resources and Hydropower Research, Beijing 100091, China; moyanSDI@cau.edu.cn or moyan@iwhr.com
[3] Scientific Research Department, Water-Saving and Equipment Limited Company of Kingland Muhe, Chifeng 024000, China; bianxy@mhjsgf.com
* Correspondence: LGYL@cau.edu.cn; Tel.: +86-10-627-38386

Received: 16 June 2017; Accepted: 21 August 2017; Published: 23 August 2017

Abstract: Given that chilling injury, which involves late spring cold and early autumn freezing, significantly affects maize growth in middle-latitude cold areas, a highly efficient cultivation technique combining suitable planting date (PD) and mulched drip irrigation is being studied to guarantee maize production. A field experiment for medium-mature variety "Xianyu 335" was conducted in 2015 to 2016 in Chifeng, Inner Mongolia, China, to explore the effects of PD on the active accumulated temperature (AAT) distribution and maize growth under mulched drip irrigation. Based on the dates (around May 1) of late spring cold occurring in the area, four PDs were designed, namely, April 20 (MD_1), May 2–3 (MD_2), May 12 (MD_3), and May 22 (MD_4), and a non-film mulching treatment ($NM-D_2$) was added on the second PD. Results indicated that: (1) the warming effect of film mulching effectively compensated for the lack of heat during the early stages of maize growth. Compared with that in $NM-D_2$, the soil temperature under mulching in MD_2 for the sowing–emergence and seedling stage increased by 14.3% and 7.6%, respectively, promoting maize emergence 4 days earlier and presenting 5.6% and 9.7% increases in emergence rate and grain yield, respectively; (2) the AAT reduction caused by PD delay was mainly observed in reproductive stage, which reached 96.6 °C for every 10 days of PD delay in this stage; (3) PD markedly affected maize growth process and yield, which were closely related to the chilling injury. The late spring cold slowed down the emergence or jointing for maize (under MD_1 and MD_2), but brought insignificant adverse effect on maize later growth and grain yield (16.1 and 15.9 Mg·ha^{-1}, respectively). While the maize in both MD_3 and MD_4 treatments suffered from early autumn freezing damage at the anthesis–maturity stages, resulting in shortening in reproductive period by 4–8 days and decrease in grain yield by 11.4–17.3% compared with those in MD_1 and MD_2; and (4) taking the typical date (May 1) of late spring cold occurring as the starting point, the grain yield penalty reached 8.5% for every 10 days of PD delay; for every 100 °C of AAT decrease during reproductive stage, the grain yield decreased by 6.1%. The conclusions offer certain reference values for maize cultivation in the same latitude areas with similar ecological environments.

Keywords: planting date; mulched drip irrigation; chilling injury; active accumulated temperature; maize growth

---

1. Introduction

Global warming significantly affects agricultural production in middle-high latitude areas. An increasing number of studies have indicated that crop varieties with a long growth period (GP)

should be selected for temperature increasing and frost-free season extension, and photo-thermal resources should be fully utilized to promote crop production [1–3]. However, the selection of planting date (PD), which corresponds to the expanded planting of crop varieties with a long GP, remains challenging [4]. In recent years, researchers have conducted extensive discussions on the changing trend of the PDs for rice, maize and other crops [4–7]. The Corn Belt in Northeast China is located in cold middle-latitude regions. Insufficient accumulated temperature caused by chilling injuries (such as late spring cold and early autumn freezing) is a crucial ecological factor that limits grain yield [8]. Therefore, film mulching has become a primary cultivation practice. However, owing to the lack of systematic research on the effects of PD on maize growth and regulation mechanism of heat factors, wide PD intervals have made the effects of improving production in cold weather imperceptible.

Planting date influences crop growth by changing the corresponding relationship between hydrothermal factors and growth stages. The main obstacles that restrict maize growth in different ecological regions should be identified to determine the optimal PD. In cold middle-latitude regions, the limitation of PD delay should be based on the required accumulated temperature for crop growth [9]. Maize growth is accelerated with PD postponement, leading to a shortening growth period and an adverse effect on biomass accumulation [10,11]. Moreover, delayed sowing increases the probability of early autumn freezing occurring at latter stages in maize growth, thereby restraining grain maturity and leading to output reduction [12,13]. Early planting with film mulching is regarded as an effective measure to increase production against the cold, because despite crop exposure to a low-temperature environment at the prophase, the growth point is not endangered; besides, no adverse effect was observed in middle-later stages, and maize can reach physiological maturity before early autumn freezing occurs [14,15].

The U.S. Corn Belt is located in the same latitude as the China Corn Belt, and the appropriate PDs for maize are from late April to early May, which tends to be advanced in recent years [2,16,17]. Maize planting season ranges from late April to mid-May in the China Corn Belt, and several reports showed that the appropriate PDs for medium-late maturing varieties are from April 25 to May 10, but with considerable yield differences [18,19].

Most of the aforementioned conclusions are drawn from studies on traditional surface irrigation. Mulched drip irrigation technology combines the characteristics of film mulching and drip irrigation. The application of this technology for maize is becoming increasingly widespread in Northeast China these years. According to the latest reports, the mulched drip irrigation area for maize in the region increased 135×10^4 ha during 2012–2015, which made a significant contribution to food security [20]. Film mulching can improve soil temperature and reduce water evaporation [21–24], whereas drip irrigation can achieve accurate and efficient management of soil moisture and nutrients [25,26]. The regulation effects of mulched drip irrigation and PD on the microclimate around the crop will perform a certain role in the maize growth process, yield formation and utilization of hydrothermal resources. However, related research under drip irrigation remains rare and unsystematic, and the reported conclusions were mainly concentrated in arid areas, wherein the primary objective was to improve yield- and water-use efficiency [27–29]. Furthermore, current studies on maize PD under mulched drip irrigation are little. Therefore, identifying a reasonable PD that corresponds to mulched drip irrigation in cold middle-latitude areas with different ecological factors under consideration is important.

This study aims to investigate the warming effect and yield-increasing potential of mulched drip irrigation for maize, to analyze maize growth response to chilling injury (that is, late spring cold and early autumn freezing) at certain growth stages caused by different PDs, quantify the accumulated temperature loss and yield reduction with PD delay, and provide a theoretical basis for improving the rational utilization of heat resources and minimizing production loss in cold middle-latitude areas.

2. Materials and Methods

2.1. Site Description

The field experiment was conducted in 2015 to 2016, Chifeng in Inner Mongolia, China (42°56′53″N, 119°4′20″E). The area has a semi-arid and continental monsoon climate with an approximately 135-day frost-free period. Early autumn freezing generally occurs from the end of September to the beginning of October, and the late spring cold always occurs around May 1. The effective accumulated temperature ($\geq$10 °C) is 2000–3200 °C, and the annual average rainfall and evaporation amounts are 350–450 and 1500–2300 mm, respectively.

Temperature and precipitation during maize GPs in 2015 and 2016 are shown in Figure 1. The late spring cold occurred in both years (the minimum daily temperature dropped to −1.8 °C in 2015 and to 1.4 °C in 2016) and the dates of the early autumn freezing in 2 years were October 2 and September 28, respectively. The weather conditions during maize GPs in 2015 and 2016 were in line with the environmental climate characteristics of years in the region [30]. Total precipitation values during maize GPs for 2 years were 180 and 250 mm, respectively. Note that the climate feature of "late spring cold" in the region is that the daily average temperature exceeding 10 °C lasts for 5–10 days at least in late Spring, but then the daily minimum temperature dropped to about 0 °C for a sudden strong cold air attack (the cold snap always continue 1–2 days), and then the temperature increases again gradually.

The soil layer of 0–60 cm in the test site belong to silt loam soil with a soil bulk density of 1.49 g·cm^{-3} and a field water-holding rate of 34.45% (volumetric water content). Soil organic matter was 10.6 g·kg^{-1}, mass fraction for total nitrogen was 0.60 g·kg^{-1}, and available potassium and phosphorus were 167 and 7.6 mg·kg^{-1}, respectively.

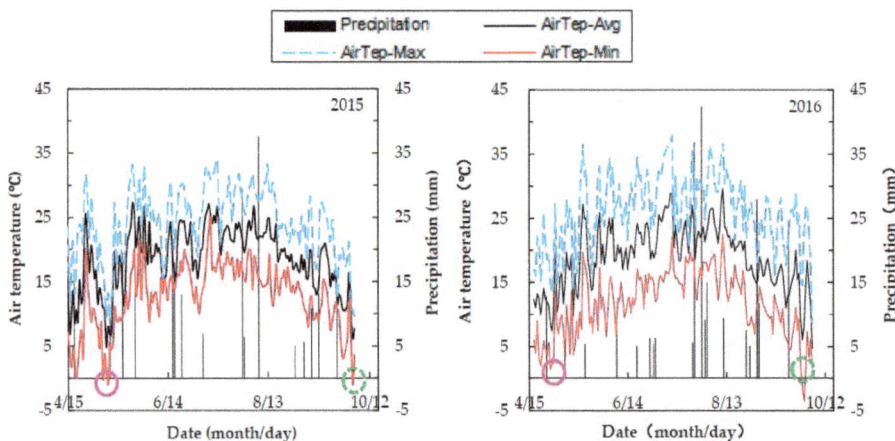

Figure 1. Daily temperature and precipitation during maize growing season in 2015 and 2016. Purple circle with full line marked position for the late spring; green broken circle marked position for the early autumn freezing.

2.2. Field Experiment

The experiment used a completely randomized design for the variable factor of PD, which involved four levels. Based on the date of late spring cold (around May 1), four PDs were set under mulched drip irrigation, namely, MD$_1$ (April 20), MD$_2$ (May 2–3), MD$_3$ (May 12), and MD$_4$ (May 22). And a treatment (NM-D$_2$) without mulching was added on the second PD under drip irrigation. Each treatment was performed in triplicate, thus, 15 plots in total were arranged randomly. The size of each plot was 40 m × 6 m.

Maize varieties need to match the light and heat resources of a certain ecological zone otherwise the yield and utilization rate of resources will be reduced [31]. For testing, we selected the medium variety "Xianyu 335", which is a high-yield plant in the Corn Belt of Northeast China [32,33]. Maize was planted in the alternate wide–narrow rows (80–40 cm) with a planting density of 83,330 plants·ha^{-1}. The drip tape with a wall thickness of 0.2 mm, a drip flow of 1.38 L·h^{-1}, and a drip spacing of 40 cm was placed in the middle of the narrow line. The maize cropping pattern and lateral layout of drip tapes under mulched drip irrigation are shown in Figure 2.

The treatments were irrigated according to the lower irrigation limit and crop water requirement, and the lower limit of irrigation at both seedling and mature stages was 70%, whereas that in other growth stages was 75%. The precipitation and irrigation amounts in each GP of maize for all treatments are shown in Figure 3. In treatments of MD_1–MD_4, the total irrigation amounts were 170, 155, 151, and 151 mm, respectively, in 2015 and 135, 135, 125, and 115 mm, respectively, in 2016. The irrigation amount in NM-D_2 treatment was same as that in MD_2 for 2 years. The total precipitation during maize growth season was 180 mm in each treatment in 2015, whereas those in MD_1–MD_4 in 2016 were 251, 242, 232, and 215 mm, respectively.

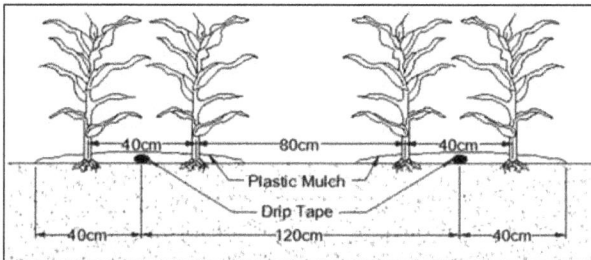

Figure 2. The schematic diagram of cropping pattern and lateral layout of drip tapes under mulch for maize.

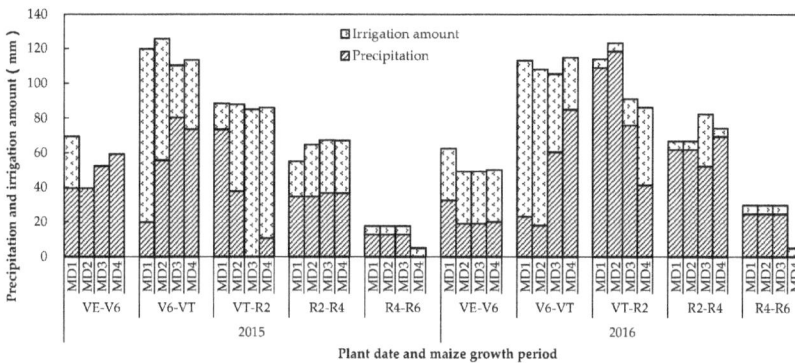

Figure 3. The precipitation and irrigation amount in various growth stages of maize for different treatments. VE—emergence, V6—sixth leaf, VT—tasseling, R2—filling, R4—dough, R6—physiological maturity.

Fertilizer schedule in all treatments was the same for 2 years, and the application amounts of N, P_2O_5, and K_2O were 285, 135, and 135 kg·ha^{-1}, respectively. Fertilizer application percentages in each GP are shown in Table 1. Seed fertilizer, which included urea (N 46%), calcium superphosphate

(P_2O_5 46%), and potassium sulfate were applied to the field by a seeder. The remaining fertilizer was applied in proportion with water by drip tapes, and the soluble fertilizers included urea (N 46%), MAP (monoammonium phosphate, P_2O_5 61%, N 12%), and potassium chloride (K_2O 62%). Other farm tasks were performed as the management in local high-yield farmlands.

Table 1. Fertilizer application percentages in each growth period.

Fertilizer	Sowing	V6–V12	VT	R2	R4
N	20%	45%	15%	15%	5%
P_2O_5	50%	30%	20%	–	–
K_2O	50%	50%	–	–	–

V6—sixth leaf, V12—twelfth leaf, VT—tasseling, R2—filling, R4—dough.

2.3. Observation Indexes and Methods

ET107 automatic meteorological station manufactured by USA Campbell Scientific was used to continuously monitor meteorological data, such as daily temperature and precipitation during the maize growing season.

Soil water content was regularly measured using the gravimetric method. The soil samples were collected by an auger boring in 20 cm increments to a depth of 80 cm, then were dried at 105 °C for 8 h. The measurement range of the electronic scale used is 500 g with an accuracy of 0.01 g. Soil moisture is calculated as the following formula.

$$\omega = \frac{m_w - m_d}{m_d} \times 100\% \tag{1}$$

ω—soil water content, %; m_w—the mass of wet soil, g; m_d—the mass of dry soil, g.

A set of curved tube thermometers were embedded in the narrow lines (under mulching) and wide lines (in bare land) of each treatment to measure the soil temperature at depths of 5, 10, 15, 20 and 25 cm. The observation times were 8:00, 14:00, and 18:00 daily, and the average value of three observations was taken as the daily soil temperature [34,35].

Dates of VE (emergence), V6 (sixth leaf), VT (tasseling), R2 (filling), and R6 (physiological maturity) were recorded according to maize growth traits. Five plants were removed from the ground except for the roots to measure height, stem diameter and dry matter (DM) accumulation (plants were separated into leaves, stems, and ears, which were heat-treated at 105 °C for 30 min and dried at 75 °C to a constant weight).

During harvest season, 10 m of maize from the middle six lines in each plot were harvested separately, grain moisture content was measured, and yield with a standard moisture content of 14% was calculated. Five maize ears from each plot were selected to measure the yield components, such as bald tip length, hundred-grain weight, and grain number per ear.

Statistical analysis and plotting were performed using Microsoft Excel 2010 and IBM SPSS Statistics 17.0.

3. Results

3.1. Effects of Planting Date and Film Mulching on Heat Distribution

3.1.1. Soil Temperature

The dynamic changes of daily soil temperature (DST) in the depth of 5–25 cm under film mulching and in bare land during maize growing season in 2015 and 2016 are shown in Figure 4. The increasing rate of the average DST under film mulching at various GPs, compared with those of bare land in different treatments, are shown in Figure 5.

From MD_1 to MD_4, the accumulated soil temperature (AST) under mulching during maize entire GP for two years averagely increased by 149.9 °C, 136.6 °C, 116.5 °C, and 99.6 °C, respectively, compared with those in bare land. The AST increments of seedling stage in MD_1–MD_4 accounted for 59.7%, 59.1%, 55.0%, and 54.8% of those during the total GP, respectively. Results indicated that the warming effect of film mulching decreased with the PD delay.

For maize plants with different PDs at two stages, namely, sowing–emergence and seeding, the average increasing rates of DST under mulching were 11.2% and 6.7%, respectively. Moreover, the warming effect gradually weakened with continuous GP progress, and the average increasing rates of DST under mulching were 2.3–3.0% during V6–harvest stage with different PDs. The results demonstrated an evident warming effect of film mulching in the prophase of maize.

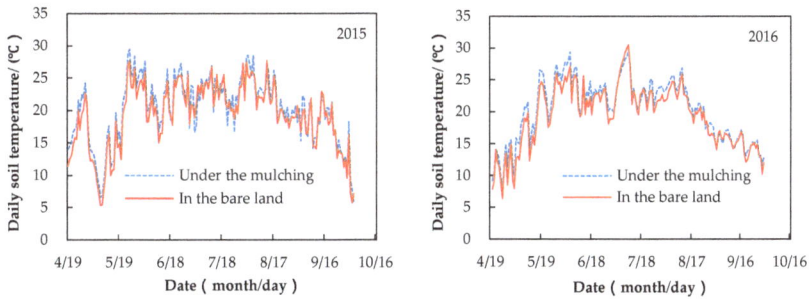

Figure 4. Changes of daily average soil temperature at 5–25 cm depth during maize growing season in 2015 and 2016.

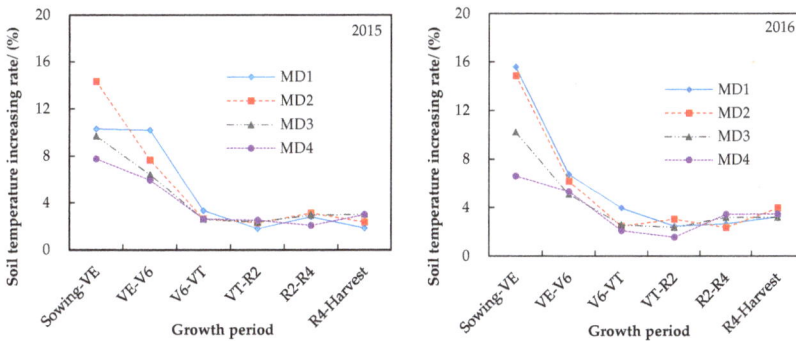

Figure 5. Daily soil temperature increasing rate under mulching compared with that in bare land of various maize growth periods in different treatments. VE—emergence, V6—sixth leaf, VT—tasseling, R—filling, R—dough.

3.1.2. Active Accumulated Temperature

The AATs (≥ 10 °C) in different treatments for various GPs of maize under mulched drip irrigation are shown in Table 2.

The total AATs exhibited a decreasing trend with the delay in PD, decreasing under MD_2–MD_4 treatments by 73 °C, 226 °C, and 365 °C, respectively, compared with those in MD_1.

The AATs at maize's different GPs were determined based on daily temperature and stage duration. The average AAT required for maize emergence was approximately 208 °C, and the annual variation was caused by fluctuation in the occurrence of late spring cold.

The AATs at both vegetative and reproductive growth stages decreased with the postponement of PD. The AATs at these two stages under MD_1–MD_4 treatments were on the average 1494–1430 °C and 1327–1022 °C, respectively. Compared with those in MD_1, AATs of the maize reproductive stage in MD_2–MD_4 decreased by 89 °C, 198 °C, and 305 °C, respectively, illustrating that PD particularly affected AATs of this stage and insufficient AATs at this stage were unfavorable for grain filling and dehydration.

Table 2. The active accumulated temperature of sowing–harvest, sowing–VE, VE–VT and VT–harvest stages for maize in different treatments (°C).

Year	Treatment	Sowing–Harvest	Sowing–VE	VE–VT	VT–Harvest
	MD_1	3050	183	1544	1323
	MD_2	2927	210	1487	1229
2015	MD_3	2793	235	1435	1122
	MD_4	2677	207	1447	1023
	Mean	2862	209	1478	1174
	MD_1	2993	219	1443	1330
	MD_2	2969	214	1508	1247
2016	MD_3	2798	214	1450	1134
	MD_4	2635	186	1413	1020
	Mean	2849	208	1454	1183
	MD_1	3021	201	1494	1327
Mean	MD_2	2948	212	1498	1238
	MD_3	2796	225	1443	1128
	MD_4	2656	197	1430	1022

VE—emergence, VT—tasseling.

3.2. Effect of Planting Date on Maize Growth Process

The durations of various maize GPs under different treatments are shown in Table 3, and no statistically significant differences occurred between years for the average parameter values of all treatments.

Table 3. The durations of sowing–harvest, sowing–VE, VE–VT and VT–harvest stages for maize in different treatments (d).

Year	Treatment	Sowing–Harvest	Sowing–VE	VE–VT	VT–Harvest
	MD_1	155d	9a	81b	65c
	MD_2	150c	17c	69a	64c
	MD_3	139b	11b	68a	60b
2015	MD_4	133a	10a	66a	57a
	F-test	89.9 **	42.9 **	52.0 **	22.7 **
	NM-D_2	153	21	67	65
	Mean	146A	14A	70A	62A
	MD_1	154d	18d	69b	67c
	MD_2	149c	14c	69b	66c
	MD_3	141b	12b	67a	62b
2016	MD_4	131a	10a	64a	57a
	F-test	200.5 **	143.9 **	16.3 **	45.2 **
	NM-D_2	152	18	67	67
	Mean	145A	14A	67A	64A
	MD_1	155d	14ab	75b	66c
	MD_2	150c	16b	69a	65c
Mean	MD_3	140b	12ab	67a	61b
	MD_4	132a	10a	65a	57a
	F-test	200.7 **	4.6 *	8.5 **	51.2 **
	NM-D_2	153	20	67	66

1 The different lowercase letters in the same column show significance at $p < 0.05$. 2 * means significant ($p < 0.05$), ** means extremely significant ($p < 0.01$). 3 Column values for the various years with same uppercase letters are insignificantly different at $p < 0.05$. 4 VE—emergence, VT—tasseling.

Compared with that under NM-D_2 treatment, the duration of total maize GP under MD_2 decreased by 3 days, and that of sowing–VE period shortened by 4 days, whereas the differences in growth durations of middle–late stages were minimal.

For treatments under mulched drip irrigation, the duration of total maize GP decreased gradually with the delay in PD, and the differences among MD_1–MD_4 reached 5–23 days. The total GP duration for maize in both MD_1 and MD_2 exceeded 150 days, which guaranteed the physiological maturity for grain. Whereas, maize in MD_3 and MD_4 experienced early autumn freezing during later growth stages, resulting in evident shortening in GP duration (10–18 days) and failing to reach physiological maturity.

A significant difference reaching 2–8 days was found in sowing–VE duration among MD_1–MD_4 treatments. When sowing was performed on April 20, maize sowing–VE duration was highly unstable with a 9-day difference within 2 years because of the fluctuation in late spring cold. In 2015, maize had entered seedling stage as late spring cold occurred, whereas the occurrence of chilling injury during sowing–VE stage delayed emergence by 9 days in 2016. In MD_2–MD_4 treatments, the sowing–VE duration decreased gradually with PD delay with insignificant inter-annual differences.

The duration of VE–VT for maize showed a decreasing trend with the PD delay. A 12-day difference in this period appeared in MD_1 treatment for 2 years because the seedlings underwent a long period of recovery after acquiring frostbite in 2015; by contrast, no chilling injury occurred during this period in 2016. The inter-annual differences of this period for 2 years were insignificant under MD_2–MD_4 treatments.

The effects of PD on the VT–Harvest duration were highly evident. MD_1 and MD_2 didn't show significant difference during this period, but shortened by 4–8 days when PD delayed to May 12 or later (MD_3 and MD_4) because of early autumn freezing.

3.3. Effects of Planting Date on Maize Growth Indexes

Planting date significantly affected maize emergence rate, plant height, stem diameter, and single plant dry weight (DW) and its allocation proportions in each organ under mulched drip irrigation in varying degrees (Table 4). No significant difference occurred in maize growth indexes under MD_1 and MD_2, but the plant height increased significantly and the stem diameter and DW per plant decreased considerably with the continuous delay in PD. And no appreciable differences existed in most growth indexes between years, other than plant height and single plant DW (with greater values in the first year).

The maize emergence rate under treatments of MD_1, MD_3 and MD_4 did not appear to be markedly different, but that in MD_2 decreased by 1.6–2.3%. The reason why maize in MD_2 obtained a lower emergence rate was that the PD (May 2–3) was near the date of late spring cold, the average temperature during maize sowing–VE period was lower than that in other treatments.

A comparison of the distribution ratios of DW in nutritive and reproductive organs of maize under mulched drip irrigation showed that the proportions of DW in stem and leaves exhibited an upward trend with the delay in PD, whereas those in the ears presented a downward trend. No significant difference was found in single-plant DW and its distribution proportion in each organ between MD_1 and MD_2; however, in MD_3 and MD_4 treatments, single-plant DW decreased by 10.0–14.5%, and the proportions of DW in stem and leaves increased by 11.7–15.4% and 16.4–22.8%, respectively, whereas those in ears decreased by 8.6–6.4%. Results indicated that the delay in PD led to "slim" plants, which are not conducive to the translocation of dry matter to ears.

In the comparison of the maize growth indexes between treatments with film mulching and NM-D_2, film mulching improved the maize emergence rate by 5.6–8.1% on average; the single-plant DW in MD_1 and MD_2 increased by 8.0–8.4%, while those in MD_3 and MD_4 decreased by 2.6–7.5%, compared with those in NM-D_2.

Table 4. Maize growth indexes of different treatments in harvest season.

Year	Item	Emergence Rate (%)	Height (cm)	Stem Diameter (cm)	Dry Matter Per Plant			
					$(g \cdot Plant^{-1})$	Proportion (%)		
						Stem	Leaves	Ear
2015	MD$_1$	94.5b	311.5a	2.6b	434.9b	19.9a	11.9a	68.2b
	MD$_2$	92.1a	314.2a	2.6b	440.2b	20.2a	12.1a	67.7b
	MD$_3$	93.5ab	317.3ab	2.4ab	400.0a	23.0ab	14.6b	62.4a
	MD$_4$	93.9ab	322.0b	2.3a	378.5a	23.9b	15.4b	60.8a
	F-test	6.6 *	4.2 *	6.8 *	11.8 **	11.5 **	16.7 **	18.7 **
	NM-D$_2$	86.7	316.2	2.4	412.5	22.7	14.1	63.2
	Mean	92.1A	316.2B	2.5A	413.2B	21.9A	13.6A	64.5A
2016	MD$_1$	93.5a	306.0a	2.5 b	426.1b	19.6a	12.4a	68.0c
	MD$_2$	93.0a	309.0b	2.5b	423.7b	20.0a	13.1ab	66.9bc
	MD$_3$	95.2a	314.4c	2.3ab	376.4a	21.5a	14.2bc	64.3ab
	MD$_4$	95.6a	319.3 c	2.1a	359.3a	22.1a	15.0c	62.9a
	F-test	3.3	6.9 **	15.2 **	39.4 **	1.9	6.6 *	7.3 *
	NM-D$_2$	88.5	311.5	2.3	384.7	21.6	13.8	64.6
	Mean	93.2A	312.0A	2.3A	394.0A	20.9A	13.7A	65.3A
Mean	MD$_1$	94.0ab	308.6a	2.6	430.5c	19.8a	12.1a	68.1b
	MD$_2$	92.5a	311.8a	2.6	432.0c	20.1a	12.6a	67.3b
	MD$_3$	94.3ab	315.9ab	2.4	388.2b	22.2b	14.4b	63.4a
	MD$_4$	94.7b	320.7b	2.2	368.9a	23.0b	15.2b	61.8a
	F-test	4.4 *	5.6 **	9.6 **	27.2 **	8.6 **	22.3 **	21.5 **
	NM-D$_2$	87.6	313.9	2.4	398.6	22.2	14	63.9

1 The different lowercase letters in the same column show significance at $p < 0.05$. 2 * means significant ($p < 0.05$), ** means extremely significant ($p < 0.01$). 3 Column values for the various years with different uppercase letters are significantly different at $p < 0.05$.

3.4. Effects of Planting Date on Maize Yield Indexes

The maize yield indexes under different treatments are shown in Table 5.

A comparison of the maize yield indexes with different PDs under mulched drip irrigation showed that the bald tip length and grain moisture content exhibited an increasing trend with a delay in PD, whereas the hundred-grain weight, grain number per ear (2016), yield, and total DM displayed a decreasing trend. The grain number per ear under MD$_2$ in 2015 was the largest in terms of relatively lower emergence rate and large ear length. And no statistically evident differences existed in most yield indexes other than the bald tip length and grain number per ear between years, with smaller bald tip length and greater grain number in the first year.

No evident difference was found in the yield indexes between MD$_1$ and MD$_2$. Under MD$_3$ and MD$_4$, the bald tip length increased by 34.3% and 54.3%, the hundred-grain weight decreased by 4.3% and 9.1%, and grain number per ear decreased by 7.5% and 9.9%, respectively, compared with those in MD$_1$ and MD$_2$. The results illustrated that delayed sowing is unfavorable for the formation of yield components. Compared with those in MD$_1$, under MD$_2$, MD$_3$, and MD$_4$ treatments, DM decreased by 1.2%, 9.5%, and 13.7%, grain moisture content increased by 1.8%, 13.5%, and 17.9%, and yield (14% moisture content) decreased by 1.4%, 11.4%, and 17.3%, respectively. The findings indicated no significant difference in total biomass and economic yield of maize sowed on April 20 and May 2, whereas delayed PD to May 12 or later resulted in a significant yield reduction.

Compared with those yield indexes under non-plastic film mulching (NM-D$_2$), the bald tip length and grain moisture of maize planted before May 3 (MD$_1$ and MD$_2$) averagely decreased by 20.5% and 3.8%, respectively, but the total DM and yield increased by 6.5% and 9.7%, respectively; while under MD$_3$ and MD$_4$ treatments, the bald tip length and grain moisture averagely increased by 9.1–22.7% and 8.0–12.3%, respectively, whereas the total DM and yield decreased by 1.5–8.0% and 2.2–6.6%, respectively. The results showed an apparent effect of film mulching on improving yield for the early sowing maize.

Table 5. Maize yield indexes in different treatments.

Year	Treatment	Bald Tip Length (cm)	Hundred- Grain Weight (g)	Grain Number Per Ear	Grain Moisture Content (%)	Yield (14%) (Mg·ha^{-1})	Total Dry Matter Accumulation (Mg·ha^{-1})
	MD$_1$	1.6a	35.5c	592b	24.9a	16.24c	34.25b
	MD$_2$	1.7a	35.1bc	604c	25.4a	16.02c	33.77b
	MD$_3$	2.2b	33.4ab	560ab	28.1b	14.45b	31.15a
2015	MD$_4$	2.4b	31.8a	548a	29.6b	13.55a	29.60a
	F-test	15.3 **	9.6 **	10.2 **	14.2 **	24.9 **	10.9 **
	NM-D$_2$	2	34.7	576	26	14.62	31.45
	Mean	2.0A	34.1A	576B	26.8A	15.20A	32.20A
	MD$_1$	1.8a	35.3b	591b	24.8a	15.91b	33.18b
	MD$_2$	1.9a	35.2b	587b	25.2a	15.68b	32.83b
	MD$_3$	2.5b	34.1b	538a	28.3b	14.04a	29.86a
2016	MD$_4$	3.0b	32.4a	521a	28.9b	13.06a	28.62a
	F-test	48.1 **	6.9 *	29.6 **	14.5 **	9.9 **	28.5 **
	NM-D$_2$	2.4	34.3	558	26.27	14.29	30.91
	Mean	2.3B	34.3A	559A	26.7A	14.53A	31.04A
	MD$_1$	1.7a	35.4c	592b	24.8a	16.08c	33.71c
	MD$_2$	1.8a	35.1c	595b	25.3a	15.85c	33.30c
Mean	MD$_3$	2.4b	33.7b	549a	28.2b	14.24b	30.50b
	MD$_4$	2.7c	32.0a	535a	29.3b	13.30a	29.11a
	F-test	30.3 **	41.8 **	21.7 **	30.4 **	59.9 **	26.9 **
	NM-D$_2$	2.2	34.5	567	26.1	14.46	31.18

1 The different lowercase letters in the same column show significance at $p < 0.05$. 2 * means significant ($p < 0.05$), ** means extremely significant ($p < 0.01$). 3 Column values for the various years with different uppercase letters are significantly different at $p < 0.05$.

According to the significance F-test based on the mean dates of 2 years, illuminating that no evident differences were found in yield indexes between MD$_1$ and MD$_2$, whereas significant differences appeared in maize bald tip length, hundred-grain weight, and grain yield among treatments of MD$_2$, MD$_3$, and MD$_4$. We use the regressions to present the relationships of maize bald tip length, hundred-grain weight, and grain yield with the days of PD delay (in Figures 6–8). When the typical date of late spring cold (around May 1) was regarded as the starting point, an evident linear relationship existed between the grain yield indexes and the days of PD delay (MD$_2$–MD$_4$). That is, for every 10 days of PD delay, bald tip length increased by 23.2%, whereas hundred-grain weight and grain yield decreased by 4.5% and 8.5%, respectively. According to the regression equation, it concluded that the yield penalty (4.2%) was less than 5.0% with a yield exceeding 15 Mg·ha^{-1} when PD delayed to May 8.

Therefore, obtaining a super high yield is possible under mulched drip irrigation when the medium-mature maize varieties with close-planting are sown from April 20 to May 8 in the area.

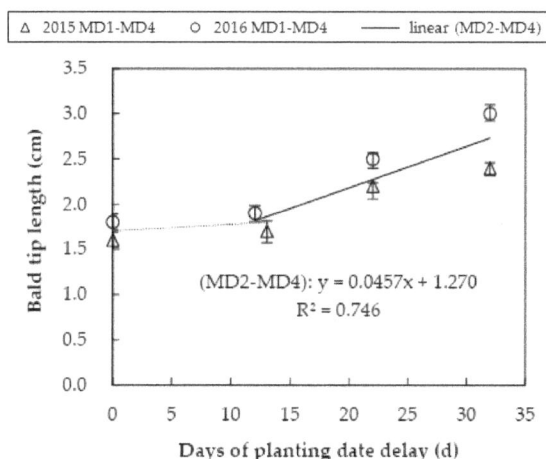

Figure 6. Effect of planting date on Bald tip length.

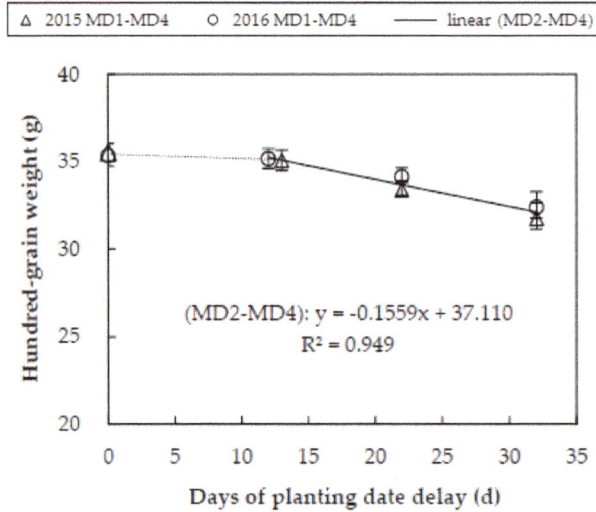

Figure 7. Effect of planting date on hundred-grain weight.

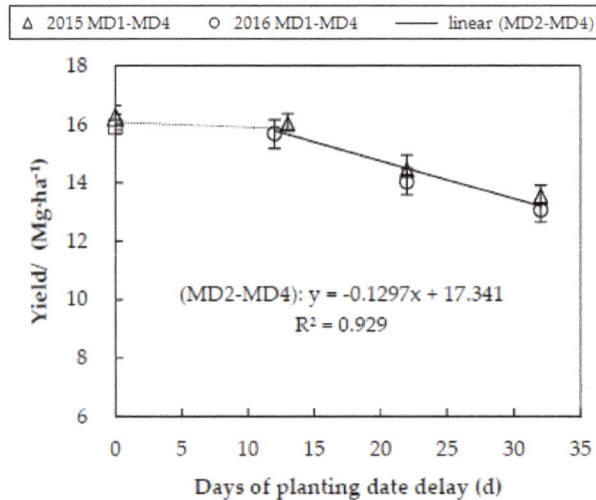

Figure 8. Effect of planting date on maize yield.

4. Discussion

The selection of an appropriate planting time for maize belongs to the category of ecological climate adaptability with regards to the degree of interaction between growth habit and climate condition [36]. Water stress is a key factor that restricts grain yield in areas with limited irrigation, and PD mainly affects yield formation by regulating the reasonable correspondence between rainfall and GP [11]. However, in areas where chilling injury frequently occurs, the selected PD should fully use photo-thermal resources for the limited accumulated temperature [13]. The test was conducted in a cold middle-latitude region with general occurrence of late spring cold from the end of April to early May, along with the regularity of spring drought. Generally, farmers tend to delay maize planting with temperature and soil moisture under consideration. Whereas, the probability of experiencing early autumn freezing (at the end of September) was increased in the later GP of maize, hence resulting in irregular grain yields. In this study, the application of mulched drip irrigation technology provided an environment with appropriate water and fertilizer levels for maize during the entire GP. Therefore, the study mainly analyzed the effects of chilling injury from certain growth stages and AAT differences caused by various PDs on maize growth.

Maize is highly sensitive to chilling injury, which mainly affects crop growth by changing crop root activity, enzyme activity, and photosynthetic rate [37,38]. The chilling injury at the seedling (MD_1) or sowing–VE (MD_2) stages delayed maize jointing or emergence, but the GP (>150 days) met the growth demand and ensured grain physiological maturity. Nevertheless, the maize sowed after May 12 (MD_3 and MD_4) suffered from early autumn freezing at latter growth stages, resulting in significant shortening in GP duration and increase in grain moisture content with failing to reach physiological maturity. The results verified the previous viewpoint that early autumn freezing is detrimental to maize growth [2,39]. On the basis of the maize growth and yield indexes with different PDs, a delay in PD was found to cause weak individual growth and reduced proportion of dry matter in ears (Table 4), which were not conducive to the formation of economic output (Table 5). The results coincided with Dong's viewpoints [40].

Planting date significantly affected the AAT distribution. The AATs in maize's entire GP decreased with the delay in PD under MD_1–MD_4, and the difference mainly manifested during the reproductive growth stage. At this stage, AATs decreased by an average of 96.6 °C for every 10 days of PD delay (Table 3). Whether the AATs in maize reproductive growth stage could meet the growth demand is a key factor in yield formation for the significant effects on grain filling and dehydration process [13,41]. The relationship between maize yield and AATs in this stage is analyzed in Figure 9, which shows that when sowing was performed before May 3 (MD_1 and MD_2), AATs were sufficient to ensure grain physiological maturity with high yield. But with continuous delay in PD (MD_2–MD_4), grain yield exhibited a negative correlation with AATs at this stage, and the amplitude of yield penalty reached 6.14–7.74% for every 100 °C of AAT decrease. Confirming the key measure to guarantee high yield is to ensure grain maturity before early autumn freezing with the maximum use of accumulated temperature [42]. Moreover, the decreasing solar radiation and light interception together with low temperature in latter growth stages exerted adverse effect on yield formation for the late sowing maize [18,43].

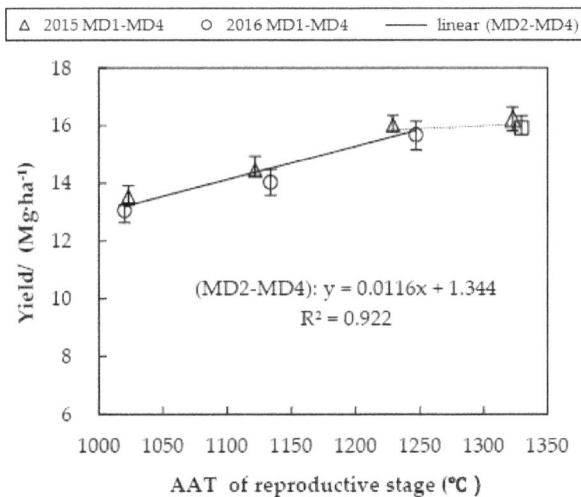

Figure 9. Effect of active accumulated temperature for reproductive stage on maize yield. AAT–active accumulated temperature.

Numerous studies have assumed that chilling injury at each growth stage will lead to varying degrees of yield reduction [13,44,45]. For the early sowing treatments in this study, the late spring cold slowed down maize early growth, but exerted no evident adverse effects on yield benefit. Reasons may be as following: the warming effect of film mulching could partly compensate for the lack of heat during the early stage of maize, thereby guaranteeing high emergence rate and nonfatal hurt from low temperature to seedlings; moreover, sufficient active accumulated temperature and solar radiation during maize growing season promoted the potential productivity. In the test, the yield from NM-D_2 treatment appears to be about the same as the yield from MD_3 treatment but less than that from MD_2 treatment, which showed an evident yield benefit from film mulching. However, early autumn freezing during the reproductive stage led to considerable yield reduction. Therefore, maize should be sown in advance with film mulching in cold middle-latitude regions to avoid early autumn freezing and ensure that GP duration and AATs can meet growth requirements.

In the test region, we suggest that the medium-mature variety of maize is sown from April 20 to May 8 with available irrigation. Compared with the conclusions of Yu et al and Cao et al. (maize cultivation without film mulching) [18,19], appropriate sowing intervals exhibited an increasing trend and were 5–7 days in advance, with 8.3–80.5% increase in grain yield. Appropriate advanced sowing gained a super-high yield in this study, also due to the technology of drip irrigation, which provided in-season supply of water and fertilizers for maize [30,46]. For the same latitude areas with similar ecological environments, it's potential to obtain high yield when the maize sowing date is in range of 10 days before or after the late spring cold, for a medium-mature variety of maize under mulched drip irrigation.

5. Conclusions

Mulched drip irrigation has shown a broad development prospect in middle-latitude spring maize planting area with low temperature and chilling injury. In this study, the effects of environment difference caused by varying PDs on maize growth under mulched drip irrigation were analyzed. The conclusions were as follows:

(1) Film mulching could significantly improve soil temperature, and the warming effect was more evident in maize early growth stages and gradually decreased with the postponement of PDs (based on MD_1). The warming effect of mulching was effective in compensating for the lack of heat caused by the low-temperature environment at the early growth stages of maize, which was beneficial to the improvement of maize emergence rate and economic yield.

(2) Planting date markedly affects AATs and maize growth process. Both AATs and maize GP duration decreased with the delay in PD (based on MD_1). When the PD was within April 20–May 2, the GP and AATs could meet the growth demand, whereas a delay of the PD until May 12 or later failed to guarantee grain physiological maturity.

(3) The effect of PD on maize growth was closely related to chilling injury (the late spring cold and early autumn freezing). The late spring cold slowed down maize early growth process (MD_1 and MD_2), but no significant adverse effect existed on later growth. As to MD_3 and MD_4, the occurrence of early autumn freezing at maize later stages led to significant decrease in AATs during the reproductive stage with adverse effect on grain filing and dehydration, and subsequently resulted in evident yield penalty (11.3–16.8%).

(4) Taking the date (May 1) of late spring cold generally occurring as the starting point, there was a significant positive correlation in maize bald tip length and the days of PD delay, while the changing tendency of hundred-grain weight and grain yield with the days of PD delay showed a linear decreasing trend; the AATs of maize reproductive stage gradually decreased with the postponement of PD, and the grain yield decreased linearly with the decrease in AATs of this stage.

Based on the threat of chilling injury in the middle-latitude area, it seems more reasonable to plant medium-mature maize varieties with close-planting in range of 10 days before and after the late spring cold to obtain a potential high yield (>15 Mg·ha^{-1}) under mulched drip irrigation.

Acknowledgments: The authors are greatly indebted to the financial support from the National Support Program of China (2014BAD12B05).

Author Contributions: Dan Wang, Yan Mo and Mingkun Cai are graduate-students of China Agricultural University, they conducted the experiments and collected relevant materials for this research. Xinyang Bian is the key working personnel in the experimental station, he participated in the design and operation of the experiment. Dan Wang analyzed the date and wrote the manuscript. Guangyong Li provided direction of this study and revised the manuscripts. All authors have read and approved the final manuscript.

Conflicts of Interest: The authors declare no conflict of interest.

References

1. Parry, M.; Canzaiani, O.; Palutikof, J.; van der Linden, P.J.; Hanson, C. Climate change 2007: Impacts, adaptation and vulnerability. *J. Environ. Qual.* **2007**, *37*, 2407.
2. Kucharik, C. A multidecadal trend of earlier corn planting in the central USA. *Agron. J.* **2006**, *98*, 1544–1550. [CrossRef]
3. Olesen, J.E.; Trnka, M.; Kersebaum, K.C.; Skjelvåg, A.O.; Seguin, B.; Peltonensainio, P.; Micale, F. Impacts and adaptation of european crop production systems to climate change. *Eur. J. Agron.* **2001**, *34*, 96–112. [CrossRef]
4. Zhao, J.; Yang, X.; Dai, S.; Lv, S.; Wang, J. Increased utilization of lengthening growing season and warming temperatures by adjusting sowing dates and cultivar selection for spring maize in northeast China. *Eur. J. Agron.* **2015**, *67*, 12–19. [CrossRef]
5. Horai, K.; Ishii, A.; Mae, T.; Shimono, H. Effects of early planting on growth and yield of rice cultivars under a cool climate. *Field Crops Res.* **2013**, *144*, 11–18. [CrossRef]
6. Sacks, W.J.; Kucharik, C.J. Crop management and phenology trends in the U.S. Corn Belt: Impacts on yields, evapotranspiration and energy balance. *Agric. For. Meteorol.* **2011**, *151*, 882–894. [CrossRef]
7. Kucharik, C.J. Contribution of planting date trends to increased maize yields in the central united states. *Agron. J.* **2008**, *100*, 328–336. [CrossRef]
8. Ma, S.Q.; Wang, Q.; Wang, C.Y.; Huo, Z.G. The risk division on climate and economic loss of maize chilling damage in Northeast China. *Geogr. Res.* **2008**, *27*, 1169–1177. (In Chinese)
9. Ma, L.L. Response of Maize Growth Process to Sowing Date and the Factors of Light and Temperature. Ph.D. Thesis, Shanxi Agricultural University, Jinzhong, China, 2014. (In Chinese)
10. Hatfield, J.L.; Prueger, J.H. Temperature extremes: Effect on plant growth and development. *Weather Clim. Extrem.* **2015**, *10*, 4–10. [CrossRef]
11. Lu, H.D.; Xue, J.Q.; Guo, D.W. Efficacy of planting date adjustment as a cultivation strategy to cope with drought stress and increase rainfed maize yield and water-use efficiency. *Agric. Water Manag.* **2016**, *179*, 227–235. [CrossRef]
12. Wang, S.Y. *Study on Low Temperature and Chilling Injury for Crops*; Meteorological Press: Beijing, China, 1995.
13. Xu, T.J. Effect of Low Temperature on Maize Senescence and Grain Filling Rate during the Grain Filling Period and PASP-KT-NAA Regulation. Ph.D. Thesis, Chinese Academy of Agricultural Sciences, Beijing, China, 2012.
14. Lauer, J.G.; Carter, P.R.; Wood, T.M.; Diezel, G.; Wiersma, D.W.; Rand, R.E.; Mlynarek, M.J. Corn hybrid response to planting date in the northern corn belt. *Agron. J.* **1999**, *91*, 834–839. [CrossRef]
15. Wu, R.X.; Liu, R.Q.; Lu, C.L.; Lu, Y.L.; Li, H.; Zhang, L.; Lu, X.Z.; Wang, X.H. The optimum sowing Time for Plastic film Corn and the Use of the Two Theories. *Sci. Agric. Sin.* **2001**, *34*, 433–438. (In Chinese)
16. Benson, G.O. Corn replant decisions: A review. *J. Prod. Agric.* **1990**, *3*, 180–184. [CrossRef]
17. Good, D. Can corn and soybean crops overcome late planting? *Corn & Soybean Digest Exclusive Insight*, 6 June 2011; 4.
18. Yu, J.L. Effects of Sowing Date and Density on Matter Production and Yield Formation in Maize. Ph.D. Thesis, Shenyang Agricultural University, Shenyang, China, 2013. (In Chinese)
19. Cao, Q.J.; Yang, F.T.; Chen, X.F.; Lamine, D.; Li, G. Effects of sowing date on growth, yield and quality of spring maize in the central area of Jilin Province. *J. Maize Sci.* **2013**, *21*, 71–75. (In Chinese)
20. Wang, Y.Q.; Gong, S.H.; Xu, D.; Zhang, Y.Q. Several aspects of the research for corn under film drip irrigation in the Northeast of China. *J. Irrig. Drain.* **2015**, *34*, 1–4.

21. Orzolek, M.D.; Murphyj, J.; Giardi, J. The effect of colored polyethylene mulch on the yield of squash. In *Tomato and Auliflower*; College of Agricultural Sciences, Pennsylvania State University: State College, PA, USA, 2003.

22. Kasperbauer, M.J.; Hunt, P.G. Mulch'Surface Color Affects Yield of Fresh-market Tomatoes. *J. Am. Soc. Hortic. Sci.* **1989**, *114*, 216–219.

23. Wang, X.K.; Li, Z.B.; Xing, Y.Y. Effects of mulching and nitrogen on soil temperature, water content, nitrate-n content and maize yield in the loess plateau of China. *Agric. Water Manag.* **2015**, *161*, 53–64.

24. Sun, G.F.; Du, B.; Qu, Z.Y.; Li, C.J.; Zhang, Z.L.; Li, X.H. Changes of soil temperature and response to air temperature under different irrigation modes. *Soils* **2016**, *48*, 581–587. (In Chinese)

25. Lamm, F.R. *Irrigation and Nitrogen Management for Subsurface Drip Irrigated Corn—25 Years of K-State's Efforts*; ASABE Paper, No. 141914980; American Society of Agricultural and Biological Engineers: St. Joseph, MI, USA, 2014.

26. Jayakumar, M.; Janapriya, S.; Surendran, U. Effect of drip fertigation and polythene mulching on growth and productivity of coconut (*Cocos nucifera* L.), water, nutrient use efficiency and economic benefits. *Agric. Water Manag.* **2017**, *182*, 87–93.

27. Sun, Y.L. Effects of Sowing Dates on Spring Corn Female Spike Differentiation and Yield of Xinjiang with High Yield of 15,000 kg/hm^2. Ph.D. Thesis, Shihezi University, Shihezi, China, 2014. (In Chinese)

28. Wei, Y.G.; Chen, L.; Jiang, J.F.; Ding, W.K.; Wang, H.L. Effects of irrigation methods and sowing date on yield and water use efficiency of mulched spring maize. *Chin. Agric. Sci. Bull.* **2014**, *30*, 203–208. (In Chinese)

29. Feyzbakhsh, M.T.; Kamkar, B.; Mokhtarpour, H.; Asadi, M.E. Effect of soil water management and different sowing dates on maize yield and water use efficiency under drip irrigation system. *Arch. Agron. Soil Sci.* **2015**, *61*, 1581–1592. [CrossRef]

30. Li, C.; Wei, X.; Hu, G.J.; Zhang, X.L.; Xu, L.Q. Variation characteristics of first frost dates of Chifeng City in recent 50 years. *Inner Mong. Agric. Sci. Technol.* **2015**, *43*, 93–95. (In Chinese)

31. Bai, C.Y.; Li, S.K.; Bai, J.H.; Zhang, H.B.; Xie, R.Z. Characteristics of accumulated temperature demand and its utilization of maize under different ecological conditions in Northeast China. *Chin. J. Appl. Ecol.* **2011**, *22*, 2337–2342. (In Chinese)

32. Zheng, H.B.; Qi, H.; Liu, W.R.; Zheng, J.Y.; Luo, Y.; Li, R.P.; Li, W.T. Yield characteristics of ZD958 and XY335 under different N application rates. *J. Maize Sci.* **2013**, *21*, 117–119, 126. (In Chinese)

33. Gao, X. Effect on Stalk Lodging-Resistance Characteristics and Photosynthetic Characteristics by Different Types of Maize Cultivar High Density. Master's Thesis, Inner Mongolia Agricultural University, Hohhot, China, 2012. (In Chinese)

34. Liu, Y.; Li, Y.F.; Li, J.S.; Yan, H.J. Effects of mulched drip irrigation on water and heat conditions in field and maize yield in sub-humid region of Northeast China. *Trans. Chin. Soc. Agric. Mach.* **2015**, *46*, 93–104. (In Chinese)

35. Yin, W.; Chen, G.P.; Chai, Q.; Zhao, C.; Feng, F.X.; Yu, A.Z.; Hu, F.L.; Guo, Y. Responses of soil water and temperature to previous wheat straw treatments in plastic film mulching maize field at Hexi Corridor. *Sci. Agric. Sin.* **2016**, *49*, 2898–2908. (In Chinese)

36. Lv, X. Studies on Effects of Ecological Factors on Growth of Maize and Establishment of Climate Ecology Model and Appraisement System. Ph.D. Thesis, Shandong Agricultural University, Jinan, China, 2002. (In Chinese)

37. Zhang, J.L.; Zhou, Y.J.; Hu, M.; Zhu, L.; Zhang, D. The effects of low-temperature stress on chilling resistance ability of maize seedings. *J. Northeast Agric. Univ.* **2004**, *35*, 129–134. (In Chinese)

38. Dong, J.W.; Liu, J.Y.; Tao, F.; Xu, X.L.; Wang, J.B. Spatio-temporal changes in annual accumulated temperature in China and the effects on cropping systems, 1980s to 2000. *Clim. Res.* **2009**, *40*, 37–48. [CrossRef]

39. Duvick, D.N. Possible Genetic Causes of Increased Variability in U.S. Maize Yields. *Plant Breed. Yield Variability* **1989**, 147–156. Available online: http://www.cabdirect.org/abstracts/19901615378.html (accessed on 22 August 2017).

40. Dong, H.F.; Li, H.; Li, A.J.; Yan, X.G.; Zhao, C.M. Relations between delayed sowing date and growth, effective accumulated temperature of maize. *J. Maize Sci.* **2012**, *20*, 97–101. (In Chinese)

41. Tsimba, R.; Edmeades, G.O.; Millner, J.P.; Kemp, P.D. The effect of planting date on maize: Phenology, thermal time durations and growth rates in a cool temperate climate. *Field Crops Res.* **2013**, *150*, 145–155. [CrossRef]

42. Dou, S.J.; Dou, S.Y. Exploration of method to determine the optimum planting date based on accumulated temperature. *Soybean Sci. Technol.* **2009**, *5*, 44–45. (In Chinese)

43. Wilson, D.R.; Muchow, R.C.; Murgatroyd, C.J. Model analysis of temperature and solar radiation limitations to maize potential productivity in a cool climate. *Field Crops Res.* **1995**, *43*, 1–18. [CrossRef]

44. Meng, Y.; Li, M.; Wang, L.M.; Wang, L.Z.; Feng, Y.J. Effects of Chilling Injury on maize and correlative research. *Heilongjiang Agric. Sci.* **2009**, *4*, 150–153. (In Chinese)

45. Martin, M.; Gavazov, K.; Körner, C.; Hättenschwiler, S.; Rixen, C. Reduced early growing season freezing resistance in alpine treeline plants under elevated atmospheric CO_2. *Glob. Chang. Biol.* **2010**, *16*, 1057–1070. [CrossRef]

46. Lv, S.Q. Study on Different Nitrogen Application Schemes under Fertigation for Maize (*Zea mays* L.). Ph.D. Thesis, Chinese Academy of Agricultural Sciences, Beijing, China, 2012. (In Chinese)

sustainability

MDPI

Article

Adaptive Effectiveness of Irrigated Area Expansion in Mitigating the Impacts of Climate Change on Crop Yields in Northern China [†]

Tianyi Zhang [1,*], Jinxia Wang [2] and Yishu Teng [3]

[1]   State Key Laboratory of Atmospheric Boundary Layer Physics and Atmospheric Chemistry,
     Institute of Atmospheric Physics, Chinese Academy of Sciences, Beijing 100029, China
[2]   School of Advanced Agricultural Sciences, Peking University, Beijing 1000871, China;
     jxwang.ccap@igsnrr.ac.cn
[3]   BICIC, Beijing Normal University, Beijing 1000875, China; tengyishu@bnu.edu.cn
*   Correspondence: zhangty@post.iap.ac.cn; Tel.: +86-10-8299-5291
[†]   This paper was presented at the Global Land Programme 3rd Open Science Meeting, Beijing, China,
     24–27 October 2016.

Academic Editors: Elaine Wheaton and Suren N. Kulshreshtha
Received: 22 February 2017; Accepted: 12 May 2017; Published: 19 May 2017

Abstract: To improve adaptive capacity and further strengthen the role of irrigation in mitigating climate change impacts, the Chinese government has planned to expand irrigated areas by 4.4% by the 2030s. Examining the adaptive potential of irrigated area expansion under climate change is therefore critical. Here, we assess the effects of irrigated area expansion on crop yields based on county-level data during 1980–2011 in northern China and estimate climate impacts under irrigated area scenarios in the 2030s. Based on regression analysis, there is a statistically significant effect of irrigated area expansion on reducing negative climate impacts. More irrigated areas indicate less heat and drought impacts. Irrigated area expansion will alleviate yield reduction by 0.7–0.8% in the future but associated yield benefits will still not compensate for greater adverse climate impacts. Yields are estimated to decrease by 4.0–6.5% under future climate conditions when an additional 4.4% of irrigated area is established, and no fundamental yield increase with an even further 10% or 15% expansion of irrigated area is predicted. This finding suggests that expected adverse climate change risks in the 2030s cannot be mitigated by expanding irrigated areas. A combination of this and other adaptation programs is needed to guarantee grain production under more serious drought stresses in the future.

Keywords: irrigated area; drought; climate; adaptation; SPEI

1. Introduction

Climate change poses serious challenges to Chinese agriculture [1–3]. In recent years, the ability to meet these challenges has been tested by several major extreme climate events. For example, the devastating drought in southwestern China in 2010 critically impaired local agriculture, resulting in an estimated loss of 317 million USD [4]. The average annual total cost of climate disasters is approximately 80 billion USD in China [5]. Climate extremes are anticipated to be aggravated and increasingly influenced by climate change. These conditions will constrain future growth in agricultural sectors; therefore, it is important to take actions to mitigate future climate risks.

Countervailing the current and future adverse climate risks will require adaptation measures. In 2011, the Chinese government announced an important policy requiring that 600 billion USD be invested in agricultural irrigation [6]. The policy set several quantitative targets for improving irrigation over 10 years, starting in 2011. The most important plan is to expand the irrigated area by 2.67 million ha (equivalent to a 4.4% increase). Although the quantitative target is clearly framed, concerns have been raised about the effectiveness of irrigated area expansion in climate risk mitigation [7].

To our knowledge, no study has evaluated the extent that the above irrigated area expansion plan [7] reduces the impacts of future climate change in China. Using process-based crop model and associated assumptions, some studies have evaluated the adaptation effectiveness of potential irrigation in facing climate change risks on Chinese agriculture. For example, assuming no crop water stress was predicted to mitigate 5–15% of the yield reduction in China under future climate scenarios [8,9]. However, the assumption of no water stress is unrealistic and difficult to link with the government plan. Several recent studies were encouraged to the integration of farming management methods into impact assessments as these methods greatly determine the degree of climate impacts on crops. For instance, fertilizer intensive farmers can largely reduce the negative effects of heat stresses in the UK, France and Italy, while the effect is small or even negative in other European countries [10]. On a global scale, vulnerability of key food crops to drought is also greatly dependent on socio-economic conditions and agricultural investments [11]. In China, recent studies have quantified the relationship among climate, crop and irrigation based on statistical data [3,12]. They employed a new data-driven approach, but the major disadvantage is a lack of socioeconomic data with fine spatial-temporal resolution. Furthermore, few of these analyses addressed potential adaptation under future climate scenarios.

Therefore, to understand the adaptation effectiveness of expanding the irrigated area in mitigating climate change impacts, the following objectives are specified in our study: (i) we quantitatively identify crop yield responses to climate and irrigated area based on county-level data during 1980–2011; (ii) we establish a statistical model to assess the adaptive effects of the irrigated areas expansion plan already underway on climate change mitigation; (iii) we explore future climate impacts on crop yields across different irrigated area scenarios.

2. Materials and Methods

2.1. Data and Pretreatment

Our study region is in northern China (Figure 1) because of the increasingly important role the region plays in grain production. The region encompasses 50% of cultivated land and produces 56% of the annual grain production in China. The major grain production areas are the northeastern, northern, and eastern parts of northwestern China. Due to low precipitation and the uneven seasonal distribution of precipitation, crop production in northern China largely depends on irrigation.

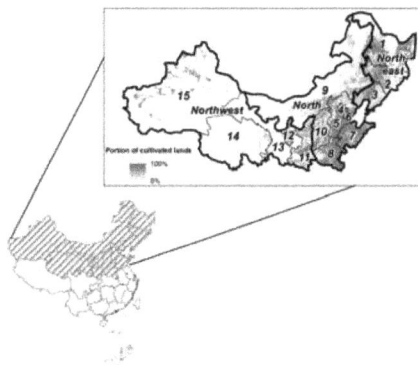

Figure 1. Illustration of the study region. The shaded area is northern China, which includes the Northeast, North and Northwest regions of China, as shown in the top-right figure. The number indicates the provinces involved in the study. 1: Heilongjiang; 2: Jilin; 3: Liaoning; 4: Beijing; 5: Hebei; 6: Tianjin; 7: Shandong; 8: Henan; 9: Inner Mongolia; 10: Shanxi; 11: Shaanxi; 12: Ningxia; 13: Gansu; 14: Qinghai; 16: Xinjiang.

Crop data used in the study were obtained from the Chinese Academy of Agricultural Sciences. These data include county-level sown areas and production data for rice, wheat, maize and soybean, which are the four major food crops in our study region, over the period 1980–2011. In addition, we considered county-level irrigated areas and cultivated land areas in our study region and period. Based on the definition by the National Standard of the People's Republic of China [13] and the Food and Agriculture Organization of the United Nations [14], "irrigated area" is the area equipped to be irrigated and it is the most often-used index to quantify irrigation level in earlier studies [3,12]. Percentage of irrigated area (PIA) was calculated based on Equation (1), which represents the irrigated areas relative to sowing areas, an index quantifying irrigated conditions, for each county in each year.

$$PIA_{c,t} = \frac{IRRI_{c,t}}{CulArea_{c,t}} \times 100\% \tag{1}$$

where $PIA_{c,t}$ is the percentage of irrigated area (%), $IRRI_{c,t}$ is the irrigated area (ha), and $CulArea_{c,t}$ is the cultivated land area (ha) of county c in year t.

As crop-specific data for irrigated area are not available, we lump data of the four crops to match the PIA data (Equation (2)).

$$Y_{c,t} = \frac{riceP_{c,t} + wheatP_{c,t} + maizeP_{c,t} + soybeanP_{c,t}}{riceA_{c,t} + wheatA_{c,t} + maizeA_{c,t} + soybeanA_{c,t}} \tag{2}$$

where $Y_{c,t}$ is the yields of the four crops weighted in each county by sown area (ton ha^{-1}); $riceP_{c,t}$, $wheatP_{c,t}$, $maizeP_{c,t}$ and $soybeanP_{c,t}$ (ton) are the production of rice, wheat, maize, and soybean, respectively; and $riceA_{c,t}$, $wheatA_{c,t}$, $maizeA_{c,t}$ and $soybeanA_{c,t}$ (ha) are the sown area of rice, wheat, maize and soybean, respectively, of county c in year t.

Daily temperature and precipitation data in 756 climate stations were downloaded from the China Meteorology Data Sharing Service System [15]. Quality controls and homogenization of these climate data have been executed by the Chinese Meteorological Administration. To derive climate data for each county, we estimated daily climate data using the algorithm presented by Thornton et al. [16]. This algorithm interpolates the abovementioned data of the 756 climate stations into 10 km grid cells and then extracts climatic information for each county from the grid data. The daily grid climatic dataset has been used in a previous study [17]. Subsequently, we calculated the daily climate data for each county by zonal averaging, and then aggregated the daily data into monthly climate data. To represent drought severity, we calculated the monthly Standardized Precipitation Evapotranspiration Index (SPEI) for each county. SPEI is a multi-scalar drought index calculated based on a climatic water balance model [18] considering the role of both precipitation and evapotranspiration. An R package, "SPEI" (https://cran.r-project.org/web/packages/SPEI/), was used to calculate the index with the lag set to 1 month to quantify the monthly moisture conditions due to the climate of the same month. Next, we calculated the mean-growing-season average temperature (Tavg) and SPEI for each crop. The growing season period for each crop was derived from the *Chinese Agricultural Phenology Atlas* [19] (Table 1). Finally, to match the PIA data, Tavg and SPEI were aggregated as weighted by the sown area of the four crops in each year (Equations (3) and (4)).

$$Tavg_{c,t} = \frac{riceTavg_{c,t} \times riceA_{c,t} + wheatTavg_{c,t} \times wheatA_{c,t} + maizeTavg_{c,t} \times maizeA_{c,t} + soybeanTavg_{c,t} \times soybeanA_{c,t}}{riceA_{c,t} + wheatA_{c,t} + maizeA_{c,t} + soybeanA_{c,t}} \tag{3}$$

where $Tavg_{c,t}$ is the mean-growing-season average temperature weighted by sown area; $riceTavg_{c,t}$, $wheatTavg_{c,t}$, $maizeTavg_{c,t}$ and $soybeanTavg_{c,t}$ are the mean growing season average temperature for rice, wheat, maize and soybean, respectively; and $riceA_{c,t}$, $wheatA_{c,t}$, $maizeA_{c,t}$ and $soybeanA_{c,t}$ are the sown area for rice, wheat, maize and soybean, respectively, of county c in year t.

$$SPEI_{c,t} = \frac{riceSPEI_{c,t} \times riceA_{c,t} + wheatSPEI_{c,t} \times wheatA_{c,t} + maizeSPEI_{c,t} \times maizeA_{c,t} + soybeanSPEI_{c,t} \times soybeanA_{c,t}}{riceA_{c,t} + wheatA_{c,t} + maizeA_{c,t} + soybeanA_{c,t}} \quad (4)$$

where $SPEI_{c,t}$ is the mean-growing-season average SPEI weighted by sown area; $riceSPEI_{c,t}$, $wheatSPEI_{c,t}$, $maizeSPEI_{c,t}$ and $soybeanSPEI_{c,t}$ are the mean growing season average SPEI for rice, wheat, maize, and soybean, respectively; and $riceA_{c,t}$, $wheatA_{c,t}$, $maizeA_{c,t}$ and $soybeanA_{c,t}$ are the sown area for rice, wheat, maize, and soybean, respectively, of county c in year t.

2.2. Statistical Model

To evaluate the relationship of climate, yield and PIA, we established a fixed-effect regression model, as given in Equation (5).

$$\log(Y_{c,t}) = \alpha_1 PIA_{c,t} + \alpha_2 PIA_{c,t}^2 + \alpha_3 Tavg_{c,t} + \alpha_4 Tavg_{c,t}^2 + \alpha_5 SPEI_{c,t} + \alpha_6 SPEI_{c,t}^2 + \alpha_7 PIA_{c,t} Tavg_{c,t}$$
$$+ \alpha_8 PIA_{c,t} SPEI_{c,t} + \alpha_{9,c} County_c + \alpha_{10,c} County_c \times Year_t + \alpha_{11,c} County_c \times Year_t^2 + \varepsilon_{c,t} \quad (5)$$

where $Y_{c,t}$ is yield for the four crops weighted by sown area in Equation (2) (ton ha^{-1}), $PIA_{c,t}$ is the percentage of irrigated area in Equation (1) (%); $Tavg_{c,t}$ and $SPEI_{c,t}$ are respectively the mean-growing-season temperature (Equation (3)) and SPEI (Equation (4)) weighted by the sown area of the four crops of county c in year t; $County$ is the dummy variable for county; $Year$ denotes time; and ε is the error term. $\alpha_1 - \alpha_{11}$ are the regression coefficients for each term.

In this model, we used quadratic terms for PIA, Tavg, and SPEI to account for the fact that crops perform best under moderate management and climate conditions and are harmed by extreme cold, hot, dry, or wet field conditions. In addition, we considered the potential interactions between irrigation and climate variables in this model, which represent the changes in climate impacts under different irrigation conditions. Unobserved possible nonlinear time trends at the county level were controlled by using county-by-year linear and quadratic terms and unobserved time-constant variations between counties using a county fixed effect. Consistent with other studies based on statistical model [20], CO_2 effects were not considered. Therefore, results in this study reflect the possible largest impacts from climate change.

The accuracy of the model was evaluated using a bootstrap analysis [21]. By constructing a number of re-samples and replacing the observations, this analysis evaluated the model accuracy defined by confidence intervals. More specifically, years were chosen randomly with replacements for 1000 iterations to estimate the regression coefficients of the model. Then, 1000 sets of regression coefficients were derived, which were then used to calculate yield changes by inputting future climate conditions. The confidence interval not spanning zero indicates a significant effect. Here, the median value and 95% confidence interval (95% CI) of those regression coefficients are reported.

2.3. Climate Scenarios

The climate change projections were taken from the Program for Climate Model Diagnosis and Inter-comparison—Coupled Model Inter-comparison Project Phase 5 for two representative concentration pathways (RCP2.6 and RCP8.5) in our study region. This ensemble climate scenarios were simulated by 26 climate models (Supplementary Materials Table S1). RCP2.6 represents a low emission pathway, i.e., greenhouse gas emissions peak between 2010 and 2020 with emissions declining substantially thereafter; RCP8.5 is a high emission pathway, i.e., emissions continue to rise throughout the 21st century.

The baseline period was set to 1980–2011, consistent with our observations, and the future period was 2020–2039 (referred to as 2030s hereafter), the target period for the abovementioned irrigated areas expansion plan. Following the steps for processing observed climate data described in Section 2.1, we derived the mean-growing-season Tavg and SPEI weighted by the sown areas of four crops for each county-year pair based on the future climate scenarios (here, we assume there is no change in the growing areas of the four crops). The difference in anticipated growing-season Tavg and SPEI relative to the baseline climate was input into our statistical model.

Table 1. Growing season of four crops for each province and mean growing season temperature.

Provinces	ID	Rice Sowing	Harvest	Tavg	Wheat Sowing	Harvest	Tavg	Maize Sowing	Harvest	Tavg	Soybean Sowing	Harvest	Tavg
Northeast													
Heilongjiang	1	1 May	30 September	17.5	1 April	31 July	14.8	1 May	30 September	17.5	1 May	30 September	17.5
Jilin	2	1 May	30 September	18.1	1 April	31 July	15.5	1 May	30 September	18.1	1 May	30 September	18.1
Liaoning	3	1 May	30 September	20.4	1 April	31 July	17.8	1 May	30 September	20.4	1 May	30 September	20.4
North													
Beijing	4	1 April	30 September	20.1	1 October	30 Jun.	6.5	1 June	30 September	22.5	1–30 June	30 September	22.5
Hebei	5	1 April	30 September	21.4	1 October	30 June	8.4	1 June	30 September	23.7	1–30 June	30 September	23.7
Tianjin	6	1 April	30 September	22.3	1 October	30 June	9.0	1 June	30 September	24.7	1–30 June	30 September	24.7
Shandong	7	1 April	30 September	22.0	1 October	30 June	10.0	1 June	30 September	24.5	1–30 June	30 September	24.5
Henan	8	1 April	30 September	22.6	1 October	30 June	11.4	1 June	30 September	24.8	1–30 June	30 September	24.8
Inner Mongolia	9	1May	30 September	17.0	1 April	31 July	14.7	1May	30 September	17.0	1May	30 September	17.0
Northwest													
Shanxi	10	1 May	30 September	19.2	1 October	30 June	5.5	1 May	30 September	19.2	1 May	30 September	19.2
Shaanxi	11	1 May	30 September	20.3	1 October	30 June	8.2	1 May	30 September	20.3	1 May	30 September	20.3
Ningxia	12	1 May	30 September	18.1	1 October	30 June	4.4	1 May	30 September	18.1	1 May	30 September	18.1
Gansu	13	1 April	30 September	16.0	1 October	30 June	3.8	1 May	30 September	16.0	1 May	30 September	16.0
Qinghai	14	NA	NA	NA	1 March	31 July	3.2	1 May	30 September	8.1	1 May	30 September	8.1
Xinjiang	15	1 May	30 September	17.3	1 October	30 June	1.5	1 May	30 September	17.3	1 May	30 September	17.3

Note: ID matches with the numbers in Figure 1; NA denotes no such crop in the province.

2.4. Irrigated Area Scenarios

To explore the effectiveness of irrigated area expansion on mitigating climate impacts, we considered four scenarios of irrigated areas: no change in PIA, 4.4% increase in PIA, 10% increase in PIA, and 15% increase in PIA. Note the maximum value of PIA is 100%. So, in cases where the PIA was greater than 100% after the addition, we reset it to 100%. No change in PIA indicates the scenario without adaptation, a 4.4% increase in PIA is consistent with the existing irrigated area expansion plan, and the last two scenarios (increased PIA by 10% and 15%) indicate potential adaptations if irrigated areas are further amplified.

3. Results

3.1. Irrigated Areas and Crop Yields under the Baseline Climate

Figure 2 demonstrates the average observed PIA and crop yields over the 1980—2011. The PIA varies by locations (Figure 2a). Better irrigation conditions are exhibited particularly in northwestern China (more than 70%) compared with northern (10–70%) and northeastern (less than 50%) China. This result is a major reflection of climatic moisture status, with the dryer climate in the Northwest requiring more irrigation to maintain the local agriculture than in northern and northeastern China. In terms of the crop yields, the spatial distribution is less clear (Figure 2b). Crop yields in most counties vary between 3.5 and 6.5 t ha^{-1}. Regions with relatively low yield include the northern region of the North and the southeastern region of northwestern China; yields vary between 2.0 and 3.5 t ha^{-1}.

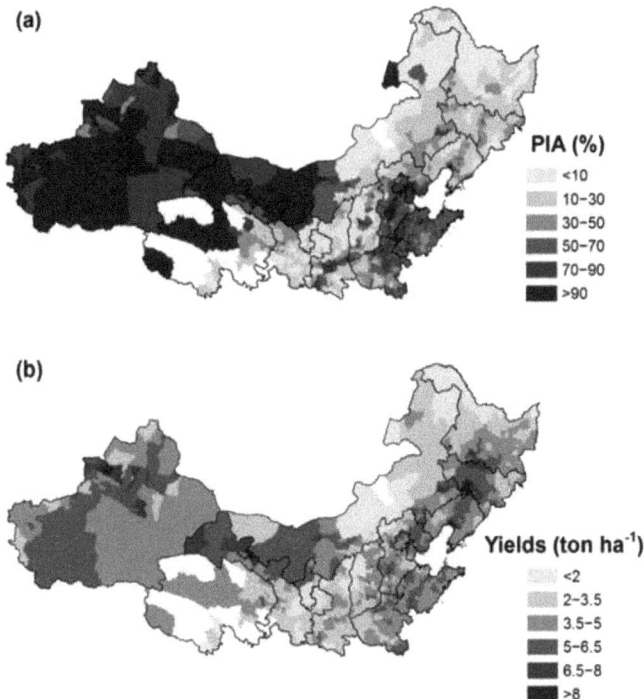

Figure 2. Percentage of irrigated area relative to the area of cultivated land (**a**); Crop yields for the four crops weighted in each county by sown area (**b**).

3.2. Effects of Climate and Irrigation on Crop Yields

A statistical model was established based on our data. Regression results of Tavg, SPEI, and PIA are presented in Table 2. We have also provided a graphical demonstration of the effects of the three individual variables on yields by artificially increasing Tavg by 1 °C, decreasing SPEI by 0.5, and increasing PIA by 10% based on above statistical model.

Table 2. Regression coefficients of the regression model, *t*-statistic and 95% confidence interval estimated using the bootstrap re-sampling approach.

Variables	Regression Coefficients	*t*-Statistic	95% CI
PIA	0.000372	0.41	(−0.0013,0.0019)
PIA^2	−4.68 × 10^{-6}	−1.07	(−1.53 × 10^{-5},5.618 × 10^{-6})
Tavg	−0.0107	−1.39	(−0.031,0.0088)
Tavg^2	−0.00052 *	−1.84	(−0.001,−0.0001)
SPEI	0.260 ***	36.2	(0.24,0.28)
SPEI^2	−0.143 ***	−21.72	(−0.16,−0.12)
PIA × Tavg	6.66 × 10^{-5} **	−25.88	(0.00002,0.00013)
PIA × SPEI	−0.00306 **	2.08	(−0.0033,−0.0027)
Sample size	28341		
R^2	0.9771		
F-value	330.9		
p-value	<0.001		

* *p*-value < 0.05; ** *p*-value <0.01; *** *p*-value < 0.001.

The full model shows a good agreement, with an R^2 of 0.9771 ($p < 0.001$). The effect of the linear Tavg term on yields is statistically insignificant ($p > 0.05$ with 95% CI between −0.031 and 0.0088). In contrast, there is a significant relationship between the Tavg quadratic term and yields ($p < 0.05$ with 95% CI between −0.001 and −0.0001), as shown in Table 2. Given the present climate, 1 °C further warming would reduce yields by 0–3% in the majority of counties when the Tavg over the growing season is greater than 0°C (Figure 3a). For SPEI, both the linear and quadratic terms on yields are statistically significant ($p < 0.001$ with 95% CI between 0.24 and 0.28 for the linear term and between −0.16 and −0.12 for the quadratic term). With SPEI reduced by 0.5, crops growing above an SPEI of approximately 0.5 in the mean-growing season tend to benefit from the drought, whereas crops grown below this threshold are likely to show a declined yield (Figure 3b).

The effects of PIA and PIA^2 on yields are both statistically insignificant ($p > 0.05$). The 95% CIs vary between −0.0013 and 0.0019 for the linear term and between −1.53 × 10^{-5} and 5.62 × 10^{-6} for the quadratic term (Table 2). CIs spanning zero suggest an inconsistent regression coefficient in the sign for each sub-sample generated using the bootstrap analysis. However, significant interaction effects of PIA on climate variables are shown ($p < 0.001$). The 95% CI for PIA × Tavg is between 0.00002 and 0.00013, while the 95% CI for PIA×SPEI is between −0.0033 and −0.0027 (Table 2). The estimated yield change is approximately 2% when the PIA increases by 10% (Figure 3c). This indicates that irrigated area expansion can alter the magnitude of climate impacts on yields.

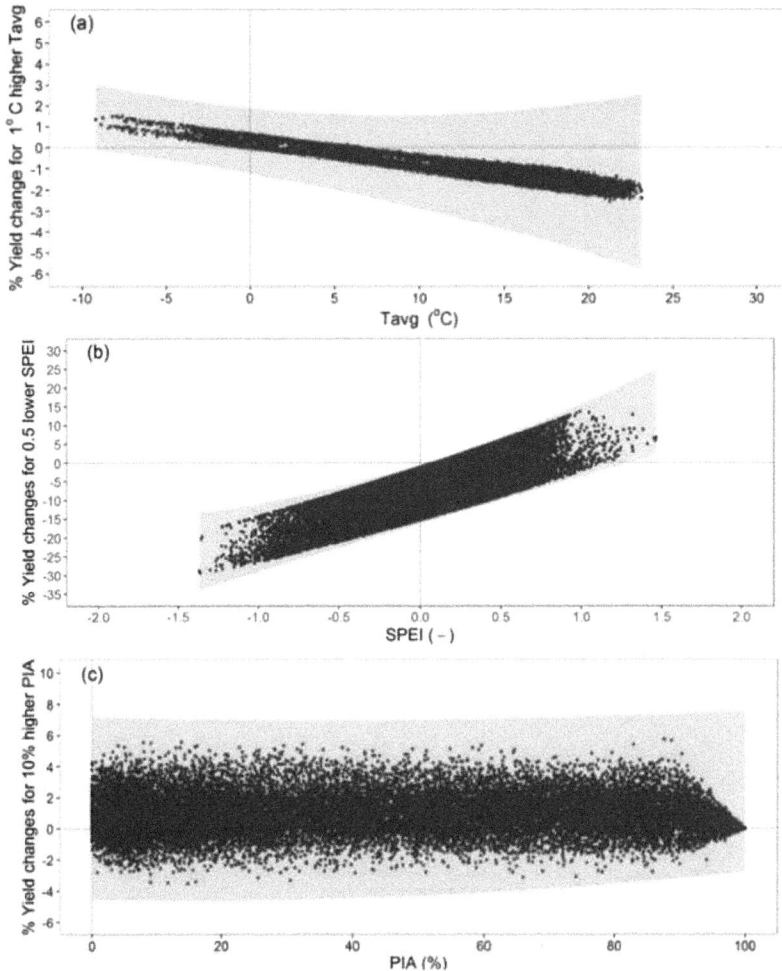

Figure 3. Model-estimated percentage yield changes for (**a**) 1 °C warmer mean-growing-season average temperature (Tavg); (**b**) 0.5 unit lower Standardized Precipitation Evapotranspiration Index (SPEI); and (**c**) 10% higher percentage of irrigated area (PIA). Each of the shaded areas shows the 95% confidence interval in the bootstrap analysis.

3.3. Future Climate Scenario

Based on climate model outputs, a warmer and dryer climate was projected in our study region (Figure 4). Under RCP2.6, it was predicted that Tavg would increase by 1–1.5 °C in northern China as well as the southern region of northeastern China and by more than 1.5 °C in other regions (Figure 4a). In addition, a dryer climate will prevail in most counties: SPEI will experience a 0.0–0.5 reduction in northeastern and most areas of northern China, and more serious decreases in SPEI (0.5–1.5) will occur in northwestern China and parts of northern China (Figure 4b). Under the RCP8.5 scenario, the increase in Tavg will be at least 1.0 °C and most counties will experience a warming with more than 1.5 °C (Figure 4c) relative to the baseline climate. The magnitude of SPEI reduction is also greater (Figure 4d) than in RCP2.6, especially in the central region of northwestern China, where SPEI is estimated to be reduced by approximately 1.0.

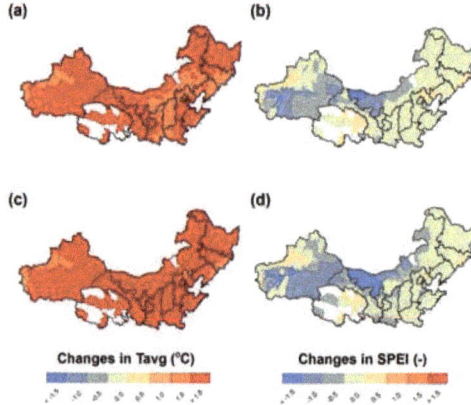

Figure 4. Changes in mean growing season Tavg and SPEI in the 2030s (2020–2039) under RCP2.6 (**a,b**) and RCP8.5 (**c,d**).

3.4. Climate Impacts on Yields under Three Irrigated Area Scenarios

Maintaining the PIA at the baseline climate is anticipated to reduce yields under future climate scenarios, with yields decreasing by 4.7% for the climate under the RCP2.6 scenario averaged over the study region (Table 3). More specifically, we predict 0–5% yield reductions in northeastern and the southern part of northern China, and certain counties in north and northwestern China would experience even more serious reductions, varying between 10% and 20% (Figure 5a). Under the RCP8.5 scenario, the reduction in yields is estimated to be approximately 7.3% (Table 3). Regions with the greatest yield reduction are predicted in the northern part of northern China and central northwestern China, where losses could exceed 20%. In the remaining areas, yields are projected to be reduced by 5–15% (Figure 5e).

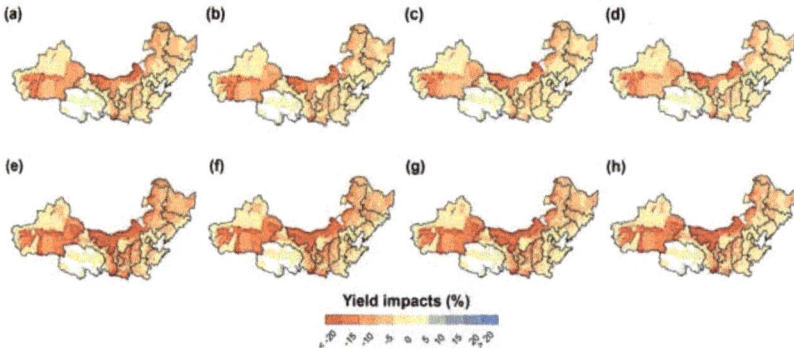

Figure 5. Model-estimated percentage changes in crop yields when (**a**) PIA is constant; (**b**) PIA was increased by 4.4%; (**c**) PIA was increased by 10% and (**d**) PIA was increased by 15% under RCP2.6. The results under RCP8.5 are shown in the bottom panel (**e–h**).

Expanding irrigated areas can alleviate the yield reductions associated with climate impacts. Our model estimates that approximately 0.7% of yields could be saved by a 4.4% increase in PIA under the RCP2.6 climate scenario, and the predicted yield improvement is 0.8% under RCP8.5 (Table 3). Further expansions in PIA are projected to result in greater yield gains. Yield increases are 1.5% under RCP2.6 and 1.8% under RCP8.5 if PIA is expanded by 10%, and the values are 2.2% under RCP2.6

and 2.7% under RCP8.5 if PIA is expanded by 15% (Table 3). However, these yield benefits are still limited relative to adverse climate impacts. The regions with the greatest yield reductions are the northern part of northern China and central northwestern China. Yield could decrease by 5–20% under the RCP2.6 scenario (Figure 5b–d) and by 10–20% under the RCP8.5 scenario (Figure 5f–h). Yield decreases are comparably lower in the northeastern and southern regions of northern China, with 0–15% decreases under RCP2.6 (Figure 5b–d) and 5–15% decreases under RCP8.5 (Figure 5f–h).

Table 3. Projected changes in temperature, SPEI and yields under the four irrigated area scenarios in the study area. The average value has been weighted by sown area.

RCP	RCP2.6	RCP8.5
Temperature change (°C)	1.6	2.0
SPEI change (−)	−0.2	−0.3
Percentage yield change with no change in irrigated area (%)	−4.7	−7.3
Percentage yield change with 4.4% increase in irrigated area (%)	−4.0	−6.5
Percentage yield change with 10% increase in irrigated area (%)	−3.2	−5.5
Percentage yield change with 15% increase in irrigated area (%)	−2.5	−4.6

4. Discussion

4.1. Yield Responses to Climate and Irrigated Areas

Our results demonstrate a significant effect of climate variables on crop yields. Increases in Tavg are harmful to yields, with a 0–2% yield reduction per additional 1 °C Tavg, because of the associated shorter growing season [14]. Lower yields will be caused by drought in most counties except under very moist climate conditions (i.e., SPEI is approximately 0.5), where yields will be increased. The inverse yield responses to SPEI are associated with less severe water logging and disease under a very wet climate, and hence higher yields, when SPEI is reduced [22,23].

Our model detected a statistically significant effect for the interaction terms of irrigation and climate variables, suggesting that expanding irrigated areas can reduce climate impacts. More specifically, with more irrigated areas, yield reductions caused by heat and drought stresses would be lower. However, our results also indicate that the effects of irrigated area expansion are still very weak on crop yields; the median magnitude is only an approximately 2% yield increase with 10% higher PIA. This finding suggests that the extent to which current irrigation practices will mitigate the negative impacts of climate are quite insufficient in China.

Two primary factors might explain the weak yield response to increased irrigated areas. First and potentially most important, expansions in irrigated areas are not associated with more irrigation water. As noted by theFood and Agriculture Organization of the United Nationsin regard to the definition of irrigated area [14]: *"Due to several reasons (e.g., crop rotation, water shortages, and damage of infrastructure) the area actually irrigated maybe significantly lower than the area equipped for irrigation"*. This means that irrigated area data do not reflect the actual accessibility of irrigation water, even though the data of irrigated area is the current primary data to quantify irrigation level. Agricultural water shortage growing in magnitude and frequency in the current [24] and future [25] climate is the main reason. Second, the household contract system was created in 1979, and the use rights of farmland were evenly distributed to the farmers by group farmland ownership [26]. These contracts encouraged farmers to work on their own farmlands but also partially shifted responsibilities previously taken care of by the government to individual farmers, which the farmers could not afford, such as irrigation infrastructure maintenance and repair [27]. Based on a survey conducted by the Ministry of Water Resources of the People's Republic of China in 2006, only 50% of household-based irrigation infrastructures are available to irrigate and 35% of areas categorized as irrigated areas cannot be irrigated [28].

4.2. Future Climate Impacts and Adaptation by Expanding Irrigated Areas

In the 2030s, a warmer and dryer climate is anticipated, posing a serious challenge to agricultural outputs and irrigation water resources over our study region. Expanding irrigated area was projected to save yields from harmful climate impacts. However, such yield benefits are quite limited compared with the negative climate impacts. Similar yield reductions were projected even when irrigated areas are increased by 15%. The scenario experiments demonstrate that yield losses are difficult to avoid under future climate no matter how the irrigated area will be increased.

The model output in our study is not consistent with earlier assessments, which investigated the potential adaptive effects of expanding irrigation using process-based models by assuming different irrigation schemes [29] or assuming no water stress [8]. In theory, there is significant potential for improving irrigation to mitigate the harmful impacts of both of heat and drought on crops [30]. However, in practice, the adaptive potential will not be fully realized. As our study has quantified, the effects of expansion of the irrigated area have little influence on yields due to the aforementioned reasons and thus cannot fundamentally countervail the expected adverse climate change impacts in the 2030s. This finding suggests that the effects of expanding the irrigated area are restricted.

4.3. Implications for the Adaptive Policies for Climate Change in China

Anticipated yield reductions across the irrigated area scenarios suggest that expanding the irrigated area alone cannot achieve our expected climate risk mitigation in the 2030s. Therefore, other solutions are needed. For example, water-saving irrigation technology has been found to reliably increase grain yields while using less water [31]. The adoption of water-saving irrigation technology in sown areas is currently very limited in China [32,33]. According to a farmers' survey across seven provinces in China, only 32% and 4% of sown areas are equipped with household-based and community-based water-saving irrigation technologies, respectively [34]. In northern China, water saving technologies have been reported to show a great potential to reduce water use and improve crop productivities. By using these new technologies, irrigated water reduces by 11.7% and water use efficiency (i.e., yield produced per unit of water) increases by 27.8% for wheat; and the irrigated water saving and water use efficiency improvements are 23.0% and 17.6% for maize, respectively [35]. Therefore, with the low application and substantial potential, water-saving irrigation technology innovation appears to be a more promising approach than establishing more irrigated areas.

Other adaptive measures helpful to improving water use efficiency should not be overlooked. For instance, due to advances in breeding technology, new rice cultivars with high-water efficiency have been bred in China [36], exhibiting a yield advantage of 31–36% under drought [37]. Therefore, policies aimed at climate stress-tolerance cultivars appear to be beneficial to adapting to climate stresses in the future. In addition, the major reason for the future drought is increased evapotranspiration associated with warming [38], thus some technologies that can reduce evapotranspiration, such as plastic sheeting and low-tillage, will be also very helpful. Linking seasonal climate forecasting with crop choice can thus provide another potential climate adaptation. Other adaptation options, including multiple rather than individual adaptive measures, appear more realistic to help reduce future climate risks and should be addressed in future studies.

Finally, it is necessary to develop a new index to represent actual irrigation and water availability at fairly fine resolution. Even though irrigated area is widely used in many earlier works [3,12], its use tends to lack adequate consideration of actual irrigation water and associated adaptive effectsas we showed in this study. Such a new index will prove critical when developing relevant agricultural water use policies.

5. Conclusions

To address the effectiveness of expanding the irrigated area in order to mitigate future climate stresses, this study used county-level data to quantify the adaptive effects of irrigated areas on crop

Sustainability **2017**, *9*, 851

yields and anticipated change in yields under future climate across different irrigated area scenarios. We concluded that expanding irrigated areas cannot countervail future adverse climate impacts on crop yields in northern China. This limitation is primarily attributed to the underutilization of the irrigated area during drought due to water shortage and impaired irrigation infrastructure. Therefore, we hypothesize that the key to improving the resilience of Chinese agriculture under climate impacts is not the size of the irrigated area but, rather, modernizing irrigation. This target change needs to be quantitatively addressed, as it has not been clearly framed in existing policy. Furthermore, the irrigated area expansion plan will require the complementation of other adaptation programs, such as crop breeding and seasonal forecasting. These practices will be particularly useful in regions facing shortages of agricultural water resources. The suitability of different adaptation programs in different regions must be identified in future investigations.

Within the limits of available data, the statistical models used here have been applied to groups of crops and irrigated areas together without crop-specific analysis. These limitations could be overcome with further work by developing an enhanced spatially intensive dataset that further separates agricultural resource inputs for individual crops. In addition, clear knowledge and integrated assessment models to inform farmers' adaptive reactions to climate extremes are in high demand. Such information would enable more accurate predictions of the adaptation potential, costs and benefits, and the agricultural system could be modified based on predicted climate change scenarios.

Supplementary Materials: The following are available online at www.mdpi.com/2071-1050/9/5/851/s1, Table S1: Climate models used to simulate ensemble climate scenarios.

Acknowledgments: This work was funded by the National Natural Science Foundation of China (41661144006; 31661143012; 41301044). We appreciated the insightful suggestions of anonymous reviewers and Zhi Chen in helping to improve this paper.

Author Contributions: Tianyi Zhang and Jinxia Wang conceived and designed the study; Tianyi Zhang and Jinxia Wang performed the analysis; Tianyi Zhang and Yishu Teng collected data; Tianyi Zhang wrote the paper and all other authors provided comments on the earlier versions of this manuscript.

Conflicts of Interest: The authors declare no conflict of interest.

References

1. Lin, E.; Xiong, W.; Ju, H.; Xu, Y.; Li, Y.; Bai, L.; Xie, L. Climate Change Impacts on Crop Yield and Quality with CO_2 Fertilization in China. *Philos. Trans. R. Soc. B* **2005**, *360*, 2149–2154.
2. Xiong, W.; Lin, E.; Ju, H.; Xu, Y. Climate Change and Critical Thresholds in China's Food Security. *Clim. Chang.* **2007**, *81*, 205–221. [CrossRef]
3. Zhang, T.; Simelton, E.; Huang, Y.; Shi, Y. A Bayesian Assessment of the Current Irrigation Water Supplies Capacity under Projected Droughts for the 2030s in China. *Agric. For. Meteorol.* **2013**, *178*, 56–65. [CrossRef]
4. The Food and Agriculture Organization of the United Nations (FAO). Drought. 2015. Available online: http://www.fao.org/docrep/017/aq191e/aq191e.pdf (accessed on 12 March 2016).
5. Asian Development Bank. *Addressing Climate Change Risks, Disasters, and Adaptation in the People's Republic of China*; Asian Development Bank: Mandaluyong City, Philippines, 2015.
6. Communist Party of China (CPC). Chinese Central Government's Official Web Portal, China's Spending on Water Conservation Doubles During 11th Five-Year Plan. 2011. Available online: http://www.gov.cn/jrzg/2011--01/29/content_1795245.htm (accessed on 12 March 2016).
7. Yu, C. China's water crisis needs more than words. *Nature* **2011**, *470*, 307. [CrossRef] [PubMed]
8. Challinor, A.; Simelton, E.; Fraser, E.; Hemming, D.; Collins, M. Increased crop failure due to climate change: Assessing adaptation options using models and socio-economic data for wheat in China. *Environ. Res. Lett.* **2010**, *5*, 3. [CrossRef]
9. Ju, H.; van der Velde, M.; Lin, E.; Xiong, W.; Li, Y. The impacts of climate change on agricultural production systems in China. *Clim. Chang.* **2013**, *120*, 313–324. [CrossRef]
10. Reidsma, P.; Ewert, F.; Oude Lansink, A.; Leemans, R. Adaptation to climate change and climate variability in European agriculture: The importance of farm level responses. *Eur. J. Agron.* **2010**, *32*, 91–102. [CrossRef]

11. Simelton, E.; Fraser, E.; Termansen, M.; Benton, T.; Gosling, S.; South, A. The socioeconomics of food crop production and climate change vulnerability: A global scale quantitative analysis of how grain crops are sensitive to drought. *Food Secur.* **2012**, *4*, 163–179. [CrossRef]

12. Simelton, E.; Fraser, E.; Termansen, M.; Forster, P.; Dougill, A. Typologies of crop-drought vulnerability: An empirical analysis of the socio-economic factors that influence the sensitivity and resilience of drought of three major food crops in China (1961–2001). *Environ. Sci. Policy* **2009**, *12*, 438–452. [CrossRef]

13. Ministry of Water Resources of China. *Technical Terminology for Irrigation and Drainage*; Ministry of Water Resources of China: Beijing, China, 1993; pp. 56–93.

14. The Food and Agriculture Organization of the United Nations (FAO). Global Map of Irrigated Areas. 2010. Available online: http://www.fao.org/nr/water/aquastat/irrigationmap/index30.stm (accessed on 12 March 2016).

15. China Meteorology Data Sharing Service. Daily climate dataset. Available online: http://cdc.nmic.cn/ (accessed on 12 March 2016).

16. Thornton, P.; Running, S.; White, M.A. Generating surfaces of daily meteorological variables over large regions of complex terrain. *J. Hydrol.* **1997**, *190*, 214–251. [CrossRef]

17. Zhang, T.; Huang, Y.; Yang, X. Climate warming over the past three decades has shortened 20 rice growth duration in China and cultivar shifts have further accelerated the 21 process for late rice. *Glob. Chang. Biol.* **2013**, *19*, 563–570. [CrossRef] [PubMed]

18. Vicente-Serrano, S.; Begueria, S.; Lopez-Moreno, J. A multi-scalar drought index sensitive to global warming: The Standardized Precipitation Evapotranspiration Index-SPEI. *J. Clim.* **2010**, *23*, 1696–1718. [CrossRef]

19. Zhang, F. *Chinese Agricultural Phenology Atlas*; Science Press: Beijing, China, 1987.

20. Liu, B.; Asseng, S.; Müller, C.; Ewert, F.; Elliott, J.; Lobell, D.; Martre, P.; Ruane, A.; Wallach, D.; Jones, J.W.; et al. Similar estimates of temperature impacts on global wheat yield by three independent methods. *Nat. Clim. Chang.* **2016**, *6*, 1130–1136. [CrossRef]

21. Lobell, D.; Burke, M.; Tebaldi, C.; Mastrandrea, M.; Falcon, W.; Naylor, R. Prioritizing climate change adaptation needs for food security in 2030. *Science* **2008**, *319*, 607–610. [CrossRef] [PubMed]

22. Deng, X.; Huang, J.; Qiao, F.; Naylor, R.; Falcon, W.; Burke, M. Impacts of El Nino-Southern Oscillation events on China's rice production. *J. Geogr. Sci.* **2010**, *20*, 3–16. [CrossRef]

23. Zhang, T.; Zhu, J.; Wassmann, R. Responses of rice yields to recent climate changein China: An empirical assessment based on long-term observations at different spatial scales (1981–2005). *Agric. For. Meteorol.* **2010**, *150*, 1128–1137. [CrossRef]

24. Shalizi, Z. *Addressing China's Growing Water Shortages and Associated Social and Environmental Consequences*; World Bank Policy Research Working Paper: No. 3895; World Bank: Washington, DC, USA, 2006.

25. Wang, S.; Zhang, Z. Effects of climate change on water resources in China. *Clim. Res.* **2011**, *47*, 77–82. [CrossRef]

26. Chen, T.; Yabe, M. *Study on the Formation of Household Management in Chinese Agriculture*; Faculty of Agriculture Publications: Fukuoka, Japan, 2009.

27. Xu, K. Why do irrigation infrastructures abandoned? *Coop. Econ. China* **2009**, *2*, 5.

28. Zhang, C.; Li, D. The concept of reinforcing rural irrigation infrastructure constructions in modern China. *China Rural Water Hydropower* **2009**, *7*, 1–3.

29. Chen, C.; Wang, E.; Yu, Q. Modeling wheat and maize productivity as affected by climate variation and irrigation supply in North China Plain. *Agron. J.* **2010**, *102*, 1037–1049. [CrossRef]

30. Zhang, T.; Lin, X.; Sassenrath, G. Current irrigation practices in the central United States reduce drought and extreme heat impacts for maize and soybean but not for wheat. *Sci. Total Environ.* **2015**, *508*, 331–342. [CrossRef] [PubMed]

31. Grassini, P.; Cassman, K. High-yield maize with large net energy yield and small global warming intensity. *Proc. Natl. Acad. Sci. USA* **2011**, *109*, 1074–1079. [CrossRef] [PubMed]

32. Blanke, A.; Rozelle, S.; Lohmar, B.; Wang, J.; Huang, J. Water saving technology and saving water in China. *Agric. Water Manag.* **2007**, *87*, 139–150. [CrossRef]

33. Liu, Y.; Huang, J.; Wang, J.; Rozelle, S. Determinants of agricultural water saving technology adoption: An empirical study of 10 provinces of China. *Ecol. Econ.* **2008**, *4*, 462–472.

34. Cremades, R.; Wang, J.; Morris, J. Policies, Economic incentives and the adoption of modern irrigation technology in China. *Earth Syst. Dyn.* **2015**, *6*, 399–410. [CrossRef]

35. Huang, Q.; Wang, J.; Li, Y. Do water saving technologies save water? Empirical evidence from North China. *J. Environ. Econ. Manag.* **2017**, *82*, 1–16. [CrossRef]

36. Zhang, Q. Strategies for developing Green Super Rice. *Proc. Natl. Acad. Sci. USA* **2007**, *104*, 16402–16409. [CrossRef] [PubMed]

37. Marcaida, M., III; Li, T.; Angeles, O.; Evangelista, G.; Fontanilla, M.; Xu, J. Biomass accumulation and partitioning of newly developed Green Super Rice (GSR) cultivars under drought stress during the reproductive stage. *Field Crop. Res.* **2014**, *162*, 30–38. [CrossRef]

38. Chen, H.; Sun, J. Changes in drought characteristics over China using the standardized precipitation evapotranspiration index. *J. Clim.* **2015**, *28*, 5430–5447. [CrossRef]

sustainability

MDPI

Article

Estimation of the Virtual Water Content of Main Crops on the Korean Peninsula Using Multiple Regional Climate Models and Evapotranspiration Methods

Chul-Hee Lim [1,2], Seung Hee Kim [2], Yuyoung Choi [1], Menas C. Kafatos [2] and Woo-Kyun Lee [1,*]

[1] Department of Environmental Science and Ecological Engineering, Korea University, Seoul 02841, Korea; limpossible@korea.ac.kr (C.-H.L.); cuteyu0@korea.ac.kr (Y.C.)

[2] Center of Excellence in Earth Systems Modeling and Observations, Chapman University, Orange, CA 92866, USA; sekim@chapman.edu (S.H.K.); kafatos@chapman.edu (M.C.K.)

* Correspondence: leewk@korea.ac.kr; Tel.: +82-2-3290-3016; Fax: +82-2-3290-3470

Received: 20 April 2017; Accepted: 1 July 2017; Published: 4 July 2017

Abstract: Sustainable agriculture in the era of climate change needs to find solutions for the retention and proper utilization of water. This study proposes an ensemble approach for identifying the virtual water content (VWC) of main crops on the Korean Peninsula in past and future climates. Ensemble results with low uncertainty were obtained using three regional climate models, five potential evapotranspiration methods, and the Environmental Policy Integrated Climate (EPIC) crop model. The productivity results of major crops (rice and maize) under climate change are likely to increase more than in the past based on the ensemble results. The ensemble VWC is calculated using three types of crop yields and fifteen consumptive amounts of water use in the past and the future. While the ensemble VWC of rice and maize was 1.18 m^3 kg^{-1} and 0.58 m^3 kg^{-1}, respectively, in the past, the future amounts were estimated at 0.76 m^3 kg^{-1} and 0.48 m^3 kg^{-1}, respectively. The yields of both crops showed a decline in future projections, indicating that this change could have a positive impact on future water demand. The positive changes in crop productivity and water consumption due to climate change suggest that adaptation to climate change can be an opportunity for enhancing sustainability as well as for minimizing agricultural damage.

Keywords: virtual water content; ensemble result; crop yield; regional climate models; PET methods

---

1. Introduction

Agriculture is the most climate-dependent production sector; thus, it is necessary to accurately assess the impacts of climate change on the agriculture sector to achieve sustainability [1,2]. The agriculture sector is also linked to the largest number of indicators of sustainable development [3]. The demand for agriculture-related predictions is expected to rise in the future with population expansion and a greater need for food security [4,5].

Agriculture and crop production are the sectors that have the highest water demand [6]. While it is necessary to manage multiple stress factors to achieve agricultural productivity, an abundant supply of water is essential. Several global studies have predicted significant changes in crop productivity and hydrological circulation in the mid-latitude regions during the 21st century [7,8]. The Korean Peninsula has a temperate, monsoon climate with a large annual variation in precipitation, and is classified as a region of water shortage by the UN (United Nations); therefore, it is necessary to study the agricultural water use in the region [9,10].

The concept of virtual water or water footprint has been presented by several international organizations and studies as an effective way to estimate water demand [11,12]. Virtual water refers to the amount of water used per unit product and is synonymous with water footprint. Many studies have estimated the water demand of the agricultural sector by calculating the virtual water of crops [13–15]. By estimating the virtual water per crop, we can also analyze the influence of regional differences and climate change on the water demand.

While estimates of the virtual water content of crops in East Asian cases can be easily found [11–16], the crop-hydrological model-based estimation approach for production and water consumption is not common. In particular, there were no cases which estimated the virtual water of a crop with a model-based approach on the Korean Peninsula, and only Zhao et al. [15] suggested the virtual water of several crops in China using a modeling approach. The information from a model-based virtual water content analysis in the Korean Peninsula can help solve problems of food and water security [4,10].

Previous model-based virtual water studies have calculated the amount of water used, crop yields, and evapotranspiration rates to estimate the virtual water content of crops [13,15,16]. The most basic data used in the estimation of evapotranspiration and crop yield is climatic data, which has significant uncertainties [10,17], depending on the climate model used [9,18].

To estimate and reduce uncertainties in a model's predictions, multiple methods have recently been employed by the crop modeling community [17,19–21]. The major source of a model's error is from the uncertainty of the input data, such as climate data. Another source is due to a model's characteristics having various approaches and parameterizations to determine the growth and the phenological development of the crop [20].

In this study, we propose a multi-input and multi-model super ensemble approach for the virtual water content of main crops based on the past and future climate data of the Korean Peninsula. Specifically, we use five evapotranspiration methods to reduce model uncertainty and three regional climate models to reduce the uncertainty in the climate data. In particular, we present virtual water content that represents the past and the future by averaging the number of virtual water contents based on multiple data and methodologies. The average of the multiple results is defined as the ensemble average, and the uncertainty reduction is analyzed by considering the difference in each value without considering the variability. Given the importance and increasing emphasis of agricultural water use under climate change, it is useful to provide the estimation of the ensemble virtual water of main crops.

2. Data and Methods

2.1. Research Area and Crops

The research area covers the whole cropland of the Korean Peninsula, which includes both South and North Korea (Figure 1). The Korean Peninsula is located on the eastern end of Asia, covers an area of about 221,000 km^2, and is located between the 33.23° N and 43.01° N latitudes and the 124.14° E and 130.93° E longitudes. The Korean Peninsula is located in the temperate zone. It is largely influenced by the temperate monsoon climate, and has large annual variations in precipitation, with high temperatures and humidity levels in the summer. Under the Köppen climate classification, a warm and dry winter climate (Cwa) and a cold and dry winter climate (Dwa) are the most common, with an average annual temperature of 10–16 °C, and an average cumulative precipitation of 1200 mm [22]. Geographically, there are high mountains in the north and east and plains in the south and west.

The cropped area of South Korea is about 20,000 km^2 and that of North Korea about 30,000 km^2, which is over 20% of the total area of the peninsula. Rice paddies and maize fields appear to be mixed, but rice is widely distributed on a broad plain and the number of maize fields increases in the north [23]. In this study, considering the resolution of the available climate data, the cropland of the peninsula was mapped with a 5 km^2 grid resolution, and all the data were processed accordingly.

In this study, rice and maize, staple crops in the Korean Peninsula, are selected for the analysis. Rice production is dominant throughout the region, where precipitation is concentrated in the summer. Maize is mainly produced in the northeastern region, where low precipitation and temperature conditions are dominant. Note that maize accounts for 35% of the total food production in North Korea [23,24].

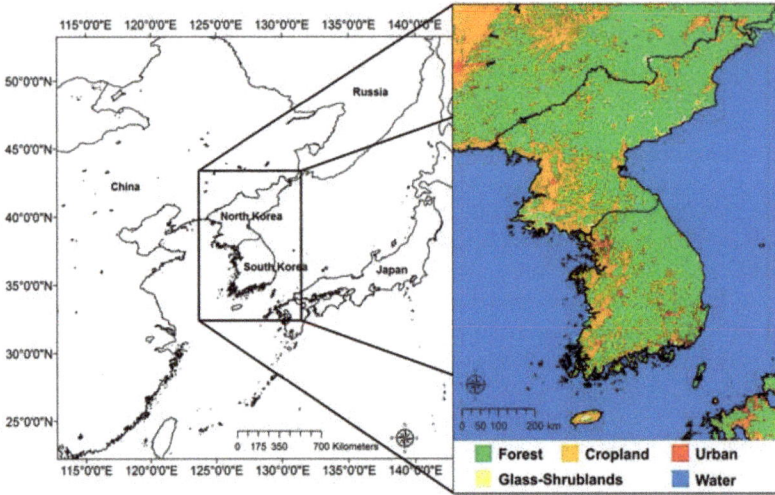

Figure 1. Land cover map of the research area.

2.2. Assessment Model Hierarchy

Figure 2 shows a data flow of this study. We have selected three regional climate models (RCM) for climate data, and each climate data are used in a simulation in an EPIC model with five different potential evapotranspiration (PET) methods. In the research framework, fifteen ensemble members are produced and are used to estimate the virtual water contents (VWC) of crops over the study region. A detailed explanation of each component is described in the following sections.

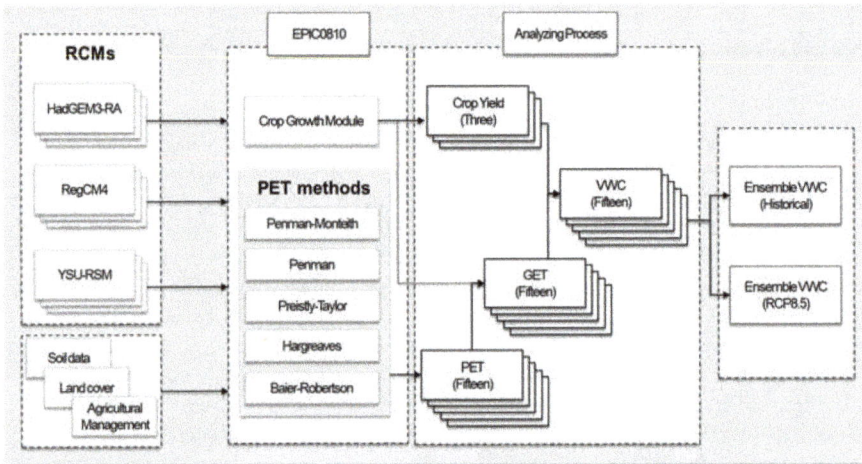

Figure 2. Schematic diagram of the applied methodology. RCM, regional climate model; PET, potential evapotranspiration; VWC, virtual water content; growing season evapotranspiration (GET).

2.3. Multiple Regional Climate Models

To reduce the uncertainty in the climate data, the data of three regional climate models (RCM) were used. The climate model data for the region are obtained through the COordinated Regional climate Downscaling EXperiment (CORDEX)-East Asia. The CORDEX initiative was launched by the Task Force for Regional Climate Downscaling (TFRCD), which was established in 2009 by the World Climate Research Program (WCRP).

The climate data of the RCMs acquired through CORDEX-East Asia are HadGEM3-RA, RegCM4, and YSU-RCM, and the three models were verified using previous studies [10,17]. These models provide dynamically downscaled data with a spatial resolution of 12.5 km^2 based on the HadGEM2-AO model, a large-scale climate model.

HadGEM3-RA is a regional version of the Hadley Center Global Environment Model, a non-hydrostatic regional climate model following the Arakawa-C horizontal grid and a terrain-following vertical coordinate [9]. RegCM4, the fourth version of the RegCM regional climate model system, is a hydrostatic, compressible model with sigma-p vertical coordinates and an Arakawa-B horizontal grid [25]. YSU-RSM, a disturbance model, is defined by a two-dimensional sine series for the perturbation of vorticity and by a two-dimensional cosine series for perturbations of pressure, divergence, temperature, and mixing ratio [26]. More detailed information about these models can be found on the CORDEX-East Asia web page (https://cordex-ea.climate.go.kr/main/aboutCordexPage.do).

We used RCMs for the representative concentration pathways (RCP) 8.5 scenario, which is the lowest greenhouse gas reduction scenario, and for the analysis of the data for 1981–2000 and 2031–2050.

2.4. EPIC Model and PET Methods

The EPIC model was developed in the 1980s to assess soil erosion and soil productivity, followed by a model on plant growth and hydrological parameters [27,28]. Since the EPIC model was first published, many components have been added, such as GLEAMS [29], Century [30], and RUSLE [31]. This model was renamed the Environmental Policy Integrated Climate model (EPIC) with the addition of environmental assessment functions for pesticides and water quality [32].

The EPIC model has a structure that converts daily energy and biomass to simulate crop growth [28]. The daily potential biomass increase is calculated using climate variables such as the solar radiation and biomass–energy conversion rates of the crop. The daily response to plant stress variables (water, nutrient, temperature, aeration, and salinity) decreases the potential biomass. Crop yields are ultimately estimated based on the crop harvest index and the actual biomass accumulation [33].

The estimation of crop productivity through the EPIC model has been successfully applied to the whole of the Korean Peninsula, as well as Eastern Asia, by previous studies [34–36]. In this study, calibration was performed to estimate the crop productivity on the Korean Peninsula. Some of the key crop parameters were modified through calibration. For rice, the biomass–energy ratio was set at 30 kg MJ^{-1}, the harvest index at 0.55 mg mg^{-1}, the optimum temperature at 25 °C, the base temperature at 10 °C, and the potential heat unit (PHU) ranged from 1300 to 1500 °C, depending on the climate of the specific grid cells. For maize, the biomass–energy ratio was set at 43 kg MJ^{-1}, the harvest index at 0.45 mg mg^{-1}, the optimum temperature at 25 °C, the base temperature at 8 °C, and the PHU range at 1000–1200 °C, depending on the climate of the specific grid cells [34,37,38].

After calibration, a statistical validation was conducted to evaluate the model's performance for estimating crop productivity using the results from 1981 to 2000 and the statistical data. The HadGEM3-RA climate model was used as a standard in the calibration and validation process. HadGEM3-RA has been used most commonly for the Korean Peninsula and shown to average between the RegCM4 and YSU-RSM [17,18,39]. Since only the amount of production by country is known and the statistics for North Korea are very limited, the model was only verified for the South Korean data. The statistical data on rice and maize production were retrieved from the Korea Statistical Information System (KOSIS). Three statistical indicators were used to validate the model's performance: (i) the Root Mean Square Error (RMSE), (ii) the Nash–Sutcliffe Efficiency Coefficient (NSEC), and (iii) the Relative Error (RE) [40,41]. These statistical measures have been described in detail in the literature [35,37,42].

Five representative methods (Penman–Monteith (PM) [43], Penman (P) [44], Preistly–Taylor (PT) [45], Hargreaves (H) [46], and Baier–Robertson (BR) [47]) were used to estimate the PET. All of the PET methods available in the EPIC model were used, and these five methods are commonly used in equations in many previous studies [12–15,21].

$$PET_{PM} = (RN \times \delta + 86.66 \times AD \times EA\,(1-RH) \times U\,/\,350)\,/\,((2.51 - 0.0022 \times T) \times (\delta + \gamma)) \quad (1)$$

$$PET_P = (RN \times \delta\,/\,(2.051 - 0.0022 \times T\,) + \gamma \times (2.7 + 1.63 \times U) \times EA\,(1-RH))/(\delta + \gamma) \quad (2)$$

$$PET_{PT} = 1.28 \times (RN \times (1.0 - AB)/(2.501 - 0.0022 \times T)) \times (\delta/(\delta + \gamma)) \quad (3)$$

$$PET_H = 0.0032 \times (RAMX\,/(2.501 - 0.0022 \times T)) \times (T + 17.8) \times (T_{max} - T_{min})^{0.6} \quad (4)$$

$$PET_{BR} = 0.288 \times T_{max} - 0.144 \times T_{min} + 0.139 \times RAMX - 4.931 \quad (5)$$

PET_{PM}, PET_P, PET_{PT}, PET_H, and PET_{BR} simulate the daily PET by each method. T_{min} and T_{max} are the daily minimum and maximum temperatures (°C), and RAMX is the solar irradiance on a clear day (MJ m^{-2} d^{-1}). T is the slope of the daily average temperature (°C), RN is the total solar irradiance (MJ m^{-2} d^{-1}), δ is the slope of the saturation vapor pressure curve (kPa °C^{-1}), γ is the psychrometric constant (kPa °C^{-1}), U is the average daily wind speed (ms^{-1}), and EA is the saturation vapor pressure at mean air temperature (kPa). RH is the average relative humidity per day, AD is air density (Kg m^{-3}), and AB is the soil albedo. The required meteorological data are different for each PET method. PET_{PM} and PET_P require the temperature, solar radiation, relative humidity, and wind speed variables; PET_{PT} requires the relative humidity and wind speed values; and PET_H and PET_{BR} require only the temperature-related variables. The climate requirement of each PET method is described in Table 1.

Table 1. Climate variables required by each PET method (◯: Required climate variables).

Method	Temperature (T, T_{min}, T_{max})	Solar Radiation	Relative Humidity	Wind Speed	Reference
Penman–Monteith	◯	◯	◯	◯	Monteith (1965)
Penman	◯	◯	◯	◯	Penman (1948)
Priestley–Taylor	◯	◯			Priestley and Taylor (1972)
Hargreaves	◯				Hargreaves and Samani (1985)
Baier–Robertson	◯				Baier and Robertson (1965)

2.5. Method to Calculate Virtual Water Content of Crops

The method proposed by Zhao et al. [15] to estimate the virtual water content (VWC) of crops was used in this study. The VWC of crops represents the amount of water used per unit of production, defined as the ratio of the Consumptive Water Use (CWU) to crop production [48]. The CWU of the crop is estimated from the actual evapotranspiration during the growing season of each crop. The VWC of the crop is an indicator of the agricultural demand, which can be used to estimate the total demand for water in the region and the supply requirements. The VWC is calculated as follows.

$$VWC = CWU/P = 10 \times GET/Y \quad (6)$$

where CWU is the volume of water (m^3) used by the crop during the growing season, P is the amount of crop (kg) produced during the same period, GET is the actual evapotranspiration in the growing season (mm), Y is the crop yield in kg ha^{-1} units, and the number 10 was used for the conversion of mm to m^3 ha^{-1}.

2.6. Other Data and Management Description

The EPIC model requires various soil-related parameters (OC (%), pH, cation exchange capacity (cmol kg^{-1}), sand (%), silt (%), bulk density (tm^{-3}), and electrical conductivity (mS cm^{-1})). The input form was completed using the Digital Soil Map of the World [49], constructed spatially using the ISRIC-WISE database [50].

We used the 2010 Global Land Cover 30 (GLC30) data to reflect the current land cover [51] (accessed from http://www.globallandcover.com). GLC30, a 30 m land cover map based on Landsat 7 satellite imagery, was extracted from the cropland of the Korean peninsula and constructed with 5 km × 5 km grid cells.

The amount of fertilizer and irrigation water required for each crop and area in the EPIC model is calculated to determine the spatially required amount. Since the planting and harvest dates for crops are different for each region, these are automatically assigned based on the temperature by setting the first farming start date to reflect agricultural activities on the Korean Peninsula. The starting date of the first farming for rice was set to 1 March and for corn, to 1 April.

3. Results

3.1. Estimation of Crop Yield

3.1.1. Evaluation of the Model's Performance

The HadGEM-EPIC results were used for evaluating the model's performance (Table 2). The statistical data for the rice yield showed an average range of 4.0–5.0 t ha^{-1} with a low standard deviation. The maize yield data showed relatively high standard deviations and an average of 3.6–5.1 t ha^{-1}. The non-main production area for maize had a very low yield, which led to high standard deviations. The estimates showed low standard deviations and were similar to the reported values. The RMSE values ranged from 0.2 to 0.9 for rice and 0.3 to 1.7 for maize, rice data showing higher accuracy (Table 2). The NSEC and RE values showed that the rice yield data were more accurate, which contributed to the high standard deviations of the maize production data (Table 2). The accuracy assessment of the statistical data for the two crops showed an overall high degree of accuracy; therefore, the data can be used in past climate models and in future climate change research.

Table 2. Evaluation of the model's performance in estimating rice and maize yield. SD, standard deviation; RMSE, root mean square error; NSEC, Nash–Sutcliffe efficiency coefficient; RE, relative error.

Year	Reported (t ha^{-1})		Estimated (t ha^{-1})		RMSE	E (NSEC)	RE (%)
	Mean	SD	Mean	SD			
			Rice Yield				
1981	4.03	0.39	3.96	0.36	0.28	−0.41	−1.12
1982	4.21	0.38	5.16	0.33	0.59	−0.71	3.35
1983	4.20	0.39	5.12	0.37	0.51	−1.01	4.28
1984	4.38	0.40	4.77	0.54	0.40	−0.01	1.79
1985	4.31	0.42	4.81	0.45	0.57	−0.65	5.21
1986	4.25	0.68	4.85	0.43	0.63	−0.72	2.56
1987	4.19	0.42	4.08	0.27	0.31	−0.08	−1.93
1988	4.64	0.45	4.98	0.31	0.44	−0.48	1.52
1989	4.56	0.34	4.67	0.41	0.32	−0.09	0.69
1990	4.34	0.42	4.87	0.56	0.56	−1.08	2.87
1991	4.28	0.34	4.69	0.27	0.48	−0.57	1.36
1992	4.44	0.34	5.44	0.63	0.87	−0.82	3.82
1993	4.08	0.58	4.84	0.33	0.74	−1.16	2.61
1994	4.45	0.32	4.56	0.45	0.27	−0.25	0.42
1995	4.35	0.28	4.90	0.29	0.59	−0.89	1.93
1996	4.94	0.26	4.56	0.36	0.41	−0.12	−1.25
1997	5.00	0.32	4.89	0.42	0.37	−0.08	−0.73
1998	4.62	0.36	5.01	0.43	0.45	−0.33	1.13
1999	4.78	0.37	3.91	0.27	0.77	−1.02	−3.27
2000	4.80	0.31	4.99	0.44	0.29	−0.04	1.11
			Maize Yield				
1981	4.38	1.58	4.32	0.18	0.42	−0.51	−0.81
1982	4.12	1.37	5.20	0.38	0.87	−1.12	3.67
1983	3.66	1.24	5.36	0.28	1.21	−2.33	6.15
1984	4.44	1.61	5.05	0.33	0.75	−0.62	2.49

Table 2. *Cont.*

Year	Reported (t ha^{-1})		Estimated (t ha^{-1})		RMSE	E (NSEC)	RE (%)
	Mean	SD	Mean	SD			
1985	5.04	1.76	5.26	0.33	0.38	−0.09	1.22
1986	4.79	1.71	5.47	0.43	0.73	−0.43	5.76
1987	4.85	1.63	4.60	0.29	0.42	−0.11	−1.93
1988	4.80	1.93	5.25	0.25	0.55	−0.91	7.32
1989	4.88	1.84	4.38	0.22	0.65	−1.15	−1.86
1990	4.61	1.78	5.07	0.29	0.71	−1.32	2.83
1991	3.41	1.34	5.57	0.37	1.63	−2.78	12.32
1992	4.40	1.52	5.63	0.38	0.88	−1.18	8.17
1993	4.18	1.54	5.15	0.29	0.91	−1.36	5.86
1994	4.09	1.53	4.39	0.27	0.37	−0.12	1.04
1995	4.25	1.54	5.19	0.35	0.82	−0.58	3.63
1996	4.03	1.60	5.83	0.29	1.44	−2.30	8.29
1997	4.11	1.48	5.19	0.27	1.01	−0.91	5.59
1998	3.98	1.31	4.68	0.30	0.71	−1.21	5.15
1999	3.94	1.37	5.30	0.30	1.09	−0.81	10.28
2000	4.06	1.16	5.06	0.24	0.91	−1.11	9.11

3.1.2. Estimation of Crop Yields Using Multiple RCMs

The estimated results of rice and maize production using three RCMs showed significant differences between the models. Based on the RCP 8.5 scenario, the productivity is predicted to increase in the future.

Rice yields showed a gradual increase from 1981 to 2000, and the RegCM4-EPIC result was the lowest (mean 3.1 t ha^{-1}) while the YSU-RSM-EPIC result was the highest (mean 4.25 t ha^{-1}) (Figure 2). The estimation results of HadGEM3-RA-EPIC were moderate (mean 4.05 t ha^{-1}), and the values of the three ensembles (mean 4.25 t ha^{-1}) were similar to those of HadGEM3-RA-EPIC (Figure 2). The results of the ensemble were similar to the agricultural statistics, and more accurate than the individual model results (Figure 2 and Table 2). The spatial distribution of productivity calculated by RegCM4-EPIC showed that the whole of North Korea and the central region of South Korea had low productivity. YSU-RSM-EPIC showed high productivity in the entirety of South Korea, while HadGEM3-RA-EPIC showed high productivity in the western plains of North Korea (Figure 3). The rice cultivation on the Korean Peninsula was more accurately expressed in the ensemble results than the individual model results, compared to the existing studies or the actual production status [52,53].

In the RCP 8.5 scenario for 2031–2050, the rice productivity estimates are predicted to increase by 15–25% compared to the historical data. In the historical data, the YSU-RSM-EPIC results were the highest (mean 5.6 t ha^{-1}), while the RegCM4-EPIC results were slightly higher than the HadGEM3-RA-EPIC results. The ensemble results showed an average rice yield of 4.7 t ha^{-1}, with annual variability, but not with a time series increase or decrease (Figure 3). Differences in spatial distribution between the models were observed, but overall, rice productivity improved in most regions except the northeast mountainous region of the Korean Peninsula (Figure 4).

The maize yield results showed no significant increase or decrease from 1981 to 2000, but the YSU-RSM-EPIC results (mean 6.4 t ha^{-1}) were the highest, similar to those for the rice data. RegCM4-EPIC showed moderate levels (mean 5.25 t ha^{-1}), while the lowest results were calculated by HadGEM3-RA-EPIC (mean 4.85 t ha^{-1}). The ensemble results showed an average value of 5.5 t ha^{-1} (Figure 3), which was similar to the existing Korean Peninsula maize productivity data [34]. Most regions, except the high plateau region of the Korean Peninsula, showed high productivity, which was more evident in the ensemble results (Figure 5).

The estimates of the future maize productivity, from 2031 to 2050, under climate change showed a slight increase in productivity compared to past productivity. The YSU-RSM-EPIC results were the highest, and the RegCM4-EPIC results were slightly higher than the HadGEM3-RA-EPIC results (Figure 3). The ensemble results showed an average maize yield of 5.75 t ha^{-1}, with annual variability, but no significant increase in the time series. There was no significant difference in spatial distribution between the models, and maize productivity was found to increase across the Korean Peninsula (Figure 5).

Three RCM and EPIC models were used to estimate the change in crop productivity, which was then used as an ensemble result to reduce the uncertainty in the climate models. This ensemble result considers the fertilization and irrigation required for crop growth, and is estimated for the whole cultivation area without distinguishing between rice paddies and maize fields.

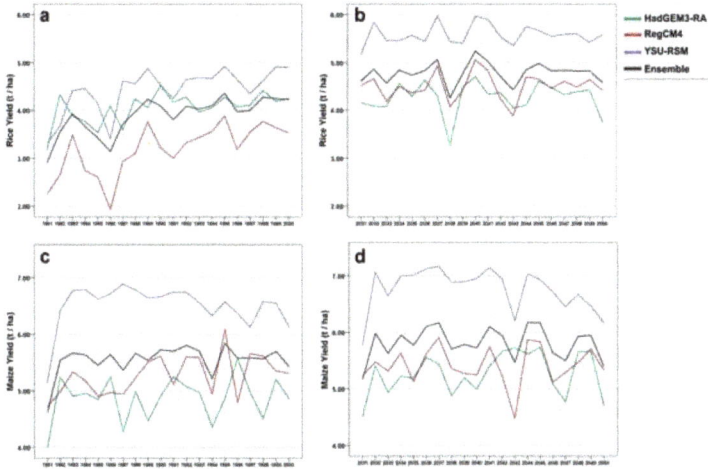

Figure 3. Estimated crop yield using multiple regional climate models (RCMs) in the past and future (**a**) Historical-rice; (**b**) representative concentration pathways (RCP) 8.5-rice; (**c**) Historical-maize; (**d**) RCP8.5-maize.

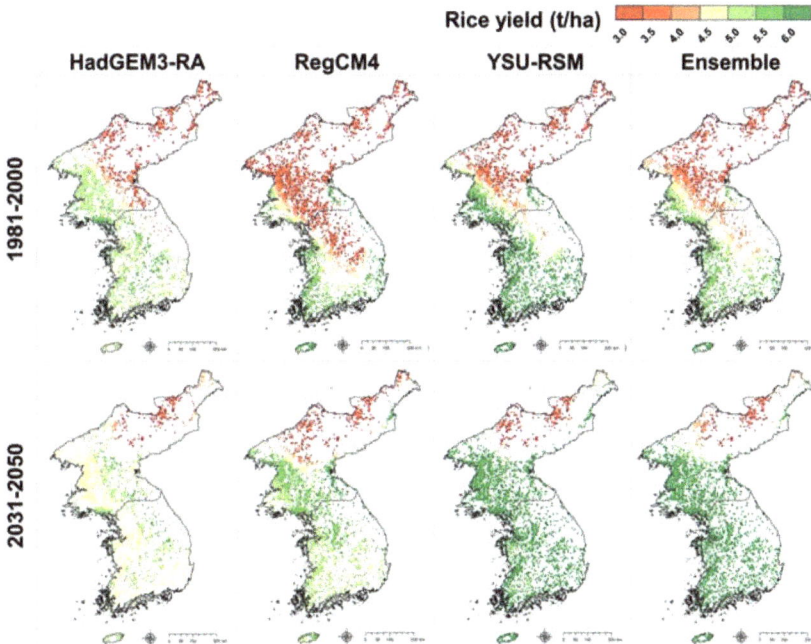

Figure 4. Spatial distribution of rice yield in the past and future for each climate model.

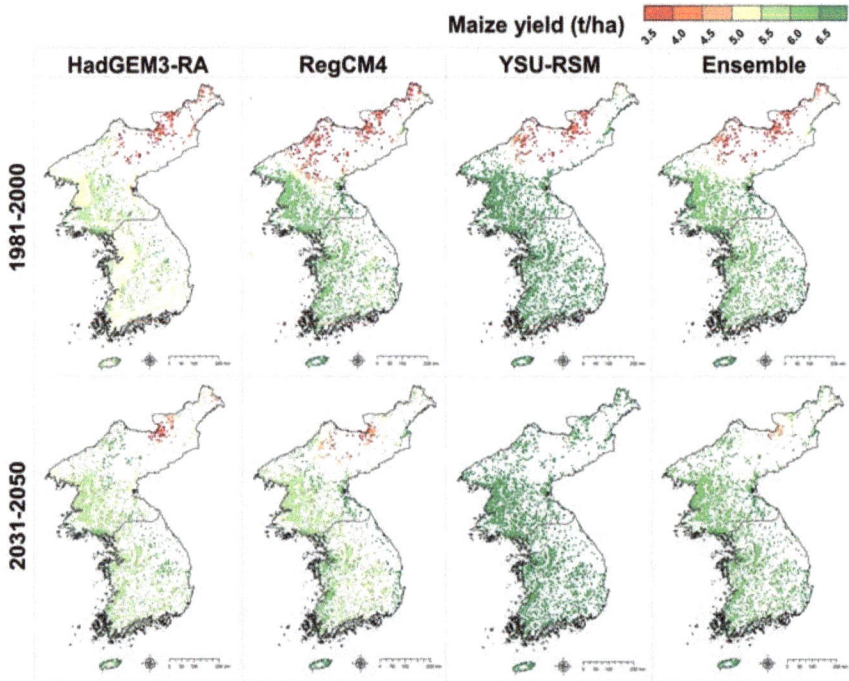

Figure 5. Spatial distribution of maize yield in the past and future for each climate model.

3.2. Estimation of PET Using Multiple Methods

Three RCMs and five PET methods were used to estimate a total of fifteen PETs in the Korean Peninsula. The PET estimation results were greater than the RCM results by over 1000 mm per year. The annual variations were minimal, and the differences between the RCM or PET method results were greater. Although all five methods were representative PET methods, the PET was largely overestimated and the BR was largely underestimated. The trends in the results for each past and future model (RCP 8.5) were similar, but the values were predicted to increase by 100 mm in the future (Figure 6). The ensemble values of the fifteen results confirmed that latitudinal differences were prevalent. In the past ensemble results, South Korea showed a PET level of 1000–1300 mm and North Korea 700–1100 mm. In the future, the PET levels for both South and North Korea are predicted to increase by an average of 100 mm (Figure 7).

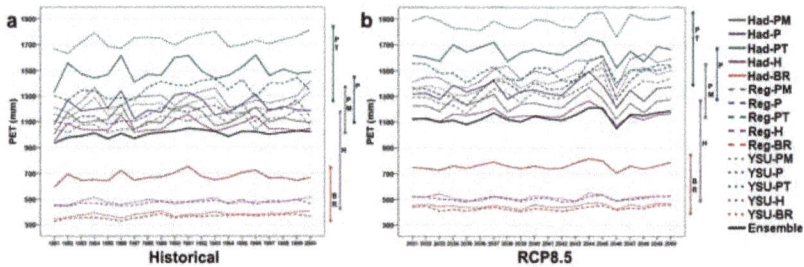

Figure 6. Estimated PET using multiple RCMs and PET methods for the past and future (**a**) Historical period (1981–2000); (**b**) RCP8.5 scenario (2031–2050).

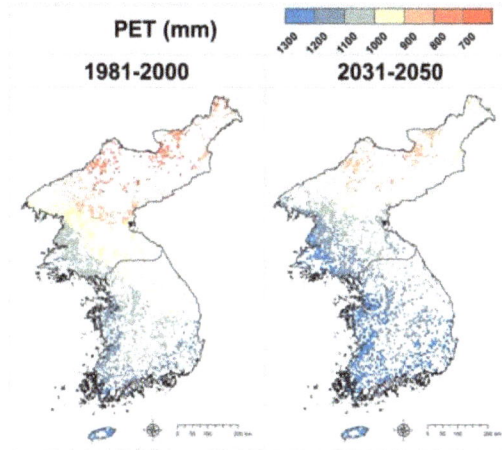

Figure 7. Spatial distribution of potential evapotranspiration for the past and future using the ensemble result.

The ensemble values were similar to the results of the existing PET studies in the Korean Peninsula or the satellite-based estimates when compared to the fifteen individual results, and the ensemble results for the future analysis period had a low level of uncertainty [41,54].

3.3. Estimation of Consumptive Water Use by Growing Season Evapotranspiration

The CWU was estimated using the growing season evapotranspiration (GET) data of each crop in the cropland. Similar to the PET estimation, three RCMs and five PET methods were applied, revealing fifteen results for the past and future periods. The GET estimation results differed largely from the RCM and PET method results, but the interval of the difference decreased as the total estimated amount decreased.

For the past data, the YSU-RSM-PT results (the most overestimated ones in PET) showed that the GET was 450 mm for both rice and maize, much higher than the other models. The RegCM4-BR results (the least underestimated PET results for rice) showed that the GET was less than 200 mm. The RegCM4-H results (least underestimated PET results for maize) showed that the GET was 165 mm on average. The average of the fifteen results was 307 mm for rice and 278 mm for maize (Figures 8 and 9). Initially, the water consumption of rice was much higher, but the results for the two crops showed a small difference in average values due to the basic latitudinal differences in GET and the relatively long growing period in North Korea.

For the future period, the estimated GET of rice was generally similar to past values, and showed a slight increase with rising temperatures (Figure 8). From 2031 to 2050, the average of the ensemble values was 312 mm. The spatial pattern was the same as for the past values, but the values were slightly decreased (Figure 10). For maize, the pattern of the estimated values according to the RCM and PET methods was similar, but the overall GET value was slightly lower than the past values (Figure 9). This lower GET is caused by the shorter growing period in North Korea following the temperature increase. The ensemble value of the fifteen GETs of maize was 271 mm. The spatial distribution results also confirmed that the GET decreased significantly in North Korea (Figure 10).

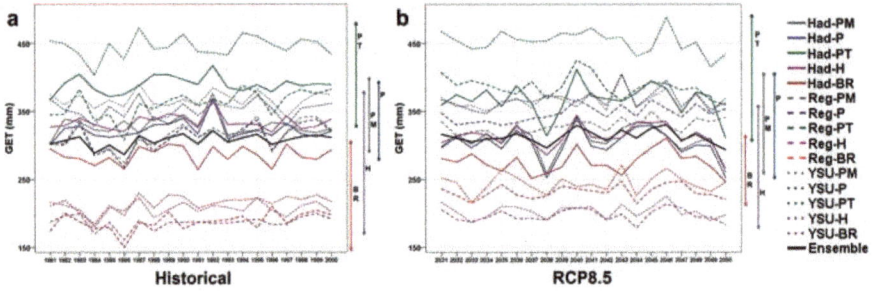

Figure 8. Estimated GET of rice using multiple RCMs and PET methods in the past and future (**a**) Historical period (1981–2000); (**b**) RCP8.5 scenario (2031–2050).

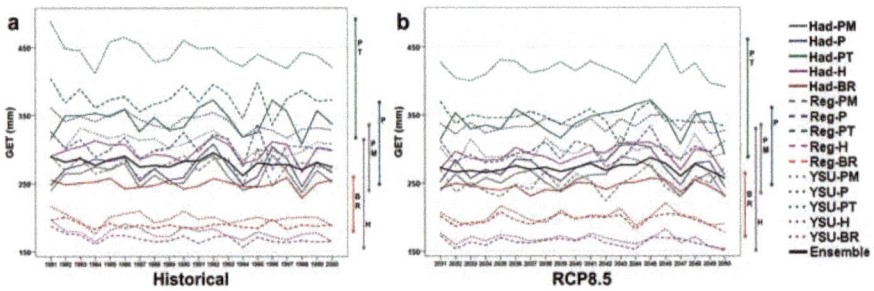

Figure 9. Estimated GET of maize using multiple RCMs and PET methods in the past and future (**a**) Historical period (1981–2000); (**b**) RCP8.5 scenario (2031–2050).

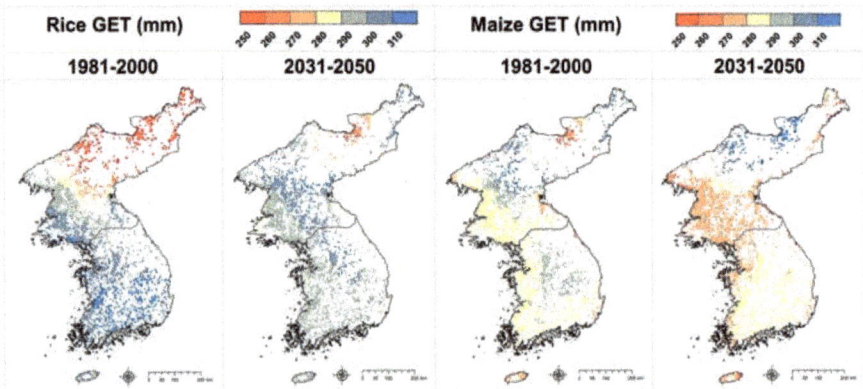

Figure 10. Spatial distribution of growing season evapotranspiration in the past and future using the ensemble result.

3.4. Virtual Water Content of Past and Future Using Multiple Data Sources

The crop yields and GET were used to calculate the VWC for each crop. Only the results from the three RCMs were estimated for the crop yield; therefore, the VWC was calculated using the five GETs and one crop yield, estimated by each RCM.

Reflected in the crop yield, the quantitative differences in the PET and the GET by the PET method significantly decreased. The VWC of rice ranged from 0.7 to 2.1 m^3 kg^{-1} and the average ensemble value was 1.18 m^3 kg^{-1}, consistent with the past data (Figure 11). In the future estimates, the range decreases to 0.4~1.1 m^3 kg^{-1} and the mean value of the ensemble is 0.76 m^3 kg^{-1}, significantly lower than the past data (Figure 11). These results indicate slight increases under the climate change scenario, but the rice yield is expected to increase greatly. The degree of change in the VWC calculated by the ratio of the two values is large. Spatially, there was a pattern of change similar to the rice yield result, with low VWC regions expanding northward (Figure 13).

The past values of the VWC of maize ranged between 0.3 and 1.0 m^3 kg^{-1}, and the ensemble average was 0.58 m^3 kg^{-1} (Figure 12). Under the climate change scenario, the range decreased to 0.2–0.7 m^3 kg^{-1} and the ensemble average to 0.48 m^3 kg^{-1}, slightly lower than the past values (Figure 12). Maize showed a smaller decrease in VWC than rice, and the maize yield is estimated to slightly increase while the GET slightly decreases. The degree of change in the VWC calculated by the ratio of the two values is relatively small. Spatial changes caused by climate change tend to be similar to the maize yield results, and the regions with lower VWCs are predicted to expand across the Korean Peninsula (Figure 13).

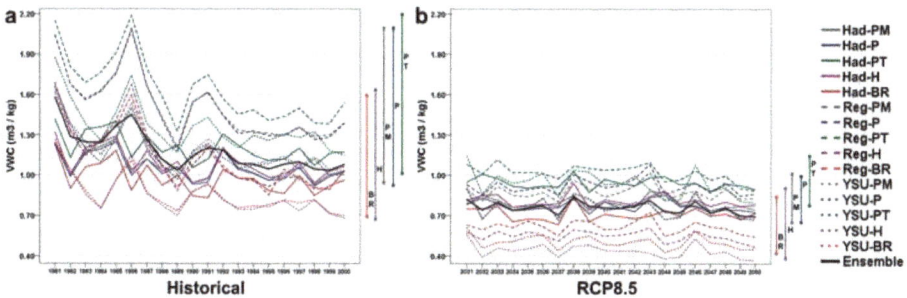

Figure 11. Estimated VWC of rice using multiple RCMs and PET methods in past and future (**a**) Historical period (1981–2000); (**b**) RCP8.5 scenario (2031–2050).

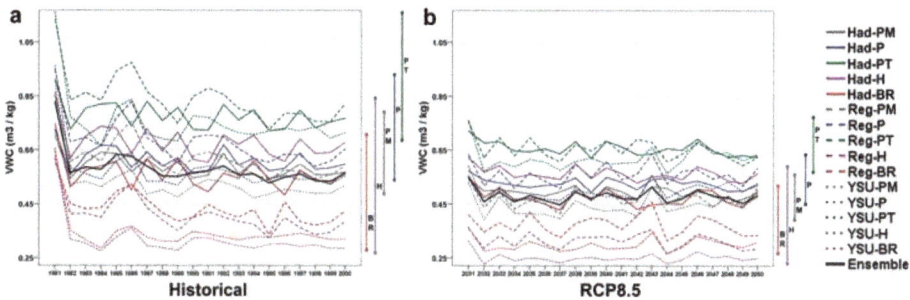

Figure 12. Estimated VWC of maize using multiple RCMs and PET methods in the past and the future (**a**) Historical period (1981–2000); (**b**) RCP8.5 scenario (2031–2050).

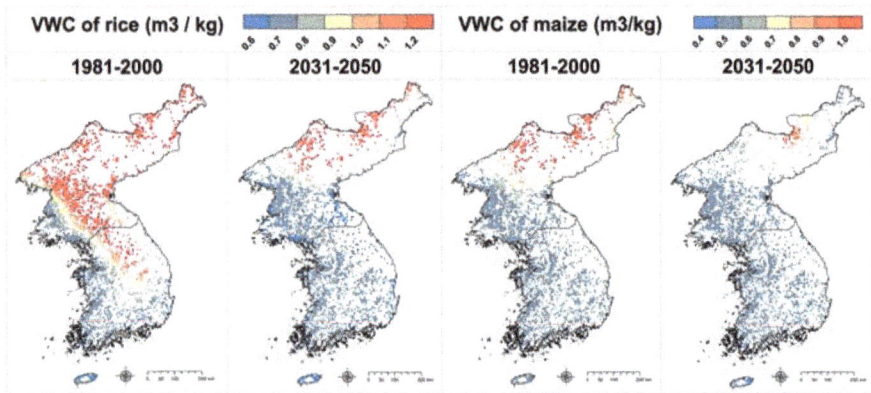

Figure 13. Spatial distribution of virtual water contents of crops in the past and future using the ensemble result.

4. Discussion

4.1. Assessing the Ensemble Result of Crop Yield, PET, GET, and VWC

The four outputs (crop yield, PET, GET, and VWC) estimated in this study were calculated using multiple methods and multiple data sources, so the ensemble results presented here have a lower level of uncertainty. This ensemble approach is also emphasized in the future climate change research outlook [19,21].

One of the uncertainties addressed in this study is the uncertainty in crop yields in the climate data. Since crop yields have a high dependence on climate, the accuracy of the climate data is very important, even if the same crop model is used [17,53]. In this study, three RCM data sets were used, derived from one global climate model (GCM). The climate values calculated by each RCM are different from the previous studies, reflected in the crop yield results of this study [10,17]. Ultimately, three crop yields from each crop and period could be used to estimate a more realistic crop productivity for the Korean Peninsula.

There are many ways to estimate the evapotranspiration required for calculating VWC and to lower the uncertainty in the results of those methods [13,21]. The data required for the five PET methods used in this study were different, and the estimated results also showed great differences. The results of the H and BR methods, estimated using only the temperature, tend to be underestimated, and the results of the PT method, estimated using temperature and solar radiation, tend to be overestimated. Hence, the methodological uncertainty is lowered by the analysis of the differences in the methodologies and the ensemble result of evapotranspiration using multiple methodologies.

In particular, we calculated fifteen VWCs for each crop, which have inherent uncertainties due to the climate models and the PET method. Presenting the average of fifteen VWCs as an ensemble value resulted in mitigating the differences in data and methods by up to 294% (0.33–0.97 m^3 kg^{-1} (in 1986)) (historical maize data, annual average). The maximum difference of 252% (0.87–2.19 m^3 kg^{-1} (in 1986)) in the historical rice data was alleviated. In the future projections, the total VWC decreased compared to the past values, but the effect of the ensemble results was similar between the past and future values. The maximum reduction in the RCP85-rice value was 262% (0.42–1.10 m^3 kg^{-1} (in 2043)) and a reduction of up to 245% (0.31–0.76 m^3 kg^{-1} (in 2031)) was observed in RCP85-maize results (Figures 10 and 11). In other words, the ensemble VWC reduced the difference by over 200% (according to data and methods) in all of the periods and crops. Although there are no VWC studies available for the Korean Peninsula, ensemble results are more similar than individual results when compared with the results of Zhao et al. [15], which calculated the VWC for China.

4.2. Implications for Agricultural Water Supply and Demand in the Korean Peninsula

Precipitation in the Korean Peninsula is concentrated in the summer monsoon season: 50–60% of annual precipitation occurs during this season. The water supply is not constant, as it is highly dependent on the river regime [10]. Most agricultural products in South Korea have been replaced by imported products due to changes at the economic level, but the self-sufficiency rate of food crops is so high that the consumption of water in agriculture is still at significant levels. In North Korea, the demand for water has increased because of the increase in food production owing to limited imports [24]. Overall, the Korean Peninsula, having a relatively high population density (South Korea: 519; North Korea: 208) and limited water supply, has a high demand for municipal, industrial, and agricultural water [55].

The average amount of water consumed by crops per unit production of major crops can be spatially determined by quantifying the agricultural water consumption using the results of the ensemble model approach employed in this study. This study also provided the future projections of the average agricultural water consumption, reflecting the impact of climate change, which has positive effects on productivity; as climate change accelerates, the amount of water consumed in the production of major crops will decrease. Climate change will have a negative impact on water supply as it affects precipitation patterns and amounts [10], but it can positively change water demand. However, this positive change in water demand due to climate change is the result of optimized adaptation modeling to water and nutrient stress. The consequences of climate change impacts without adaptation can vary significantly.

5. Conclusions

Multiple data sources and methods were employed in this study to estimate the past and future VWC of each crop in the Korean Peninsula. The EPIC crop model was used, and three RCMs and five PET methods were applied to reduce uncertainties in the data and methods. The rice and maize productivity varied significantly in the RCM results, confirming the increased potential production of both crops in the future. Positive changes in the northern part of the Korean Peninsula are noticeable, and maize is predicted to have high productivity in the entire peninsula in the future. The fifteen PETs and GETs from the RCM and PET methods were significantly different, and the water consumption of the crops was estimated by minimizing the errors by calculating the average values. The VWC was calculated for past and future crop yields and consumptive water use, with over 200% difference between them according to the RCM and PET methods. Computing the ensemble VWC for each period and crop by averaging fifteen VWCs could reduce the errors. The VWCs of the crops in the future projections were lower than those using the past data, which reflected a positive change in productivity and a decline in the length of the growth period. The past and future ensemble VWCs presented in this study provide quantitative data to shape the overall water demand for agriculture in the Korean Peninsula. We conclude that these results can be useful to improve agricultural sustainability, including food and water security, in the Korean Peninsula.

Acknowledgments: This study was supported by the Korean Ministry of Environment as part of the "Public Technology Development Project based on Environmental Policy" (Project Number: 2016000210001) and "Climate Change Correspondence Program" (Project Number: 2014001310008). We also thank Hanbin Kwak for technical support.

Author Contributions: Chul-Hee Lim designed this research, analyzed the result and wrote the paper; Seung Hee Kim and Yuyoung Choi participated in the data work. All authors, including Menas C. Kafatos and Woo-kyun Lee, gave comments and approved the final manuscript.

Conflicts of Interest: The authors declare no conflict of interest.

References

1. Sachs, J.D. From millennium development goals to sustainable development goals. *Lancet* **2012**, *379*, 2206–2211. [CrossRef]

2.   IPCC. *Climate Change 2014: Impacts, Adaptation, and Vulnerability. Part A: Global and Sectoral Aspects. Contribution of Working Group II to the Fifth Assessment Report of the Intergovernmental Panel on Climate Change*; Cambridge University Press: Cambridge, UK, 2014.

3.   Griggs, D.; Stafford-Smith, M.; Gaffney, O.; Rockström, J.; Öhman, M.C.; Shyamsundar, P.; Noble, I. Policy: Sustainable development goals for people and planet. *Nature* **2013**, *495*, 305–307. [CrossRef] [PubMed]

4.   Wheeler, T.; Von Braun, J. Climate change impacts on global food security. *Science* **2013**, *341*, 508–513. [CrossRef] [PubMed]

5.   Godfray, H.C.J.; Garnett, T. Food security and sustainable intensification. *Philos. Trans. R. Soc. B* **2004**, *369*, 20120273. [CrossRef] [PubMed]

6.   Elliott, J.; Deryng, D.; Müller, C.; Frieler, K.; Konzmann, M.; Gerten, D.; Glotter, M.; Flörke, M.; Wada, Y.; Best, N.; et al. Constraints and potentials of future irrigation water availability on agricultural production under climate change. *Proc. Natl. Acad. Sci. USA* **2014**, *111*, 3239–3244. [CrossRef] [PubMed]

7.   Calzadilla, A.; Rehdanz, K.; Betts, R.; Falloon, P.; Wiltshire, A.; Tol, R.S. Climate change impacts on global agriculture. *Clim. Chang.* **2013**, *120*, 357–374. [CrossRef]

8.   Haddeland, I.; Heinke, J.; Biemans, H.; Eisner, S.; Flörke, M.; Hanasaki, N.; Stacke, T. Global water resources affected by human interventions and climate change. *Proc. Natl. Acad. Sci. USA* **2014**, *111*, 3251–3256. [CrossRef] [PubMed]

9.   Huang, B.; Polanski, S.; Cubasch, U. Assessment of precipitation climatology in an ensemble of CORDEX-East Asia regional climate simulations. *Clim. Res.* **2015**, *64*, 141–158. [CrossRef]

10.   Park, C.; Min, S.K.; Lee, D.; Cha, D.H.; Suh, M.S.; Kang, H.S.; Kwon, W.T. Evaluation of multiple regional climate models for summer climate extremes over East Asia. *Clim. Dyn.* **2016**, *46*, 2469–2486. [CrossRef]

11.   Hoekstra, A.Y.; Hung, P.Q. Globalisation of water resources: International virtual water flows in relation to crop trade. *Glob. Environ. Chang.* **2005**, *15*, 45–56. [CrossRef]

12.   Zhuo, L.; Mekonnen, M.M.; Hoekstra, A.Y. The effect of inter-annual variability of consumption, production, trade and climate on crop-related green and blue water footprints and inter-regional virtual water trade: A study for China (1978–2008). *Water Res.* **2016**, *94*, 73–85. [CrossRef] [PubMed]

13.   Liu, J.; Zehnder, A.J.; Yang, H. Global consumptive water use for crop production: The importance of green water and virtual water. *Water Resour. Res.* **2009**, *45*. [CrossRef]

14.   Liu, J.; Folberth, C.; Yang, H.; Röckström, J.; Abbaspour, K.; Zehnder, A.J. A global and spatially explicit assessment of climate change impacts on crop production and consumptive water use. *PLoS ONE* **2013**, *8*, e57750. [CrossRef] [PubMed]

15.   Zhao, Q.; Liu, J.; Khabarov, N.; Obersteiner, M.; Westphal, M. Impacts of climate change on virtual water content of crops in China. *Ecol. Inform.* **2014**, *19*, 26–34. [CrossRef]

16.   Zhuo, L.; Mekonnen, M.M.; Hoekstra, A.Y. Consumptive water footprint and virtual water trade scenarios for China—With a focus on crop production, consumption and trade. *Environ. Int.* **2016**, *94*, 211–223. [CrossRef] [PubMed]

17.   Chun, J.A.; Li, S.; Wang, Q.; Lee, W.S.; Lee, E.J.; Horstmann, N.; Park, H.; Veasna, T.; Vanndy, L.; Pros, K.; et al. Assessing rice productivity and adaptation strategies for Southeast Asia under climate change through multi-scale crop modeling. *Agric. Syst.* **2016**, *143*, 14–21. [CrossRef]

18.   Baek, H.-J.; Lee, J.; Lee, H.-S.; Hyun, Y.-K.; Cho, C.; Kwon, W.-T.; Marzin, C.; Gan, S.-Y.; Kim, M.-J.; Choi, D.-H. Climate change in the 21st century simulated by HadGEM2-AO under representative concentration pathways. *Asia-Pac. J. Atmos. Sci.* **2013**, *49*, 603–618. [CrossRef]

19.   Rosenzweig, C.; Elliott, J.; Deryng, D.; Ruane, A.C.; Müller, C.; Arneth, A.; Boote, K.J.; Folberth, C.; Glotter, M.; Khabarov, N. Assessing agricultural risks of climate change in the 21st century in a global gridded crop model intercomparison. *Proc. Natl. Acad. Sci. USA* **2014**, *111*, 3268–3273. [CrossRef] [PubMed]

20.   Wallach, D.; Rivington, M. A framework for assessing the uncertainty in crop model predictions. *FACCE MACSUR Rep.* **2014**, *3*, 1–5.

21.   Liu, W.; Yang, H.; Folberth, C.; Wang, X.; Luo, Q.; Schulin, R. Global investigation of impacts of PET methods on simulating crop-water relations for maize. *Agric. For. Meteorol.* **2016**, *221*, 164–175. [CrossRef]

22.   Choi, I.H.; Woo, J.C. Developmental process of forest policy direction in Korea and present status of forest desolation in North Korea. *J. For. Sci.* **2007**, *23*, 14.

23.   Jeon, Y.; Kim, Y. Land reform, income redistribution, and agricultural production in Korea. *Econ. Dev. Cult. Chang.* **2000**, *48*, 253–268. [CrossRef]

24. Korea Rural Economic Institute (KREI). *KREI Quarterly Agriculture Trends in North Korea*; Korea Rural Economic Institute: Seoul, Korea, 2014.

25. Lee, J.W.; Hong, S.Y.; Chang, E.C.; Suh, M.S.; Kang, H.S. Assessment of future climate change over east Asia due to the RCP scenarios downscaled by GRIMs-RMP. *Clim. Dyn.* **2014**, *42*, 733–747. [CrossRef]

26. Giorgi, F.; Coppola, E.; Solmon, F.; Mariotti, L.; Sylla, M.B.; Bi, X.; Elguindi, N.; Diro, G.T.; Nari, V.; Giuliani, G.; et al. RegCM4: Model description and preliminary tests over multiple CORDEX domains. *Clim. Res.* **2012**, *52*, 7–29. [CrossRef]

27. Williams, J.R.; Jones, C.A.; Dyke, P.T. A modeling approach to determining the relationship between erosion and soil productivity. *Trans. ASABE* **1984**, *27*, 129–144. [CrossRef]

28. Williams, J.R.; Jones, C.A.; Kiniry, J.R.; Spanel, D.A. The EPIC crop growth model. *Trans. ASAE* **1989**, *32*, 497–511. [CrossRef]

29. Knisel, W.G. *GLEAMS: Groundwater Loading Effects of Agricultural Management Systems 2006*; Version 2.10 (No. 5); University of Georgia: Athens, GA, USA, 2006.

30. Izaurralde, R.; Williams, J.R.; McGill, W.B.; Rosenberg, N.J.; Jakas, M.Q. Simulating soil C dynamics with EPIC: Model description and testing against long-term data. *Ecol. Model.* **2006**, *192*, 362–384. [CrossRef]

31. Renard, K.G.; Foster, G.R.; Weesies, G.A.; Porter, J.P. Rusle: Revised Universal Soil Loss Equation. *J. Soil Water Conserv.* **1991**, *46*, 30–33.

32. Rinaldi, M. Application of EPIC model for irrigation scheduling of sunflower in Southern Italy. *Agric. Water Manag.* **2001**, *49*, 185–196. [CrossRef]

33. Williams, J.R. The EPIC model. In *Computer Models of Watershed Hydrology*; Singh, V.P., Ed.; Water Resources Publications: Littleton, CO, USA, 1995.

34. Xiong, W.; Balkovič, J.; van der Velde, M.; Zhang, X.; Izaurralde, R.C.; Skalský, R.; Lin, E.; Mueller, N.; Obersteiner, M. A calibration procedure to improve global rice yield simulations with EPIC. *Ecol. Model.* **2014**, *273*, 128–139. [CrossRef]

35. Lim, C.H.; Lee, W.K.; Song, Y.; Eom, K.C. Assessing the EPIC model for estimation of future crops yield in South Korea. *J. Clim. Chang. Res.* **2015**, *6*, 21–31. [CrossRef]

36. Lim, C.H.; Kim, M.; Lee, W.K.; Folberth, C. Spatially Explicit Assessment of Agricultural Water Equilibrium in the Korean Peninsula. *Agric. Water Manag.* **2017**, under review.

37. Balkovič, J.; van der Velde, M.; Schmid, E.; Skalský, R.; Khabarov, N.; Obersteiner, M.; Xiong, W. Pan-European crop modelling with EPIC: Implementation, up-scaling and regional crop yield validation. *Agric. Syst.* **2013**, *120*, 61–75. [CrossRef]

38. Lim, C.H.; Choi, Y.; Kim, M.; Jeon, S.W.; Lee, W.K. Impact of deforestation on agro-environmental variables in cropland, North Korea. *Sustainability* **2017**, under review.

39. Yoo, B.H.; Kim, K.S. Development of a gridded climate data tool for the Coordinated Regional climate Downscaling Experiment data. *Comput. Electron. Agric.* **2017**, *133*, 128–140. [CrossRef]

40. Nash, J.E.; Sutcliffe, J.V. River flow forecasting through conceptual models part I—A discussion of principles. *J. Hydrol.* **1970**, *10*, 282–290. [CrossRef]

41. Niu, X.; Easterling, W.; Hays, C.J.; Jacobs, A.; Mearns, L. Reliability and input-data induced uncertainty of the EPIC model to estimate climate change impact on sorghum yields in the US Great Plains. *Agric. Ecosyst. Environ.* **2009**, *129*, 268–276. [CrossRef]

42. Balkovič, J.; van der Velde, M.; Skalský, R.; Xiong, W.; Folberth, C.; Khabarov, N.; Smirnov, A.; Mueller, N.D.; Obersteiner, M. Global wheat production potentials and management flexibility under the representative concentration pathways. *Glob. Planet. Chang.* **2014**, *122*, 107–121. [CrossRef]

43. Monteith, J. Evaporation and environment. *Symp. Soc. Exp. Biol.* **1965**, *19*, 205–234. [PubMed]

44. Penman, H.L. Natural evaporation from open water, bare soil and grass. *Proc. R. Soc. Lond. Ser. A* **1948**, *193*, 120–145. [CrossRef]

45. Priestley, C.; Taylor, R. On the assessment of surface heat flux and evaporation using large-scale parameters. *Mon. Weather Rev.* **1972**, *100*, 81–92. [CrossRef]

46. Hargreaves, G.H.; Samani, Z.A. Reference crop evapotranspiration from temperature. *Appl. Eng. Agric.* **1985**, *1*, 96–99. [CrossRef]

47. Baier, W.; Robertson, G.W. Estimation of latent evaporation from simple weather observations. *Can. J. Plant Sci.* **1965**, *45*, 276–284. [CrossRef]

48. Liu, J.; Williams, J.R.; Zehnder, A.J.; Yang, H. GEPIC—Modelling wheat yield and crop water productivity with high resolution on a global scale. *Agric. Syst.* **2007**, *94*, 478–493. [CrossRef]

49. Food Agriculture Organization. *FAO Digital Soil Map of the World*; FAO: Rome, Italy, 1995.

50. Batjes, N.H. *ISRIC-WISE Derived Soil Properties on a 5 by 5 Arc-Minutes Global Grid*; ISRIC—World Soil Information: Wageningen, The Netherlands, 2006.

51. Chen, J.; Chen, J.; Liao, A.; Cao, X.; Chen, L.; Chen, X.; Zhang, W. Global land cover mapping at 30 m resolution: A POK-based operational approach. *ISPRS J. Photogramm. Remote Sens.* **2015**, *103*, 7–27. [CrossRef]

52. Kim, H.Y.; Ko, J.; Kang, S.; Tenhunen, J. Impacts of climate change on paddy rice yield in a temperate climate. *Glob. Chang. Biol.* **2013**, *19*, 548–562. [CrossRef] [PubMed]

53. Shin, Y.; Lee, E.J.; Im, E.S.; Jung, I.W. Spatially distinct response of rice yield to autonomous adaptation under the CMIP5 multi-model projections. *Asia-Pac. J. Atmos. Sci.* **2017**, *53*, 21–30. [CrossRef]

54. Liaqat, U.W.; Choi, M. Accuracy comparison of remotely sensed evapotranspiration products and their associated water stress footprints under different land cover types in Korean peninsula. *J. Clean. Prod.* **2017**, *155*, 93–104. [CrossRef]

55. Chung, E.S.; Won, K.; Kim, Y.; Lee, H. Water resource vulnerability characteristics by district's population size in a changing climate using subjective and objective weights. *Sustainability* **2014**, *6*, 6141–6157. [CrossRef]

![sustainability logo] *sustainability*

MDPI

Article

The Impact of Climatic Change Adaptation on Agricultural Productivity in Central Chile: A Stochastic Production Frontier Approach

Lisandro Roco [1,*]**, Boris Bravo-Ureta** [2,3]**, Alejandra Engler** [3,4] ⓘ **and Roberto Jara-Rojas** [3,4]

[1] Department of Economics and Institute of Applied Regional Economics (IDEAR),
    Universidad Católica del Norte, Antofagasta 1240000, Chile
[2] Department of Agricultural and Resource Economics, University of Connecticut, Storrs 06269, CT, USA;
    boris.bravoureta@uconn.edu
[3] Department of Agricultural Economics, Universidad de Talca, Talca 3460000, Chile;
    mengler@utalca.cl (A.E.); rjara@utalca.cl (R.J.-R.)
[4] Center for Socioeconomic Impact of Environmental Policies (CESIEP), Talca 3460000, Chile
* Correspondence: lisandro.roco@ucn.cl; Tel.: +56-55-235-5770

Received: 23 June 2017; Accepted: 13 September 2017; Published: 16 September 2017

Abstract: Adaptation to climate change is imperative to sustain and promote agricultural productivity growth, and site-specific empirical evidence is needed to facilitate policy making. Therefore, this study analyses the impact of climate change adaptation on productivity for annual crops in Central Chile using a stochastic production frontier approach. The data come from a random sample of 265 farms located in four municipalities with different agro-climatic conditions. To measure climate change adaptation, a set of 14 practices was used in three different specifications: binary variable, count and index; representing decision, intensity and quality of adaptation, respectively. The aforementioned alternative variables were used in three different stochastic production frontier models. Results suggest that the use of adaptive practices had a significant and positive effect on productivity; the practice with the highest impact on productivity was irrigation improvement. Empirical results demonstrate the relevance of climate change adaptation on farmers' productivity and enrich the discussion regarding the need to implement adaptation measures.

Keywords: climate change; adaptation; agricultural systems; productivity; technical efficiency; Chile

---

1. Introduction

Agriculture represents a relevant economic sector for the analysis of climate change, given that it is situated at the interface between ecosystems and society, and it is highly affected by changes in environmental conditions [1,2]. Climate change is affecting food prices, food security, land use [3] and raising uncertainty for crop managers [4]. According to Kahil [5], the severity of climate change impact depends on the degree of adaptation at the farm level, farmers' investment decisions and policy choices, and these factors are interrelated. Thus, it is necessary to recognize the effect that limitations in natural resources will have on agriculture to build resilience to climate change at the farm level [6].

On the other hand, as natural resources available for food production become more constraining, crop productivity is essential for fostering the growth and welfare of the agricultural sector [7]. To relax these constraints, farmers have been modifying their practices to cope with climatic variability for centuries; however, climate change is now threatening their livelihoods with increasing unpredictability, including frequent and intense weather extremes such as droughts, floods and frosts [8]. According to Zilberman et al. [9], adaptation is the response of economic agents and societies to major shocks such as climate change. Adaptation practices are adjustments intended to enhance resilience or reduce vulnerability to observed or expected changes in climate [10]. Nelson et al. [11] claim that adaptation is imperative for three reasons: (i) many future environmental risks are now more apparent and predictable than ever; (ii) even where risks are not quantifiable, environmental changes may be very significant; and (iii) environmental change, although often the outcome of multiple drivers, has indisputable human causes. Changes in food production affect all consumers; however, it is producers that need to adapt to insure adequate supplies and who bear the costs involved in improving efficiency [12].

There is a wide range of methodological approaches that have been developed over the years to generate multiple measures of productivity and efficiency [13]. A relevant measure of productivity for management recommendations is technical efficiency (TE) [14]. This indicator evaluates the difference between frontier or maximum attainable output and observed output given an input bundle and technology. Given that TE is an important component in overall productivity, the development and implementation of public policies can be more effective if the TE of any given farming system is known [15]. Several studies have investigated factors associated with agricultural productivity across the globe, but the literature linking TE with climate change adaptation is scanty. One exception is the study by Mukherjee et al. [16], which finds that heat stress in the southeastern U.S. has a significant and negative impact on milk production, while adaptation through a fairly simple cooling technology has a positive and significant effect on efficiency. In addition, in the same analysis, when climate change is factored into the production function (frontier) specification, the resulting estimates are more accurate, because they avoid possible parameter bias stemming from the omitted variable problem.

It is thus important to model the full range of interactions that might exist between productivity and climate change [17]. Most of the scientific information related to climate change and its effects on agriculture comes from case studies in developed countries. In developing countries, where there are high levels of uncertainty and vulnerability to climate change, there is need to target policy instruments to adapt the productive systems, particularly considering the lack of articulation between climate change adaptation and agricultural policy [18].

In this work, we investigate whether adaptive practices can increase productivity in different agricultural production systems based on annual crops in Central Chile. Major adaptation practices in farming systems include: conserving soil, using water efficiently, planting trees, changing planting dates and using improved varieties [19–23]. It is expected that farmers who are more aware of and better adapted to climate change will be able to make more efficient use of their resources and thus cope with any adversities. This study adds valuable information for agricultural policy design, as it provides evidence of the impact of alternative adaptation strategies to climate change. Additionally, farmer and agricultural system characteristics are linked to productivity to inform agricultural policy.

The rest of the paper is organized as follows: Section 2 gives a description of the study area, the methodological approach and the empirical models; Section 3 presents and discusses the empirical results; and Section 4 summarizes and concludes.

2. Materials and Methods

2.1. Study Area and Data

The study area covers 8,958 farms in four municipalities of the Maule Region, in Central Chile, a Mediterranean transition zone between the arid north and the rainy south. Projections for the study area comprise a decrease in precipitation of up to 40% and a rise in temperatures between 2 °C and 4 °C in the next 40 years [24,25]. This region is a major contributor to the agricultural output of the country and, despite rapid technological progress in recent years, the cultivation of annual crops, fruits and vegetables is not changing fast enough to counteract the predicted adverse effects of climate change [26,27]. Specific adverse effects expected in the near future concern losses in the quality of the environment for agricultural production [28].

The four municipalities selected for the study were: Pencahue, San Clemente, Cauquenes and Parral. Pencahue and Cauquenes are dryland areas; San Clemente is primarily composed of irrigated land near the Andes Mountains; and Parral is in the central irrigated valley. San Clemente has a total of 226,826 hectares (ha) dedicated largely to the production of forage, cereals and seeds. Cauquenes and Parral have 128,017 and 125,630 ha, respectively, with a significant area devoted to vineyards, cereals and forage. Pencahue is the smallest municipality, with 65,118 ha dedicated mostly to vineyards, orchards and cereals [26]. Table 1 presents some key characteristics of the four municipalities and the main cropping systems for each one.

Table 1. General information for the study area.

Municipality	Area	Rainfall (mm/Year)	Farms	Farms Interviewed	Main Crop System (%)				
					Wheat and Oat	Spring Crops [a]	Spring Vegetable [b]	Rice	Others Crops [c]
Pencahue	Irrigated dryland	709	1129	40	12.5	35.0	52.5	0.0	0.0
Cauquenes	Non-irrigated dryland	670	3026	81	97.5	2.5	0.0	0.0	0.0
San Clemente	Irrigated Andean foothill	920	2990	89	40.4	42.6	12.4	0.0	4.6
Parral	Irrigated central valley	900	1813	89	54.5	7.3	1.8	36.4	0.0
	Total		8958	265	56.6	77.4	12.5	7.5	1.5

[a] Spring crops are: maize, beans and potatoes. [b] Spring vegetables are: peas, onion, tomato, melon, watermelon, cucumber and squash. [c] Other crops are: tobacco and cabbage.

During August and November of 2011, a random survey was conducted that involved 274 interviews, representing 3.06% of the farmers in the study area. This survey targeted farmers that specialized in annual crops. The surveys with missing information were excluded from the analysis, leaving 265 valid surveys. Previous work in the study area inquired about the perception of and adaptation to climate change [24,26]; however, this article goes further by linking adaptation to climate change and productivity at the farm level.

Table 2 shows a description of the variables used in the study. The mean crop production value is US\$66,383 (MM\$31.2 where MM\$ is equivalent to millions of Chilean pesos; and the prevailing exchange rate was 470 Chilean pesos per U.S. dollar when the data were collected). Farms range in size from 0.5–595 hectares, with a mean of 55.5 hectares. The average cultivated land area is 17.1 hectares. The mean value of purchased inputs (seeds, fertilizers, pesticides and hired machinery) is MM\$11.4, and the mean investment in labor for crop production is MM\$2.2. Crop diversification is measured using a variant of the Herfindahl index (H) calculated for each farm as: $H = \left(1 - \sum_{i=1}^{n} \left(\frac{c_i}{T}\right)^2\right) \times 100$, where c_i is the area under the i-th crop and T is the total cropped area [29]. The H index for the sample is 23.7%, ranging from 0–96.4%.

The average age for farmers is 55.5 years, while the average level of formal education is 7.2 years. The majority (82.6%) of farmers claimed that agriculture is their main income source, accounting, on average, for 62.1% of their total income. Eighty-one farms are in dryland areas. Meteorological information from the Internet and mass media (radio, TV and newspaper) is used by 93.2% of the farmers, and 52.4% of them participate in farmer associations. The mean distance from the farms to the city of Talca, the regional capital, is 77.4 kilometers.

Table 2. Description of the variables used in the stochastic production frontier (SPF) and inefficiency models. MM$, millions of Chilean pesos.

Variable	Name	Unit	Definition	Mean	SD
Production Function Variables					
y	Agricultural production	MM$	Crop production value in Chilean pesos [a]	31.2	14.0
L	Cultivated land	Ha	Hectares with crops	17.1	53.3
C	Capital	MM$	Value of seeds, fertilizers, pesticides and machinery contracted in Chilean pesos	11.4	51.7
W	Labor	MM$	Value of family and hired labor	2.2	6.8
D	Dryland	%	Dummy variable = 1 if the farm is located in a dryland area and 0 otherwise	30.6	46.2
H	Diversification	%	Crop diversification index	23.7	27.5
A_1	Climate change adaptation	Decision	%	Dummy variable = 1 if there are at least one practice adopted and 0 otherwise	56.6	49.7
A_2		Intensity	Number	Number of climate change adaptation practices adopted in the farm	1.8	2.2
A_3		Quality	%	Index of adaptation based on experts' opinion	12.6	15.4
Inefficiency Model Variables					
z_1	Age	Years	Age of the head of the farm in years	55.5	14.1
z_2	Schooling	Years	Years of schooling of the head of the farm	7.2	4.1
z_3	Dependence	%	Dummy variable = 1 if agriculture is the main source of income for the household and 0 otherwise	82.6	37.9
z_4	Specialization	%	Percent of total income that corresponds to income from crops	62.1	32.0
z_5	Use of meteorological information	%	Dummy variable = 1 if the farmer is a user of meteorological information and 0 otherwise	93.2	25.2
z_5	Membership	%	Dummy variable = 1 if the farmer is a member of an association and 0 otherwise	52.4	50.0
z_7	Farm size	Ha	Total farm size in hectares	56.4	122.3
z_8	Distance to market	Km	Distance to the regional capital city in kilometers	77.4	43.8

[a] Four hundred seventy Chilean pesos = US$1 for the study period.

2.2. Practices Considered for Climate Change Adaptation

In recent studies, adaptive practices are identified as investment in technologies such as irrigation, the use of drought- and heat-tolerant and early-maturing varieties [19,30] and the adoption of strategies such as changing planting and harvesting dates, crop diversification, agroforestry and soil and water conservation practices [20–22]. Tambo and Abdoulaye [23] highlight the relevance of adaptation and its intensity regarding climate change. The authors just mentioned use as a first hurdle the decision to adopt a drought-resistant variety of maize and then intensity as the degree to which they will invest in adaptation measured as the area cultivated with the resistant variety.

A panel of experts was consulted to determine the most appropriate climate change adaptation strategies for the farming systems of Central Chile. This expert panel was composed of 14 national experts in agricultural systems and climate change. These experts were asked to assign a score from 0–3, where 0 is no impact and 3 is high impact, to 14 practices according to the importance of each practice for adaptation. These practices, described in Table 3, fall into three main categories: (1) water and soil conservation practices (WSC); (2) changes in cropping schedule and varieties (Cr); and (3) improvement of irrigation systems (I). These practices have been used previously in the literature [19,20,31]. We used this list of practices in the producers' survey to learn about what practices are being used by them. In several quantitative studies, the adaptation to climate change has been measured as the adoption of strategies, practices and technologies to increase the capacity of a farm to cope with changing climate and variability ([19–23] and others), and in most studies, the adaptation variable is defined as a binary decision. To carry out a more comprehensive analysis of adaptation, we include alternative measures of adaptation, from a simple binary variable to a more complex adaptation quality index. Each measure accounts for different interpretations of adaptation described as follows:

- Binary decision: a dichotomous variable indicating that at least one practice was adopted (A_1). In this case, the aim is to analyze the impact of being able to carry out a basic strategy.
- Intensity: measured as the number of practices or technologies adopted on the farm (A_2). Compared to A_1, this measure analyzes the impact of passing the first hurdle, i.e., the decision to adapt.
- Quality: an index calculated as the sum of adaptation practices weighted by the experts' score (A_3). The objective here is to estimate the impact of adopted practices that are more effective to face climate change. The weights were estimated by normalizing the average scores (0–3) given by the panel of experts to each practice, to generate a scale. The quality adaptation index (A_3) was constructed considering the sum of all the practices on a given farm multiplied by the weight assigned by experts (W_{ij}), divided by the sum of all weights (W_i). The formula used is as follows:

$$A_{3_i} = \left[\frac{\sum_{j=1}^{14} W_{ij}}{\sum_{j=1}^{14} W_j} \right] \times 100,$$ where i are the farms (from 1–265) and j are the practices (from 1–14).)

The value of A_3 ranges from 0–100% where 100% implies that the practice presents the highest valuation assigned by the experts.

The number of farmers who have decided to adopt at least one of the practices is 150, representing 56.6% of the sample. The intensity in the number of practices adopted by farmers ranges from 0–11, with a mean of 1.8. The quality of adaptation (average index) is 12.6%, ranging from 0–79.3% (as can be seen in Table 2).

Table 3. Climate change adaptation practices according to the recommendation by experts.

Practice	Type [a]	Weight %	Farmers (n = 265)	
			No. of Respondents	% of Total
Incorporation of crop varieties resistant to droughts	Cr	85.7	2	0.7
Use of drip and sprinkler	I	83.3	31	11.7
Incorporation of crops resistant to high temperatures	Cr	80.9	2	0.7
Changes in planting and harvesting dates	Cr	78.6	110	41.5
Afforestation	WSC	76.2	5	1.9
Zero tillage	WSC	69.0	3	1.1
Use of water accumulation systems	I	66.7	38	14.3
Use of green manure	WSC	66.0	33	12.4
Use of mulching	WSC	61.9	24	9.0
Use of cover crops	WSC	61.9	16	6.0
Other WSC practices	WSC	61.9	16	6.0
Use of hoses and pumps for irrigation	I	59.5	52	19.6
Implementation of infiltration trenches	WSC	57.1	19	7.1
Cleaning of canals	WSC	54.8	60	22.6

[a] Cr: changes in crops, I: improvement of irrigation systems, WSC: water and soil conservation practices.

2.3. Analytical Framework and Empirical Model

The stochastic production frontier (SPF) model developed by Battese and Coelli [32] was used to estimate the following Cobb–Douglas frontier:

$$lny_i = \beta_0 + \beta_1 lnL_i + \beta_2 lnC_i + \beta_3 lnW_i + \beta_4 D_i + \beta_5 H_i + \beta_6 A_i + (v_i - u_i) \tag{1}$$

where y_i is the value of agricultural production of the i-th farm, including the value of the output marketed, as well as the value of home consumption; L is the number of hectares assigned to annual crops by the farmer; C represents capital and is the sum of seeds, fertilizers, pesticides purchased and machinery contracted; W is the value of family and hired labor; D is a dichotomous variable that indicates if a farm is located in a dryland area and is thus expected to have lower production; H is the crop diversification index used to control for the intensity of agricultural activity and land use on the farm; A is the climate change adaptation measured as explained in Section 2.2; βs are the parameters to be estimated; and $v - u = \varepsilon$ is the composed error term.

The term v is a two-sided random error with a normal distribution ($v \sim N [0, \sigma_v^2]$) that captures the stochastic effect of factors beyond the farmer's control and statistical noise. The term u is a one-sided

($u \geq 0$) component that captures the TE of the producer; in other words, u measures the gap between observed production and its maximum value given by the frontier. This error can follow various statistical distributions including half-normal, exponential or gamma [33–35]. A high value of u implies a high degree of technical inefficiency; conversely, a value of zero implies that the farm is completely efficient. According to Battese and Coelli [32], the TE of the i-th farm is given by:

$$TE_i = exp(-u_i) \tag{2}$$

where u is the efficiency term specified in (1). TE for each farm is calculated using the conditional mean of $exp(-u)$, given the composed error term for the stochastic frontier model [36]. The maximum-likelihood method developed by Battese and Coelli [32] allows for a one-step estimation of u and v, and u can be expressed in terms of a set of explanatory variables Z_{nj} as:

$$u_j = \delta_0 + \sum_{n=1}^{k} \delta_n Z_{nj} + e_j \tag{3}$$

where δ_n are unknown parameters to be estimated.

The variables that affect technical inefficiency in our study (Table 2) are related to human capital (age, schooling, dependence, specialization and the use of meteorological information); social capital variables (membership in associations or organizations); and structural factors (distance to regional capital and farm size).

The adoption of climate change adaptation practices is a choice variable and, as in studies related to soil conservation adoption and credit access (e.g., [37–39]), might be correlated with the error term in Equation (1). Instrumental variables are commonly used to address endogeneity biases, and the Durbin–Wu–Hausman test (DWH) [40] is often the approach employed to statistically evaluate if this is indeed a problem. This test is based on the difference between the ordinary least square (OLS) and instrumental variables estimators [41]. The idea of the DWH test is to check whether the dissimilarity across these estimators is significantly different from zero given the data from the available sample. Under the null hypothesis that the error terms are uncorrelated with all the regressors against the alternative that they are correlated with at least some of the regressors, an F-test is performed [42]. The instrumental variables approach has been used in several recent studies of agricultural production analysis [43–47].

Therefore, to resolve the potential endogeneity of the variables A_1, A_2 and A_3, an instrumental variable approach was used to obtain their predicted values in a first-step regression, where A_1', A_2', and A_3' are the predicted values for A_1, A_2 and A_3, respectively. In the first step regression, the predicted values were generated as follows: A_1' was estimated using a logistic regression model; A_2' was assumed to have a zero-inflated negative binomial distribution; and for A_3', a truncated regression was applied. The models used to estimate the first step are shown in the Tables A2–A4, respectively.

To identify possible differences in TE across various technologies, we performed a Student's t-test comparing the mean of the expected TE for producers that did and did not adopt the following: (a) at least one irrigation improvement, (b) change in planting and harvesting schedule, and (c) at least two conservation practices. This simple procedure allows one to compare two independent groups by testing the null hypothesis of equal means.

3. Results and Discussion

3.1. Production Frontiers

Table 4 shows the estimations of the three SPF models. The parameter gamma is significant at the 1% level for the three models, with values of 0.42 for the Intensity model and 0.54 for the Decision and Quality models. In addition, the null hypothesis that sigma is equal to zero is rejected, confirming that the stochastic model is superior to the model that would result from using OLS. The presence of endogeneity is confirmed according to the DWH test implemented (as detailed in Table A1).

For the three models, the parameter for *L*, *C* and *W* are positive and statistically significant at the 1% level presenting also similar values across models. Capital (*C*) represents the most important production factor, with estimated coefficients around 0.60. Other studies reveal that capital is also important in the production function, with estimated parameters between 0.3 and 0.5 [39,48,49]. The size of the area under cultivation has an estimated parameter close between 0.23 and 0.29, consistent with those reported in other studies [50–52]. The lowest values are related to labor, *L*, around 0.11, consistent with the results from Rahman et al. [53] and Mariano et al. [52].

As expected, *D* is significant and negative, indicating that farms located in areas with lower quality soils and without irrigation are relatively less productive. Various agricultural production studies have shown that less-favored areas in terms of soil fertility or irrigation have lower productivity levels [52,54] and that this condition tends to be associated with high levels of inefficiency [48,49].

On the other hand, it is expected that crop diversification helps farmers to increase output, *ceteris paribus*, by allowing the continuous and more intensive use of the available soil and labor, and other resources. Crop diversification is one of the strategies used by farmers to minimize agricultural risk and to stabilize income [55]. Based on the *H* index, our results are consistent with expectations, revealing that higher diversification is positively associated with productivity. The Herfindahl index has been used in several studies to measure crop concentration or diversification [29,56]. Manjunatha et al. [57] incorporated this index in a production function for crops in India; Rahman [51] used it as a variable explaining crop efficiency in Bangladesh, demonstrating that crop diversification is associated with high levels of TE; and Kassali et al. [55] established a positive relation between crop diversification and efficiency among farmers in Nigeria.

The adoption of climate change adaptation technologies, for the three specifications (A_1, A_2, A_3) resulted in a positive and significant effect on productivity, evidencing the importance of adaptation in farming. As envisioned by Sauer et al. [58], over the next two decades, there will be pressing need for new agricultural responses in the face of population and economic growth, and these responses include increases in irrigated area and in water use intensity. Adaptation measures will need to play an increasingly important role to equilibrate food supply and demand in a global context [11,17].

The sum of the coefficients associated with *L*, *C* and *W* (partial elasticities of production) is close to one, an indication of nearly constant returns to scale for all models. This finding is consistent with those of Nyemeck et al. [48], Karagiannis and Sarris [50], Sauer and Park [59], and Reddy and Bantilan [49], but differs from that of Jaime and Salazar [60], who found increasing returns to scale in a sample of Chilean wheat farmers.

3.2. Technical Efficiency

Table 4 (bottom) shows that the average values of TE for the three models accounting for endogeneity are 67.8% (Decision), 76.4% (Intensity) and 72.3% (Quality). The mean TEs for models of decision are statistically the same. Table 5 shows that the range of TE for the 30% most efficient farms (the last three intervals) ranges from 53.9% to 74.1%. The average TE value is consistent with other studies done in Latin America using SPF models. Solís et al. [39] reported an average TE of 78%, and Bravo-Ureta et al. [14] reported a value of 70%. Table 5 also reveals high correlation coefficients between TE levels across the various models with values exceeding 0.95. In addition, Table 5 shows that the estimated TE values tend to be higher for models acknowledging endogeneity, indicating the relevance of considering this issue in the analysis.

Now we go back to Table 4 to examine the results concerning the Inefficiency Model. According to Gorton and Davidova [61], variables affecting farm efficiency can be divided into agency and structural factors. Agency factors, such as age, experience, education, specialization and training (i.e., human and social capital), represent the capacity of individuals to act independently and to make their own free choices. By contrast, structural factors, such as access to markets and credit, land tenure and farm size, influence or limit an agent in his or her decisions.

Table 4. Cobb–Douglas parameters for the stochastic production frontiers estimated considering endogeneity and three different specifications to measure climate change adaptation.

Variables	Climate Change Adaptation Measurement		
	Decision	Intensity	Quality
Constant (β_0)	4.1356 (0.9463) ***	4.7996 (0.9253) ***	4.7690 (0.9894) ***
Land (β_1)	0.2284 (0.0849) ***	0.2876 (0.0850) ***	0.2726 (0.0877) ***
Capital (β_2)	0.6184 (0.0739) ***	0.5950 (0.0710) ***	0.6041 (0.0779) ***
Labor (β_3)	0.1224 (0.0278) ***	0.1044 (0.0276) ***	0.1140 (0.0275) ***
Dryland (β_4)	−0.3485 (0.1303) ***	−0.4280 (0.1222) ***	−0.3882 (0.1350) ***
Diversification (β_5)	0.5670 (0.1312) ***	0.5933 (0.1373) ***	0.6074 (0.1361) ***
Climate change adaptation (β_6)	0.1092 (0.3012) ***	0.1656 (0.0546) ***	0.0075 (0.0052) *
Inefficiency Model			
Constant (δ_0)	0.2005 (0.6762)	0.3462 (0.6594)	−0.3082 (0.6554) ***
Age (δ_1)	0.0124 (0.0083) *	0.0171 (0.0080) **	0.0212 (0.0084) **
Schooling (δ_2)	0.0200 (0.0175)	0.0147 (0.0270)	0.0107 (0.0296)
Dependence (δ_3)	−0.7099 (0.1738) ***	−0.8436 (0.1797) ***	−0.7310 (0.1800) ***
Specialization (δ_4)	−0.0085 (0.0034) ***	−0.0099 (0.0034) ***	−0.0112 (0.0031) ***
Use of meteorological information (δ_5)	−0.6258 (0.2770) **	−0.8279 (0.2463) ***	−0.7480 (0.2556) ***
Membership (δ_6)	0.2027 (0.1698)	0.2533 (0.1742) *	0.1915 (0.1884)
Farm size (δ_7)	−0.0036 (0.0008) ***	−0.0028 (0.0029)	−0.0035 (0.0026) *
Distance to market (δ_8)	0.0085 (0.0033) ***	0.0038 (0.0031) *	0.0057 (0.0031) **
Returns to scale	0.9692	0.9870	0.9907
Maximum Likelihood Function	−209.18	−209.60	−212.76
Sigma²	0.4209 (0.0731) ***	0.4203 (0.0693) ***	0.4828 (0.0747) ***
Gamma	0.5363 (0.1043) ***	0.4247 (0.1111) ***	0.5411 (0.0989) ***
TE	67.8	76.4	72.3
TE difference with models without correcting endogeneity	ns	***	***

Climate change adaptation (A) is estimated through a logit regression (A_1') in the model for Decision, a zero-inflated negative binomial regression (A_2') in the model for Intensity and using a truncated regression (A_3') in the model for Quality (see the Appendix A). Numbers in parentheses are standard errors. * $p < 0.1$; ** $p < 0.05$; *** $p < 0.01$; ns: not significant. Estimations using Frontier Version 4.1 and STATA 11.1.

Most of the literature on TE uses human capital as the main source for explaining inefficiency [61]. Studies show that the relation of the age of farmers and TE levels varies according to geographic region and context. A negative and significant relation was described by Jaime and Salazar [60] for Chilean farmers; similar results were found by Mariano et al. [52] for rice producers in The Philippines and by Bozoğlu and Ceyhan [62] for vegetable farms in Turkey. Conversely, a positive relation is described by other authors [51,54,63]. In our study, the positive sign for age indicates that older farmers are less efficient.

It is expected that schooling has a negative effect on inefficiency levels, as noted by Jaime and Salazar [60], because education improves access to information, facilitates learning and the adoption of new processes and promotes forward-looking attitudes. Other studies support this conclusion [39,48,51,54,63–65]. However, in our study, schooling, measured by the number of years of formal instruction, has a negative, though not significant relationship with TE.

Our study found that the farmers who depend on agriculture as a primary source of income tend to be more efficient than those who do not. Similarly, Jaime and Salazar [60] report that the degree of dependence of Chilean wheat farmers on agriculture has a significant and positive relation with efficiency. Along this same line, Melo-Becerra and Orozco-Gallo [66] found that Colombian households that are dedicated exclusively to agricultural production are more efficient.

A similar relationship was found between specialization and TE; producers who specialize in crop production are more efficient than those who do not. Karagiannis et al. [67] showed that TE depends on specialization for both organic and conventional milk farms. Guesmi et al. [68], using the proportion of vineyard revenue to total agricultural revenue as a measure for specialization, also observed a positive relation between specialization and TE.

The use of meteorological information also shows a positive and significant relation with TE; farmers with access to meteorological information can be more alert about changes in weather and, in this way, minimize negative effects on productivity at the farm level. It is to be expected that access to information can have a positive effect on farm management and on the adoption of technologies related to farm productivity improvements. The use of meteorological information can represent a way to reduce uncertainty in productive operations. However, Lemos et al. [69] and Roco et al. [26] argue that the use of forecasts in decision-making is not straightforward and that much work is required to narrow the gap between producers and users of this kind of information.

Table 5. Distribution of TE and the correlation matrix for fitted models.

		Farms in Interval (%)					
Interval TE		Not-Correcting Endogeneity			Correcting Endogeneity		
		Decision	Intensity	Quality	Decision	Intensity	Quality
0–29		2.6	3.0	3.0	6.4	2.6	3.4
30–39		9.1	5.3	5.3	7.9	3.0	4.5
40–49		7.2	6.8	6.4	6.4	4.9	6.0
50–59		10.6	6.4	6.4	9.1	6.0	6.4
60–69		16.6	13.3	13.7	10.9	9.4	12.1
70–79		25.6	23.0	23.0	22.7	16.7	23.4
80–89		23.8	35.8	34.7	30.6	45.7	35.9
>90		4.5	6.4	7.5	6.0	11.7	8.3
Average TE		67.5	71.3	71.5	67.8	76.4	72.3
Correlation Matrix for TE Values							
Not-correcting for endogeneity	Decision	1	-	-	-	-	-
	Intensity	0.9872	1	-	-	-	-
	Quality	0.9876	0.9999	1	-	-	-
Correcting for endogeneity	Decision	0.9666	0.9779	0.9766	1	-	-
	Intensity	0.9532	0.9842	0.9841	0.9569	1	-
	Quality	0.9874	0.9967	0.9969	0.9741	0.9839	1

Social capital is another important factor to be considered in efficiency analyses. Membership in farmers' organizations can help to reduce inefficiency. Dios et al. [70] relate technical efficiency to innovation among farmers in Spain. Jaime and Salazar [60] note that in the Bío Bío Region in Chile, producers with higher levels of participation in organizations had higher levels of efficiency. Similar results were found by Nyemeck et al. [48] among producers in Cameroon. While in general, we found a positive relation between membership in organizations and TE levels, our results are not conclusive.

Intra- and inter-organizational arrangements are relevant for farm efficiency [61]. Our analysis, reveals a positive association between farm size and TE levels. There is evidence supporting the notion that large farms have higher levels of efficiency, due to advantages derived from economies of scale [49,53,54,60,63,66,71,72]. Considering the high percentage of small farms in the area under study, 28.6% according to ODEPA, which is the Chilean National Service for Agricultural Policy (the acronym stands for *Oficina de Estudios y Políticas Agrarias*) [72], this factor is likely a barrier to improve productivity levels in the region.

As expected, our results indicate that distance from the regional capital city has a negative and significant effect on TE levels. Proximity to markets, extension agencies and information coming from the regional capital tend to enhance farmers' TE. Tan et al. [54] claim that distance to a major city has a negative effect on TE levels for rice producers in China. Nyemeck et al. [48] highlight the importance of accessibility and find that TE is higher for farmers located near main roads.

In fact, Henderson et al. [73] found a strong and statistically-significant relationship between market participation and performance for crop-livestock smallholders in Sub-Saharan Africa.

3.3. Efficiency and Climate Change Adaptation

The analysis of efficiency in agriculture has been widely used to propose improvements in the management of farm systems. Areal et al. [74] argue that if the information received by policy makers concerning farm efficiency levels is harmonized with policy aims, policy measures may be targeted to support the targeted farms. This deserves further consideration given that the literature that links efficiency and climate change adaptation is limited.

Various *t*-tests were performed to relate efficiency levels and climate change adaptation (Table 6). We found a positive relation between TE and adopting at least one irrigation technology, i.e., farmers that adopt irrigation improvements exhibit a higher TE. In this regard, Yigezu et al. [75] argue that the use of modern irrigation methods yields an improvement of 19% in TE for wheat farmers in Syria.

However, a comparison across municipalities shows considerable geographical variability. In San Clemente, TE and the implementation of at least one irrigation alternative is evident regardless of the crops involved. For Pencahue, no differences are found between groups probably because most of the farmers in the sample (62.5%) have adopted at least one irrigation technology. In Cauquenes and Parral, we also find no significant difference and this is probably due to the low number of adopters. These results demonstrate the importance of climate change adaptation through the improvement of irrigation at the farm level to increase resource use efficiency. Kahil et al. [5] argue that water management policies, such as irrigation subsidies and efficient water markets, are key to face climate change in agriculture. Policy measures include supply enhancements to remove the threat of immediate water scarcity along with demand management measures and improved governance [76].

In general, changes in planting and harvesting dates show no relation with TE levels; however, in San Clemente, where the crops are highly diversified, farmers who have changed their planting calendars appear to have higher efficiency. Thus, it appears that this strategy that a priori could be expected to play a significant role for climate change adaptation, does not have a clear direct effect on efficiency. Additional information is required, to understand in a deeper way, the effects of a climate change practices portfolio on productivity and efficiency of agricultural systems.

The higher TE values detected for the groups who have more intensive adaptation strategies and with higher quality (number of practices and quality index) substantiate the importance of further research focusing on adaptation. It is not only necessary to adapt, but is also relevant to determine what and how much to adapt. Therefore, it is essential to foster effective adaptation and to improve the design of relevant programs to promote the adaptation capacity across farming systems. In Pencahue, 65% of the sample has adopted at least one adaptation practice, and 60% is above 25% in the adaptation index. In contrast, only 3.7% of the sample for Cauquenes has implemented at least one adaptation practice, and none of the farmers interviewed show an adaptation index over 25%. Based on this analysis, it seems clear that climate change adaptation in agriculture requires a complex set of actions including technical and managerial dimensions to reduce vulnerability and improve farmer productivity.

Table 6. *t*-tests for average TE levels grouped into various categories.

Average TE	Model	Adoption of at Least One Irrigation Improvement			Changes in Planting and Harvesting Schedules			Adoption of at Least Two Adaptation Practices			Value of Adaptation Index ≥ 25%		
		Yes	No	Sig	Yes	No	Sig	Yes	No	Sig	Yes	No	Sig
Complete sample	Decision	73.5	65.3	***	64.9	70.0	**	81.4	65.1	***	86.5	65.4	***
	Intensity	80.9	74.4	***	75.1	77.3	ns	86.3	74.5	***	88.8	74.8	***
	Quality	77.5	69.9	***	70.9	73.3	ns	84.4	69.9	***	87.3	70.3	***
	%	54.7			42.6			16.2			11.3		

Table 6. *Cont.*

Average TE	Model	Grouping Criteria											
		Adoption of at Least One Irrigation Improvement			Changes in Planting and Harvesting Schedules			Adoption of at Least Two Adaptation Practices			Value of Adaptation Index ≥ 25%		
		Yes	No	Sig	Yes	No	Sig	Yes	No	Sig	Yes	No	Sig
Pencahue	Decision	86.1	85.0	ns	85.7	85.7	ns	85.1	86.0	ns	86.0	85.2	ns
	Intensity	88.2	88.5	ns	87.8	88.7	ns	88.2	88.5	ns	88.1	88.5	ns
	Quality	86.7	86.7	ns	86.3	87.0	ns	86.8	86.6	ns	86.7	86.7	ns
	%	62.5			37.5			65.0			60.0		
Cauquenes	Decision	45.2	50.2	ns	47.2	50.9	ns	41.7	49.4	ns	-	-	
	Intensity	62.0	62.8	ns	62.0	63.3	ns	63.6	62.6	ns	-	-	
	Quality	55.7	58.2	ns	56.3	59.0	ns	59.6	57.6	ns	-	-	
	%	21.0			48.1			3.7			0.0		
San Clemente	Decision	82.2	77.0	***	83.3	77.1	***	81.6	78.2	ns	86.3	78.3	**
	Intensity	86.3	81.7	***	87.2	81.8	***	87.6	82.5	**	90.0	82.8	**
	Quality	83.2	77.4	***	84.4	77.6	***	85.0	78.4	**	87.8	78.9	**
	%	31.5			24.7			13.5			4.5		
Parral	Decision	66.5	64.1	ns	64.0	65.8	ns	79.5	64.0	*	93.1	63.5	***
	Intensity	80.0	76.3	ns	76.7	77.6	ns	89.0	76.6	*	94.5	76.4	**
	Quality	75.7	70.9	ns	71.2	71.6	ns	86.1	71.2	*	92.9	71.0	***
	%	20.0			67.3			3.6			3.6		

* $p < 0.1$; ** $p < 0.05$; *** $p < 0.01$, ns: not significant.

4. Concluding Remarks

This study analyzes the impact of climate change adaptation in productivity and efficiency for producers of annual crops in Central Chile. We used three measures of adaptation: a binary choice of adopting at least one adaptation practice or technology; an intensity measure given by the number of practices or technologies adopted; and a quality index measure. A positive association between productivity and climate change adaptation was observed for the three measures. The fitted stochastic production frontier models revealed that climate change adaptation is endogenous. Incorporation of instrumental variables allowed us to check the robustness of our results and improved the TE estimations. The fitted models showed important levels of inefficiency, suggesting the potential for increasing crop production using the current level of inputs and available technology.

Our results also show that factors such as dependence on annual crop production for income and high levels of specialization in production are associated with elevated TE levels. The use of meteorological information is also positively related with TE. In addition, our results indicate that farm size is positively related to efficiency while distance to a major city exhibits a negative relationship.

Farmers who have adopted irrigation technologies have higher TE levels. These results suggest that climate change adaptation is significant for agricultural production, especially for the intensity of climate change adaptation. Our results validate the importance, of incorporating climate change adaptation in agricultural policies designed to promote productivity growth. Our analysis also sheds light on the relevance of using meteorological information by farmers given the positive link between the latter variable and technical efficiency.

The connection between productivity with the implementation of specific farm-level adaptive practices, as well as with actions that ease adoption barriers deserves additional analyses. These analyses are essential to generate information required by policy makers to formulate robust action plans across differing cultural, economic and agricultural environments.

Acknowledgments: This work was supported by a research grant from The Latin American and Caribbean Environmental Economics Program (LACEEP). The authors thank the farmers who courteously answered our survey and the Excellence Program of Interdisciplinary Research: Adaptation of Agriculture to Climate Change (A2C2) of The University of Talca.

Author Contributions: All authors contributed extensively to the work presented in this paper. L.R., B.B-U. and A.E. designed the study. L.R. and R.J-R. implemented the fieldwork. L.R. wrote the paper. L.R. and R.J-R. conducted the analysis. B.B-U. and A.E. contributed extensively to the revision of several drafts of the manuscript.

Conflicts of Interest: The authors declare no conflict of interest.

Appendix A

Table A1. Cobb–Douglas parameters for stochastic production frontiers estimated considering three different specifications for the measurement of climate change adoption and without considering endogeneity.

Variables	Climate Change Adaptation Measurement		
	Decision	Intensity	Quality
Constant (β_0)	4.0218 (0.9857) ***	4.6682 (0.9741) ***	4.6090 (0.9891) ***
Land (β_1)	0.2314 (0.0887) ***	0.2654 (0.0869) ***	0.2602 (0.0857) ***
Capital (β_2)	0.6828 (0.0764) ***	0.6206 (0.0754) ***	0.6255 (0.0758) ***
Labor (β_3)	0.1043 (0.0283) ***	0.1112 (0.0283) ***	0.1110 (0.0270) ***
Dryland (β_4)	−0.4204 (0.1334) ***	−0.3578 (0.1270) ***	−0.3614 (0.1314) ***
Diversification (β_5)	0.5990 (0.1381) ***	0.5957 (0.1357) ***	0.6054 (0.1349) ***
Climate change adaptation (β_6)	0.0331 (0.0735)	0.0035 (0.0017) ***	0.0046 (0.0024) **
Inefficiency Model			
Constant (δ_0)	0.2035 (0.6194)	0.2166 (0.5937)	0.1591 (0.7713)
Age (δ_1)	0.0189 (0.0072) ***	0.0177 (0.0075) ***	0.0185 (0.0096) **
Schooling (δ_2)	0.0097 (0.0250)	0.0129 (0.0259)	0.0130 (0.0262)
Dependence (δ_3)	−0.4878 (0.1697) ***	−0.7657 (0.1776) ***	−0.7480 (0.2117) ***
Specialization (δ_4)	−0.0099 (0.0031) ***	−0.0098 (0.0034) ***	−0.0099 (0.0032) ***
Use of meteorological information (δ_5)	−0.7010 (0.2326) ***	−0.7406 (0.2423) ***	−0.7420 (0.2981) ***
Membership (δ_6)	0.0877 (0.1701)	0.2591 (0.1773) *	0.2663 (0.1970) *
Farm size (δ_7)	−0.0040 (0.0026) *	−0.0035 (0.0010) ***	−0.0034 (0.0009) ***
Distance to market (δ_8)	0.0056 (0.0029) **	0.0053 (0.0029) **	0.0051 (0.0030) **
Returns to scale	1.0185	0.9972	0.9967
MLF	−218.13	−211.14	−211.51
Sigma^2	0.4588 (0.0611) ***	0.4516 (0.0652) ***	0.4493 (0.0725) ***
Gamma	0.5632 (0.0996) ***	0.5222 (0.1044) ***	0.5178 (0.1144) ***
TE	67.52	71.34	71.50
Endogeneity (F value)	4.868 ***	14.266 ***	13.012 ***

Climate change adaptation (A) is measured as: the adoption of at least one practice (A_1) in the model for decision; the number of practices adopted (A_2) in the model for intensity; the number of practices weighted according to experts' opinion (A_3) in the model for quality. Numbers in parentheses are standard errors. * $p < 0.1$; ** $p < 0.05$; *** $p < 0.01$. Estimations using Frontier Version 4.1 and STATA 11.1.

Table A2. Logit regression estimation.

Variable Name	Description	Coefficient
A_1	**Dependent Variable**	
ExpAgIndep	Years of independent experience in agriculture.	−0.0153 * (0.0087)
SanClemente	Dummy variable = 1 if the farm is located in San Clemente and 0 otherwise	−0.9189 *** (0.2886)
TTPropia	Dummy variable = 1 if the farmer is owner and 0 otherwise	0.3590 (0.2821)
Internet	Dummy variable = 1 if the farmer has access to meteorological information principally form the Internet and 0 otherwise	0.9667 *** (0.3290)
Constant		0.5849 ** (0.3012)
	Log-likelihood	−170.73
	N	265
	Pseudo R^2	5.86
	Correctly classified values by Logit (%)	62.2

Numbers in parenthesis are standard errors. * $p < 0.1$; ** $p < 0.05$; *** $p < 0.01$.

Table A3. Zero inflated negative binomial regression estimation.

Variable Name	Description	Coefficient
A_2	**Dependent Variable**	
ExpAgIndep	Years of independent experience in agriculture.	−0.0121 *** (0.0034)
RXP	Dummy variable = 1 if the farmer has adopted any irrigation improvement and the location is in Pencahue municipality and 0 otherwise	0.7731 *** (0.1324)

Table A3. *Cont.*

Variable Name	Description	Coefficient
A_2	Dependent Variable	
SupProd	Surface designated to production in hectares	0.0003 (0.0003)
Internet	Dummy variable = 1 if the farmer has access to meteorological information principally form the Internet and 0 otherwise	0.2233 * (0.1329)
Constant		1.0172 *** (0.1362)
	Log-likelihood	−411.76
	N	265
	Correlation of predicted values (A_1') with A_1 (%)	53.51

Numbers in parenthesis are standard errors. * $p < 0.1$; *** $p < 0.01$.

Table A4. Truncated linear regression estimation.

Variable Name	Description	Coefficient
A_3	Dependent Variable	
ExpAgIndep	Years of independent experience in agriculture.	−0.2518 *** (0.0893)
RXP	Dummy variable = 1 if the farmer has adopted any irrigation improvement and the farm location is Pencahue and 0 otherwise	18.445 *** (4.2773)
SupProd	Surface designated to production in hectares	0.0173 * (0.0105)
Internet	Dummy variable = 1 if the farmer has access to meteorological information principally form the Internet and 0 otherwise	3.4477 (3.1870)
Constant		20.1456 *** (2.8415)
	Log-Likelihood	−574.27
	N	265
	Correlation of predicted values (A_2') with A_2 (%)	51.35

Regression was truncated in values with 0 as the lower limit and 100 as the upper limit. Numbers in parenthesis are standard errors. * $p < 0.1$; *** $p < 0.01$.

References

1. Oelesen, J.; Bindi, M. Consequences of climate change for European agricultural productivity, land use and policy. *Eur. J. Agron.* **2002**, *16*, 239–262. [CrossRef]
2. IPCC. *Managing the Risk of Extreme Events and Disasters to Advance Climate Change Adaptation*; Special Report; Cambridge University Press: Cambridge, UK, 2012; p. 582.
3. Lobell, D.B.; Field, C.B. Global scale climate-crop yield relations and the impacts of recent warming. *Environ. Res. Lett.* **2007**, *2*, 1–7. [CrossRef]
4. Pathak, H.; Wassmannn, R. Quantitative evaluation of climatic variability and risk for wheat yield in India. *Clim. Chang.* **2009**, *93*, 157–175. [CrossRef]
5. Kahil, M.T.; Connor, J.D.; Albiac, J. Efficient water management policies for irrigation adaptation to climate change in Southern Europe. *Ecol. Econ.* **2015**, *120*, 226–233. [CrossRef]
6. Jackson, T.M.; Hanjra, M.; Khan, S.; Hafeez, M.M. Building a climate resilient farm: A risk based approach for understanding water, energy and emissions in irrigated agriculture. *Agric. Syst.* **2012**, *104*, 729–745. [CrossRef]
7. Fuglie, K.; Schimmelpfennig, D. Introduction to the special issue on agricultural productivity growth: A closer look at large, developing countries. *J. Product. Anal.* **2010**, *33*, 169–172. [CrossRef]
8. Clements, R.; Haggar, J.; Quezada, A.; Torres, J. *Technologies for Climate Change Adaptation—Agriculture Sector*; Zhu, X., Ed.; UNEP Risø Centre: Roskilde, Denmark, 2011.
9. Zilberman, D.; Zhao, J.; Heirman, A. Adoption versus adaptation, with emphasis on climate change. *Annu. Rev. Resour. Econ.* **2012**, *4*, 27–53. [CrossRef]
10. IPCC. *Climate Change 2007: Impacts, Adaptation and Vulnerability*; Intergovernmental Panel on Climate Change, Fourth Assessment Report; Cambridge University Press: Cambridge, UK, 2007.
11. Nelson, D.; Adger, W.N.; Brown, K. Adaptation to environmental change: contributions of a resilience framework. *Annu. Rev. Resour. Econ.* **2007**, *32*, 395–419. [CrossRef]

12. AGRIMED. *Impactos Productivos en el Sector Silvoagropecuario de Chile Frente a Escenarios de Cambio Climático*; U. de Chile, CONAMA, ODEPA, FIA Report; Universidad de Chile: Santiago, Chile, 2008.

13. Paul, C. Productivity and efficiency measurement in our "New Economy": Determinants, interactions, and policy relevance. *J. Product. Anal.* **2003**, *19*, 161–177. [CrossRef]

14. Bravo-Ureta, B.; Solís, D.; Moreira, V.; Maripani, J.; Thiam, A.; Rivas, T. Technical efficiency in farming: A meta-regression analysis. *J. Product. Anal.* **2007**, *27*, 57–72. [CrossRef]

15. Coelli, T.; Rao, D.S.; O'Donnell, C.; Battesse, G. *An Introduction to Efficiency and Productivity Analysis*, 2nd ed.; Springer: New York, NY, USA, 2005; p. 350.

16. Mukherjee, D.; Bravo-Ureta, B.; de Vries, A. Dairy productivity and climatic conditions: Econometric evidence from South-eastern United States. *Aust. J. Agric. Resour. Econ.* **2013**, *57*, 123–140. [CrossRef]

17. USDA. *Climate Change and Agriculture in the United States: Effects and Adaptation*; United States Department of Agriculture Technical Bulletin 1935; USDA: Washington, DC, USA, 2013; p. 186.

18. Roco, L.; Poblete, D.; Meza, F.; Kerrigan, G. Farmers' options to address water scarcity in a changing climate: Case studies from two basins in Mediterranean Chile. *Environ. Manag.* **2016**, *109*, 958–971. [CrossRef] [PubMed]

19. Deressa, T.T.; Hassan, R.M. Economic impact of climate change on crop production in Ethiopia: Evidence from cross-section measures. *J. Afr. Econ.* **2009**, *18*, 529–554. [CrossRef]

20. Gbetibouo, G.A. *Understanding Farmers' Perceptions and Adaptations to Climate Change and Variability: The Case of the Limpopo Basin, South Africa*; IFPRI Discussion Paper No. 849; International Food Policy Research Institute: Washington, DC, USA, 2009; p. 36.

21. Di Falco, S.; Veronesi, M.; Yesuf, M. Does adaptation to climate change provide food security? A micro-perspective from Ethiopia. *Am. J. Agric. Econ.* **2011**, *93*, 829–846. [CrossRef]

22. Sofoluwe, N.; Tijane, A.; Baruwa, O. Farmers' perception and adaptation to climate change in Osun State, Nigeria. *Afr. J. Agric. Resour. Econ.* **2011**, *6*, 4789–4794.

23. Tambo, J.A.; Abdoulaye, T. Climate change and agricultural technology adoption: The case of drought tolerant maize in rural Nigeria. *Mitig. Adapt. Strateg. Glob. Chang.* **2012**, *17*, 277–292. [CrossRef]

24. Roco, L.A.; Engler, B.; Bravo-Ureta, B.E.; Jara-Rojas, R. Farm level adaptation decisions to face climatic change and variability: Evidence from Central Chile. *Environ. Sci. Policy* **2014**, *44*, 86–96. [CrossRef]

25. Chilean Ministry of Agriculture. Plan de Adaptación al Cambio Climático del Sector Silvoagropecuario. 2012. Available online: http://www.mma.gob.cl/1304/articles-52367_PlanAdaptacionCCS.pdf (accessed on 16 September 2017).

26. Roco, L.A.; Engler, B.; Bravo-Ureta, B.E.; Jara-Rojas, R. Farmers' perception of climate change in Mediterranean Chile. *Reg. Environ. Chang.* **2015**, *15*, 867–879. [CrossRef]

27. FIA. *El Cambio Climático en el Sector Silvoagropecuario de Chile*; Fundación para la Innovación Agraria, Ministerio de Agricultura de Chile: Santiago, Chile, 2010; p. 16.

28. Hannah, L.; Ikegami, M.; Hole, D.G.; Seo, C.; Butchart, S.H.; Peterson, A.T.; Roehrdanz, P.R. Global climate change adaptation priorities for biodiversity and food security. *PLoS ONE* **2013**, *8*, e72590. [CrossRef] [PubMed]

29. Malik, D.P.; Singh, I.J. Crop diversification—An economic analysis. *Indian J. Agric. Res.* **2002**, *36*, 61–64.

30. Moniruzzaman, S. Crop choice as climate change adaptation: Evidence from Bangladesh. *Ecol. Econ.* **2015**, *118*, 90–98. [CrossRef]

31. Bryan, E.; Deressa, T.T.; Gbetibouo, G.A.; Ringler, C. Adaptation to climate change in Ethiopia and South Africa: Options and constraints. *Environ. Sci. Policy* **2009**, *12*, 413–426. [CrossRef]

32. Battese, G.E.; Coelli, T.J. A model for technical inefficiency effects in stochastic frontier production function for panel data. *Empir. Econ.* **1995**, *20*, 325–332. [CrossRef]

33. Aigner, D.J.; Lovell, C.A.K.; Schmidt, P. Formulation and estimation of stochastic frontier production function models. *J. Econom.* **1977**, *6*, 21–37. [CrossRef]

34. Greene, W.H. Maximum likelihood estimation of econometric frontier functions. *J. Econom.* **1980**, *13*, 27–56. [CrossRef]

35. Meeusen, W.; van den Broeck, J. Efficiency estimation from Cobb–Douglas production function with composed error. *Int. Econ. Rev.* **1977**, *18*, 435–444. [CrossRef]

36. Battese, G.E.; Coelli, T.J. Prediction of firm-level technical efficiencies with a generalized frontier production function and panel data. *J. Econom.* **1988**, *38*, 387–399. [CrossRef]

37. Jones, S. A framework for understanding on-farm environmental degradation and constraint to the adoption of soil conservation measures: Case studies from Highland Tanzania and Thailand. *World Dev.* **2002**, *30*, 1607–1620. [CrossRef]

38. Chavas, J.P.; Ragan, P.; Roth, M. Farm household production efficiency: Evidence from The Gambia. *Am. J. Agric. Econ.* **2005**, *87*, 160–179. [CrossRef]

39. Solís, D.; Bravo-Ureta, B.; Quiroga, R. Technical efficiency among peasant farmers participating in natural resource management programs in Central America. *J. Agric. Econ.* **2009**, *60*, 202–219. [CrossRef]

40. Davidson, R.; MacKinnon, J.G. *Estimation and Inference in Econometrics*; Oxford University Press: New York, NY, USA, 1993.

41. Cameron, C.; Trvedi, P. *Microeconometrics Using Stata*, Revised ed.; Stata Press: College Station, TX, USA, 2010; p. 706.

42. Davidson, R.; MacKinnon, J.G. *Econometric Theory and Methods*; Oxford University Press: New York, NY, USA, 2004.

43. Di Falco, S.; Yesuf, M.; Kohlin, G.; Ringler, C. Estimating the impact of climate change on agriculture low-income countries: Household level evidence from the Nile Basin, Ethiopia. *Environ. Resour. Econ.* **2012**, *52*, 457–478. [CrossRef]

44. Weber, J.; Key, N. How much do decoupled payments affect production? An instrumental variable approach with panel data. *Am. J. Agric. Econ.* **2012**, *94*, 52–66. [CrossRef]

45. Stifel, D.; Fafchamps, M.; Minten, B. Taboos, agriculture and poverty. *J. Dev. Stud.* **2011**, *47*, 1455–1481. [CrossRef]

46. Mishra, A.K.; El-Osta, H.S.; Shaik, S. Succession decisions in US family farm business. *J. Agric. Resour. Econ.* **2010**, *35*, 133–152.

47. Kilic, T.; Carletto, C.; Miluka, J.; Savanasto, S. Rural nonfarm income and its impact on agriculture: Evidence from Albania. *Agric. Econ.* **2009**, *40*, 139–160. [CrossRef]

48. Nyemeck, J.B.; Tonyé, J.; Wandi, N.; Nyambi, G.; Akoa, M. Factors affecting the technical efficiency among smallholder farmers in the slash and burn agriculture zone of Cameroon. *Food Policy* **2004**, *29*, 531–545.

49. Reddy, A.A.; Bantilan, M.C. Competitiveness and technical efficiency: Determinants in the groundnut sector of India. *Food Policy* **2012**, *37*, 255–263. [CrossRef]

50. Karagiannis, G.; Sarris, A. Measuring and explaining scale efficiency with the parametric approach: The case of Greek tobacco growers. *Agric. Econ.* **2005**, *33*, 441–451. [CrossRef]

51. Rahman, S. Women's labour contribution to productivity and efficiency in agriculture: Empirical evidence fron Bangladesh. *J. Agric. Econ.* **2010**, *61*, 318–342. [CrossRef]

52. Mariano, M.J.; Villano, R.; Fleming, E. Technical efficiency of rice farms in different agroclimatic zones in the Philipines: An application of a stochastic metafrontier model. *Asian Econ. J.* **2011**, *25*, 245–269. [CrossRef]

53. Rahman, S.; Wiboopongse, A.; Sriboonchitta, S.; Chaovanapoonphol, Y. Production efficiency of jasmine rice producers in Northern and North-eastern Thailand. *J. Agric. Econ.* **2009**, *60*, 419–435. [CrossRef]

54. Tan, S.; Heerink, N.; Kuyvenhoven, A.; Qu, F. Impact of land fragmentation on rice producers' technical efficiency in South-East China. *NJAS Wagening. J. Life Sci.* **2010**, *57*, 117–123. [CrossRef]

55. Kassali, R.; Ayanwale, A.B.; Idowu, E.O.; Williams, S.B. Effect of rural transportation systems on agricultural productivity in Oyo State, Nigeria. *J. Agric. Rural Dev. Trop. Subtrop.* **2012**, *113*, 13–19.

56. Sarris, A.; Savastano, S.; Christiaensen, L. *The Role of Agriculture in Reducing Poverty in Tanzania: A Household Perspective from Rural Kilimanjaro and Ruvuma*; FAO Commodity and Trade Policy Research Working Paper No. 19; FAO: Rome, Italy, 2006; p. 30.

57. Manjunatha, A.V.; Anikc, A.R.; Speelmand, S.; Nuppenaua, E.A. Impact of land fragmentation, farm size, land ownership and crop diversity on profit and efficiency of irrigated farms in India. *Land Use Policy* **2013**, *31*, 397–405. [CrossRef]

58. Sauer, T.; Havlik, P.; Scheneider, U.A.; Schmidt, E.; Kindermann, G.; Obersteiner, M. Agriculture and resource availability in a changing world: The role of irrigation. *Water Resour. Res.* **2010**, *46*, W06503. [CrossRef]

59. Sauer, J.; Park, T. Organic farming in Scandinavia—Productivity and market exit. *Ecol. Econ.* **2009**, *68*, 2243–2254. [CrossRef]

60. Jaime, M.; Salazar, C. Participation in organizations, technical efficiency and territorial differences: A study of small wheat farmers in Chile. *Chil. J. Agric. Res.* **2011**, *71*, 104–113. [CrossRef]

61. Gorton, M.; Davidova, S. Farm productivity and efficiency in the CEE applicant countries: A synthesis of results. *Agric. Econ.* **2004**, *30*, 1–16. [CrossRef]
62. Bozoğlu, M.; Ceyhan, V. Measuring the technical efficiency and exploring the inefficiency determinant of vegetable farms in Samsun province, Turkey. *Agric. Syst.* **2011**, *94*, 649–656. [CrossRef]
63. Külekçi, M. Technical efficiency analysis for oilseed sunflower farms: A case study in Erzurum, Turkey. *J. Sci. Food Agric.* **2010**, *90*, 1508–1512. [CrossRef] [PubMed]
64. Phillips, J. Farmer education and farmer efficiency: A meta-analysis. *Econ. Dev. Cult. Chang.* **1994**, *43*, 149–165. [CrossRef]
65. Phillips, J.; Marble, R. Farmer education and efficiency: A frontier production approach. *Econ. Educ. Rev.* **1996**, *5*, 257–264. [CrossRef]
66. Melo-Becerra, L.A.; Orozco-Gallo, A.J. Technical efficiency for Colombian small crop and livestock farmers: A stochastic metafrontier approach for different production systems. *J. Product. Anal.* **2017**, *47*, 1–16. [CrossRef]
67. Karagiannis, G.; Salhofer, K.; Sinabell, F. *Technical Efficiency of Conventional and Organic Farms: Some Evidence for Milk Production*; OGA Tagungsband: Wien, Austria, 2006.
68. Guesmi, B.; Serra, T.; Kallas, Z.; Roig, M.G. The productive efficiency of organic farming: The case of grape sector in Catalonia. *Span. J. Agric. Res.* **2012**, *10*, 552–566. [CrossRef]
69. Lemos, M.C.; Kirchhoff, C.J.; Ramprasad, V. Narrowing the climate information usability gap. *Nat. Clim. Chang.* **2012**, *2*, 789–794. [CrossRef]
70. Dios, R.; Martínez, J.M.; Vicario, V. Eficiencia versus innovación en explotaciones agrarias. *Estud. Econ. Apl.* **2003**, *21*, 485–501.
71. Yang, Z.; Mugera, A.M.; Zhang, F. Investigating yield variability and inefficiency in rice production: A case study in Central China. *Sustainability* **2016**, *8*, 787. [CrossRef]
72. ODEPA. Caracterización de la Pequeña Agricultura en Chile. 2011. Available online: http://www.odepa.gob.cl/odepaweb/servicios-informacion/publica/Pequena_agricultura_en_Chile.pdf (accessed on 16 September 2017).
73. Henderson, B.; Godde, C.; Medina-Hidalgo, D.; van Wijkb, M.; Silvestri, S.; Douxchamps, S.; Stephenson, E.; Power, B.; Rigolot, C.; Cacho, O.; et al. Closing system-wide yield gaps to increase food production and mitigate GHGs among mixed crop-livestock smallholders in Sub-Saharan Africa. *Agric. Syst.* **2016**, *143*, 106–113. [CrossRef] [PubMed]
74. Areal, F.J.; Tiffin, R.; Balcombe, K.G. Provision of environmental output within a multi-output distance function approach. *Ecol. Econ.* **2012**, *78*, 47–54. [CrossRef]
75. Yigezu, Y.; Ahmed, M.; Shideed, K.; Aw-Hassan, A.; El-Shater, T.; Al-Atwan, S. Implications of a shift in irrigation technology on resource use efficiency: A Syrian case. *Agric. Syst.* **2013**, *118*, 14–22. [CrossRef]
76. Vargherse, S.K.; Veettil, P.C.; Speelman, S.; Buysse, J.; van Huylenbroeck, G. Estimating the causal effect of water scarcity on the groundwater use efficiency of rice farming in South India. *Ecol. Econ.* **2013**, *86*, 55–64. [CrossRef]

sustainability

MDPI

Article

Participatory Sustainability Assessment for Sugarcane Expansion in Goiás, Brazil

Heitor Luís Costa Coutinho [1,†], Ana Paula Dias Turetta [1,*], Joyce Maria Guimarães Monteiro [1], Selma Simões de Castro [2] and José Paulo Pietrafesa [3]

[1] Embrapa Solos, Rua Jardim Botânico 1024, Rio de Janeiro-RJ 22460-000, Brazil; heitor.coutinho@embrapa.br (H.L.C.C.); joyce.monteiro@embrapa.br(J.M.G.M.)
[2] Laboratory of Geomorphology, Pedology and Physical Geography—LABOGEF, IESA, UFG, Campus Samambaia, Goiânia, GO 74.690-900, Brazil; selma@ufg.br
[3] Education Faculty of UFG—Program of Pos Graduation in Education (PPGE). Rua 235, s/n-Setor Universitário, Goiânia, GO 74.605-050, Brazil; jppietrafesa@gmail.com
* Correspondence: ana.turetta@embrapa.br; Tel.: +55-21-2179-4579
† Deceased.

Received: 26 June 2017; Accepted: 28 August 2017; Published: 5 September 2017

Abstract: The sugarcane expansion in Brazil from 1990 to 2015 increased crop area by 135.1%, which represents more than 10 million hectares. Brazilian ethanol production hit a record high in 2015, reaching 30 billion liters, up 6% compared to 2014. In 2009, the Sugarcane Agroecology— ZAE-CANA—was launched to be a guideline to sustainable sugarcane production in Brazil. However, although it aims at sustainable production, it only considered natural aspects of the country, such as soil and climate. It is still necessary to develop instruments for studies on sustainability in all pillars. The aim of this study is to present the results regarding the application of the FoPIA (Framework for Participatory Impact Assessment) methodology in the Southwestern Goiás Planning Region (SGPR). FoPIA is a participatory methodology designed to assess the impacts of land use policies in regional sustainability, and the results showed the capacity of FoPIA to assess the impacts of land use change of the sugarcane expansion in that area. The major advantage of FoPIA is its participatory method feature, as it is possible to join stakeholders to debate and define sustainability guidelines.

Keywords: sugar cane expansion; sustainability assessment; FoPIA methodology

1. Introduction

Bio-ethanol from Brazil is an attractive type of biofuel because of its low price and relatively large greenhouse gas emissions reduction potential [1,2].

In the late 1970s, Brazil's National Bioethanol Program (PROALCOOL) ordered the mixture of anhydrous bioethanol (BE) in gasoline (blends up to 25%) and encouraged automakers to produce engines running on pure hydrated ethanol (100%) [3]. The Brazilian adoption of mandatory regulations to determine the amount of BE to be mixed with gasoline was essential to the success of the program [4]. The goal was to reduce oil imports which consumed one half of the total hard currency from exports. Although it was a decision made by the federal government during a military regime, it was well accepted by civil society, the agricultural sector, and car manufacturers [5].

Taking advantage of all the learning and experience of that period, the Brazilian government undertook some responsibilities against the international scenario related to climate change. In Brazil's Nationally Determined Contribution (NDC), submitted during COP21 in December 2015 and ratified in September 2016, the country agreed to reduce greenhouse gas (GHG) emissions by 37% by 2025 and 43% by 2030, with the 2005 emissions as a reference. To do so, the government agreed to increase biofuel (biodiesel and ethanol) participation by 18% in the energy matrix by 2030 [6]. Other commitments were related to actions to reduce GHG emissions by some 37% by 2020 [7]. The sugarcane expansion in Brazil increased 135.1% in crop area from 1990 to 2015, which represents more than 10 million hectares [8].

In 2009, with Decree 6961 [9], the Sugarcane Agroecology Zoning—ZAE-CANA [10] was created to guide the sustainable sugarcane expansion in Brazil. ZAE-CANA's main goals are to provide technical subsidies to policy makers to direct sugarcane expansion into legally recommended areas and sustainable production in Brazil. To achieve these goals, the study followed the guidelines that will allow the expansion of production: indication of areas with agricultural potential for sugarcane harvesting without environmental restrictions; exclusion of areas with original vegetation and indication of areas currently under anthropic use; exclusion of areas for cultivation in the Amazon, Pantanal biomes and the Upper Paraguay basin; reduction of direct competition with food production areas; indication of areas with agricultural potential (soil and climate) for the cultivation it means with slopes below 12%, by mechanical harvesting.

However, although it aims at sustainable production, the zoning only considered natural aspects of the country, such as soil, climate and relief. Also, the governance continues through contracts, to guarantee the productive supply and effectiveness of the productive chain, thus enhancing uncertainty regarding the sustainability of the Brazilian biodiesel production program [11].

Since the 1990s, environmental studies identified a wide range of reflections on sustainability and agricultural production systems. These reflections converged to the idea that economic growth, environmental preservation and social equity should be considered together to achieve a satisfactory development level [12–15].

Also at that time, the concept of sustainable development (SD), also in the agro-energy sector, was widespread, despite shortcomings in making SD operational. Therefore, policymakers are increasingly demanding comprehensive and reliable analyses of policy impacts on the economic, social and environmental dimensions of SD [16,17].

The use of criteria on sustainability allowed the assessment of the impacts caused by development processes, both in urban and rural areas. This process has contributed to regional assessments such as the implementation of public policies aimed at developing measures to mitigate social and environmental liabilities [18,19]. For instance, the European Union and some of its countries have specific directives to access the sustainability of biofuels such as the Renewable Energy Directive 2009/28/EC [20] and the Fuel Quality Directive 2009/30/EC [21], which established sustainability criteria to meet EU targets and to be eligible for financial support.

The development of instruments for studies on sustainability criteria and indicators, as well as those on the impacts on land use, is quite recent and shows gaps that are still under analysis. However, they are important instruments to understand changes that take place in social, environmental or economic phenomena. They can drive a particular need or even resources indicating trends that are undetectable in the processes. Therefore, it is necessary to develop instruments to help land managers to assess the social, economic and environmental impacts caused by land use-related public and corporate policies. These instruments may be quantitative, based on indicators' response models and functions built according to scientific knowledge and census databases; as well, they may be qualitative instruments based on technical knowledge integrated to that of the local stakeholders.

Particularly, the FoPIA (Framework for Participatory Impact Assessment) methodology has been useful to prepare for the participatory assessment of significant changes in land use and in the possibility of sustainability. The FoPIA is designed to enable assessments of policy impacts that are sensitive to national, regional and local sustainability priorities by harnessing the knowledge and expertise of national, regional and local stakeholders who play a central role in the analytical process. The analysis of specific sustainability problems gives rise to realistic national and regional policy and land use change scenarios [17].

The FoPIA was originally developed for application in the European Union to conduct stakeholder-based impact assessments of alternative land use policies, for example, to assess the policy options for biodiversity conservation in Malta [17]. This approach has been adapted for the assessment of land use policies in developing countries, with experiences in Indonesia, Tunisia and China [22–24].

Considering the assessment of impacts on the sustainability of sugarcane expansion policies in the southwestern region of Goiás state, Brazil, a participatory consultation was held to promote a structured interdisciplinary discussion about the sugarcane expansion in the region, to select public policy instruments for the construction of sugarcane expansion scenarios, as well as to define land use functions and indicators to be used in FoPIA. Hence, the aim of this study is to present the results regarding the application of this methodology.

2. Material and Methods

2.1. Study Area

The Southwestern Goiás Planning Region (SGPR) was chosen as the focus area because of its prominent expansion of sugar cane. In 2012, the sugarcane planted area in the SGPR was 286,512 ha and, in 2015, it was nearly twice that area (412,466 ha) [8]. The region spreads across an area of 61,498,463 km^2 and 26 municipalities (Figure 1).

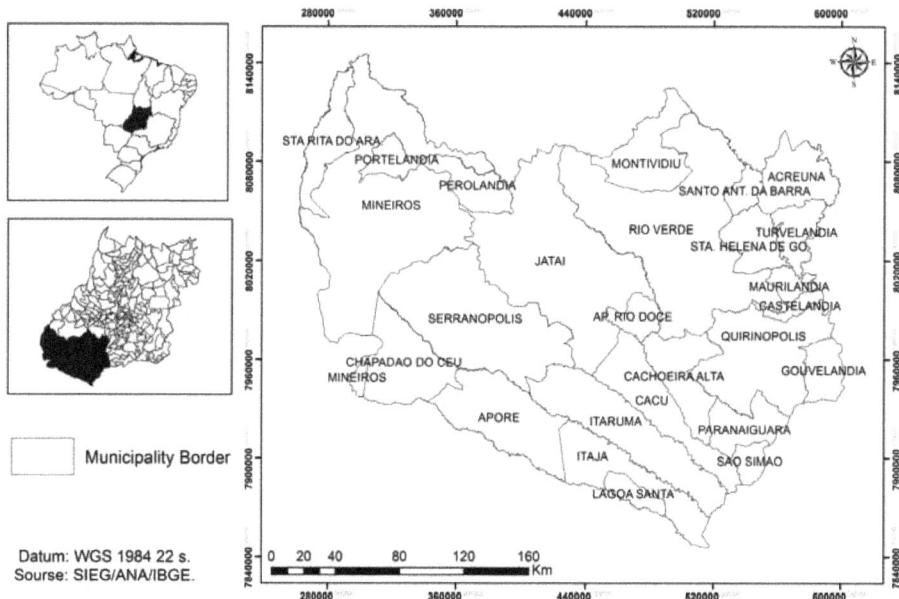

Figure 1. The Southwestern Goiás Planning Region and its municipalities. Source: SIEG—GO. Prepared by Trindade [25].

The sugarcane expansion in the large southern Goiás mesoregion is considered recent, as it started in 2004, mainly due to the advance of sugarcane agribusiness. The expansion of agribusiness in Goiás is characterized by high competition for land, favoring the leasing of large plots to harvest sugarcane for the sugar industry. This shows land concentration, mainly in parts of Southwestern Goiás involved in soybean and sugar cane production, and the exclusion of crops like rice, beans and, more recently, corn [26].

The SGPR has regional economic importance in the state, as well as consolidated logistics. However, it needs to be recovered and expanded to support the sugarcane expansion. At first

(2004–2008), much of the sugarcane expansion replaced soybean plantations, which, in turn, have shifted to pasture areas.

2.2. Methodology—FoPIA Background

The FoPIA methodology was developed as part of the EU-SENSOR project ("Sustainability Impact Assessment: Tools for Environmental, Social and Economic Effects of Multifunctional Land Use in European Regions") and applied to assess the impact of changing land use on different socioenvironmental situations of the European continent [17]. Subsequently, the FoPIA was adapted and used in the participatory impact assessment of different decision-making contexts and of environmental problems associated with land use and management in different Asian and African countries [22–24].

The FoPIA is a participatory methodology designed to assess the impacts of land use policies on regional sustainability. Its conceptual model, in which instrument users and public policy makers indicate policy scenarios to be assessed, considers variables such as driving forces, pressure, states, impacts, and responses (DPSIR). Each scenario generates different economic, fiscal or legislative conditions that, in turn, become driving forces of changes in land use. The pressures are changes in land use and management resulting from the implementation of policies. The pressures act on the states, as well as on the social, economic and environmental features of the regions subjected to changes, and they are represented by indicators. The impacts on sustainability are assessed through changes in the values of indicators as a response to the pressures, and through its relation with the sustainability limits or goals set to the region. Then, the decisions (responses) concerning the mitigation of or the adaptation to the impacts are then taken by the instrument users. However, due to limitations during the project, it was not possible to apply the entire methodology (Figure 2).

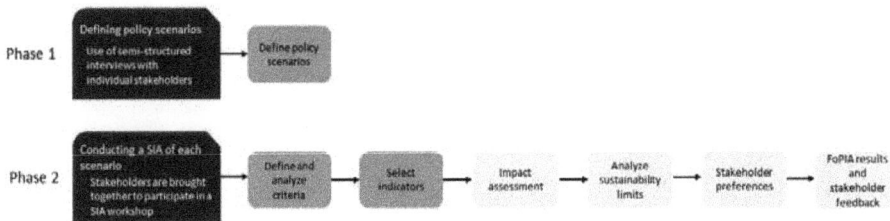

Figure 2. Framework for Participatory Impact Assessment (FoPIA) methodology steps [17]. Light gray boxes indicated what was not applied in this study.

In this study the FoPIA methodology comprises three stages: (1) The development of policy implementation scenarios and the consequent changes in land use, as well as the preliminary assessment of sustainability issues in the case study; (2) The definition of the Land Use Functions [27] suitable to the case study on a regional scale, by structuring the sustainability issue in the social, economic and environmental dimensions similarly balanced; (3) The definition of indicators for each Land Use Function, their responses to each scenario presented, followed by the integrated analysis of the results.

2.3. The Methodological Construction of FoPIA for Sugarcane Expansion in SGPR

In order to apply the FoPIA methodology to assess the impacts of land use change due to sugarcane expansion in the southwestern region of Goiás, a workshop for the participatory assessment of impacts on the sustainability of sugarcane expansion policies in Southwestern Goiás, was held at the Institute of Social and Environmental Studies (IESA—Instituto de Estudos Sócio-Ambientais) of the Federal University of Goiás (UFG), in Goiania, Goiás - Brazil, on 12 December 2012. The goals of this consultation were to promote a structured interdisciplinary discussion on the sugarcane expansion in SGPR, to select public policy instruments for the construction of sugarcane occupation scenarios, and to define land use functions and indicators to be used in the FoPIA. It condensed phase 1 and two steps of phase 2 of the FoPIA methodology (Figure 2). These steps concern topics to be discussed by experts, who have experience and knowledge in the study subject, to promote the technical base for the furthers steps. The goal was the establishment of the sugarcane expansion drivers and their potential indicators. The further steps—unfortunately, not considered in this paper—would include other stakeholders' consultation, such as government representatives, farmers and practitioners.

The workshop was a consultation activity to regional experts to subsidize the construction of policy scenarios, as well as the selection of Land Use Functions and impact indicators, by taking into consideration the case study of the expansion of sugarcane in the SGPR.

The workshop for the participatory assessment of impacts on the sustainability of sugarcane expansion policies in Southwestern Goiás included 32 experts. The number of participants the research team considered ideal to the required dynamic activities of the FoPIA workshop. The criteria to define the experts was their performance in research projects at the SGPR or their experience in studies on the impacts generated by land use changes of sugarcane expansion in the region, favoring researchers, professors, Master's and PhD students experts in geography, agronomy, ecology, climatology, soil and rural sociology.

The workshop structure was based on guiding lectures, followed by group works and plenary discussions. The workshop was divided into three study sessions: (1) Public policy scenarios; (2) land use functions; and (3) impact indicators. These steps followed the FoPIA methodology [17] (Figure 2). A previous material contend technical information to support discussions during the workshop sessions was prepared and distributed to the participants. Thus, they could work on this material and upgrade it with their knowledge. The following sections describe each session and their outcome are described in the results.

2.3.1. Session I: Public Policy Scenarios

Many factors are driving the increase in sugar cane production to meet the increase in ethanol demand in the national and international markets, such as government incentives (laws, decrees, public plans, programs, and policies) and foreign investment and market pressure (demand × supply). Based on past trends and on the experts' opinions it is possible to draw up reference scenarios. These scenarios represent developments without interference—in the absence of policy changes—they are the counterfactual, the background against which the impact of a policy can be evaluated. These scenarios are needed to know what the situation in the target year would be if policies did not change [28].

In the workshop one of the objectives was to identify the driving forces (key trends) that expand sugarcane in the SGPR, and select which would be considered in the baseline scenario. Also, the main public policies that promote sugarcane expansion sustainability were identified and selected to elaborate public policy scenarios.

Thus, six work groups were randomly formed for this session in order to discuss and summarize their findings around four guiding questions. These questions intended to exchange knowledge on sustainability concepts and how this knowledge applies to the case study and to the sugarcane expansion in the SGPR. These questions also aimed to point out the most appropriate public policies to the assessment, namely:

1. What are the main issues related to sugarcane expansion sustainability?
2. What are the main factors driving the sugarcane expansion in SGPR?
3. What are the main public policies promoting this process?
4. Consensually define which public policy would have greater influence on the sugarcane crop expansion and spatial distribution in the SGPR. Present two options for the implementation of this policy to build scenarios.

After the discussion, each group wrote down their answers in cards, which were presented at a plenary sitting and posted on a board. The session ended with a debate including all participants.

2.3.2. Session II: Land Use Functions

The definition and use of Land Use Functions (LUFs) contribute to the aggregation and prioritization of indicators according to different social, economic and environmental functions performed by land use [29], that is, the LUFs summarize the relationship between the sustainability dimensions and the indicators to be built.

The SENSOR project considered nine LUFs, three for each sustainability dimension, namely: (a) the social dimension: Labor supply; human health and recreation; cultural (landscape identity, scenic beauty, cultural heritage); (b) the economic dimension: Industrial activities and construction; rural production and mining; transport; and (c) the environmental dimension: Supply and conservation of abiotic resources; support, provision and conservation of biotic resources; maintenance of ecosystem processes (Table 1). However, the LUFs are flexible and allow changes to better meet the goals of each case study. Therefore, the original LUFs were presented and discussed by the audience. It was reviewed in terms of relevance and suitability considering the investigated object of study—the sugarcane expansion in SGPR.

Table 1. Land Use Functions (LUF) defined by the SENSOR Project team based on the LUFs suggested to the European Union [26].

Sustainability Dimension	LUF
SOCIAL	Labor supply Quality of life Human health and recreation
ECONOMIC	Industrial activities and construction Rural production Infrastructure
ENVIRONMENTAL	Conservation of abiotic resources Conservation of biotic resources Maintenance of ecosystem processes

The working groups were reorganized into two groups for knowledge and experience representation in each sustainability dimension (social, economic and environmental). A panel was set up and the groups presented their results in plenary by posting cards containing land use functions defined by them, along with their supporting justifications.

2.3.3. Session III: Impact Indicators

The experts participating in the workshop used the following criteria to select the sustainability indicators, built according to the recommendations of the Organization for Economic Cooperation and Development [30,31].

- Relevance to the formulation of policies
- Simplicity, conciseness, and ease of interpretation
- Analytical robustness
- Measurability
- Operability
- Availability (spatial and temporal).

The same work groups formed in the previous session applied the above-mentioned criteria to point out the most appropriate indicators to represent each LUF. Each group was asked to select three indicators per LUF. The results of each group were presented in plenary and the indicators were grouped—eliminating redundancies—and systematized for further analysis.

3. Results and Discussion

The activities carried out in the three sessions produced a summary of results obtained from the experts' work. Subsequently, a set of aspects that should compose the policy scenarios for participatory evaluation was presented by taking into consideration the feasibility of implementing the FoPIA methodology to assess social, environmental and economic impacts on the sugarcane expansion process in the SGPR.

3.1. Session I: Public Policy Scenarios

To build reference scenarios, it was necessary to draw up guiding questions based on the local reality considering the development of the biofuel industry in the region. The responses given by the work groups were summarized for each question, as follows.

- What are the main issues related to the sustainability of sugarcane expansion?

There was strong transversality between the responses and comments of the work groups, and this shows the importance of the sustainability issues in their three dimensions (social, economic and environmental).

Regarding the social aspect, the factor "concentration of lands controlled by large companies (sugar and alcohol plants) and landowners for cattle breeding and soybean crops, at the expense of small/medium farmers and family farmers" was pointed out by nearly all groups. One group reported that the process started through land leases (for soybean, and later, sugarcane), as the price was attractive, especially for small producers. During the successive sugar plantation renewals, for example, the lease price fell due to the reduction in the sucrose content linked to the lease price. As the situation continued, the plants made a proposal to purchase the lands and thus consolidate the concentration of rural properties in the studied region. There was strong impact on the land ownership structure. It mainly affected small farmers, who lost their identity as rural producers and it significantly changed the local agricultural profile.

This reality has given rise to several issues related to the growth of the area cultivated with sugarcane monocultures in the SGPR: The transition from manual to mechanized harvesting systems without burning led to sugarcane cutters losing their jobs. These workers had no training and thus were not employed by the sugar and alcohol industry. These sugarcane cutters, and the small farmers who sold their lands, migrated to cities with insufficient infrastructure to absorb the new population. Another critical factor was the inadequate working conditions for the sugarcane cutters who remained in the non-mechanized plantations. As many of these workers are required to cut a very large amount of sugarcane per day, the physical strain often leads to exhaustion and occupational diseases [32]. There is also labor supply seasonality, a characteristic of the conventional sugarcane production system.

As for the economic aspect, the issue of increasing land prices was highlighted, since the sale of small properties caused strong impacts. Such impacts also occur in the urban real estate sector, as the demand for housing also increases due to the arrival of new workers trained to work in plants and the rural exodus of farmers who lease or sell their properties.

Tax evasion was reported as relevant to the local economy, as most of the income generated from the sugar and sugarcane production does not stay in the municipality. It is transferred to the plants' centers of origin installed in the SGPR, which are mostly located in the Brazilian southeastern and northeastern regions. Income concentration, characterized by the low equity in the distribution of the economic benefits generated by sugarcane production, was identified as an important sustainability issue, despite the large tax collection increase in the state and municipalities.

It was reported that the local food production has been strongly impacted by sugarcane expansion due to reduced family production or land use change of these families' lands for sugarcane production.

The study groups presented some environmental aspects. They highlighted soil compaction, worsened by the standardization of mechanized crop management techniques that replaced the extensive pastures and the annual crops of family farmers previously managed through manual techniques. Mechanical harvesting includes heavy machinery and its successive use in the fields, which leads to soil detachment contributing to its compaction. In addition to soil quality loss, water infiltration and retention capacity is reduced. This leads to increased rainwater runoff, as well as increased runoff of the water used in irrigation systems. The outcome is soil erosion and transportation of nutrients and pesticides to the beds of streams, creeks and rivers [33].

The change in sugarcane field drainage dynamics increases the risk of groundwater and aquifers contamination with pesticides and industrial wastes such as vinasse and heavy metals, dumped into the soil through fertirrigation. This is worse in soils presenting sandier texture, and the experts showed concern regarding this topic. The strong water footprint resulting from high evapotranspiration—typical of the sugarcane culture—is another relevant issue to the overall environmental impacts, since it threatens water availability to humans, fauna and flora [34].

In addition to these facts, the experts mentioned that burning-based sugarcane harvest systems are still used in the SGPR, and this leads to severe air quality issues.

The issue of biodiversity loss, which becomes more evident when livestock areas are converted into sugarcane fields, was also mentioned. Although these areas are already deforested, livestock coexists with small forest fragments and scattered trees, as they provide shade and shelter for farm animals, as well as landing to several bird species, especially in livestock systems with low technological input. These systems are quite evident in large areas that have not been converted into sugarcane crops or grain monocultures.

- What are the main factors driving the sugarcane expansion in Southwestern Goiás?

The agribusiness model in the SGPR and in the state of Goiás is similar to that used in the São Paulo countryside—the largest national sugarcane producer—facilitating crop expansion. The high domestic and international demand for energy sources as alternatives to fossil fuels was considered an important sugarcane expansion factor. This expansion was supported by public policies and international agreements that favored the transformation of ethanol into an agricultural commodity. The other expansion factors mentioned were the technological development of the sugarcane and bioethanol productive chain and land availability.

Infrastructure and logistics such as agricultural flow paths and the ethanol pipeline installation plan, were considered the main sugarcane expansion factors in the SGPR. In addition, the agroecological suitability (soil, climate, topography) of most of the territory for sugarcane cultivation was considered average. This scenario qualified owners and entrepreneurs to receive government economic incentives for such purpose Manzatto et al. [10] (Figure 3).

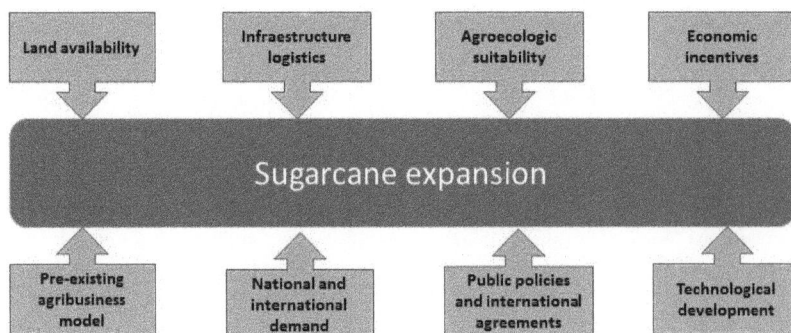

Figure 3. Sugarcane crop expansion drivers in the Southwestern Goiás Planning Region (SGPR).

- What are the main public policies promoting this process?

Some information on macro and micro state intervention policies were identified to reflect on intervention policies in the sugarcane industry and to simultaneously develop mechanisms to assess the industry impacts. The participants noticed that a set of policies should be highlighted in the SGPR: the Kyoto Protocol (at the global level); the National Agro-Energy Plan and the ZAE Cana (at the federal level); the Goiás Industrial Development Program ("Programa Produzir") and Grants and Taxes on services (ISS) at the state level.

The Kyoto Protocol (KP) is an international agreement linked to the United Nations Framework Convention on Climate Change (UNFCCC), which commits developed countries by setting internationally emission reduction targets. Brazil does not have a mandatory GHG emission reduction target; however, it has participated in the KP through the Clean Development Mechanism (CDM) projects. Clean energy generation projects, biofuels and other renewable sources could generate carbon credits to be traded on the carbon market. In this context, the National Plan for Climate Change (NPCC) aimed, among other things, "to foster the sustainable increase in the share of biofuels in the national transport matrix and work to structure an international market of sustainable biofuels".

One of the goals of the National Agro-Energy Plan [35] was to create conditions to internalize and regionalize the development based on agro-energy expansion and on the value added to the supply chain. The guidelines for this expansion were provided by the Agro-Ecological Sugarcane Zoning—ZAE-CANA [10].

Goiás state programs, such as *Programa Produzir* [36], which focuses on to the implementation and expansion of industries, also have a strong influence on the sugarcane expansion and reduce the due value-added tax (VAT) installments.

Finally, municipal policies were mentioned, including conveniences such as land grants and taxes on services (ISS) to implement new industries in Goiás. In addition, policies that have restricted sugarcane advancement, such as those found in the Rio Verde, one of the municipalities of the SGPR, which limited the sugarcane occupation to 10% of the municipal territory to protect areas planted with soybean, as well as the local food industry supply.

- Policy proposals that could be implemented.

The last item discussed in the sessions to create scenarios concerned proposals for public policy instruments to be implemented. These instruments would result in the effective transition of the production chains and allow the creation of sustainable logic-related actions.

Among public policy instruments, it was suggested that the ZAE-CANA should be further detailed to condition the government incentive contributions under the National Agro-Energy Plan, as this mechanism is considered critical to the process of sustainability. The agroecological zonings of the main crops in Goiás (scale 1:100,000) would enable a specific regulatory benchmark for Goiás and for the SGPR through the definition of sugarcane production priority areas. According to Manzatto et al. [10], many municipalities in the SGPR have large areas with medium/high suitability for sugarcane production, for example, Rio Verde (72%) and Quirinópolis (73%).

It was suggested that the environmental planning should be considered when formulating tax incentive policies.

Also, the development of an environmental education policy to encourage better understanding of the use and impacts related to soil, water resources and biodiversity was suggested. The ecological VAT could be used to reward municipalities that encourage land management changes. Accordingly, the development of a certification policy with economic incentives to certified systems was suggested. Figure 4 summarizes the main proposals built in public policies discussions.

Figure 4. Strategies and public policy themes suggested for the sustainable development of sugarcane expansion in the SGPR.

3.2. Session II: Land Use Functions

The work groups presented their views on LUFs initially adopted by Sensor project (Table 1) and suggested some modifications. A plenary consensus was obtained and a set of three LUFs was proposed for each sustainability dimension, as well as the potential indicators to achieve each LUF (Table 2).

Regarding the social dimension, the experts showed concern about the quality of the manual labor found in the SGPR agricultural production systems. These activities were often considered inhospitable, especially in the case of crops using sugarcane burning before the harvest. The proposal for the inclusion of the health and quality of life aspects in order to compose LUFs to assess this issue was accepted unanimously. The inclusion of the local socio-cultural development LUF was also suggested as LUFs hold recreational, educational, religious, scientific and cultural land use functions. The SGPR is notable for its natural beauty and its rich, exuberant landscape, mainly due to the heterogeneity of its elements.

Regarding the economic dimension, rural production along with the local consumption expansion issue were considered important to comprise another LUF as a way to reflect on the local impacts of the sugarcane expansion, and about the difficulty to produce and consume other food products. The income evasion was pointed out in Session I. If only the agricultural production is assessed, a positive result might hide a local negative impact, such as the reduced consumption of locally produced agricultural products. A similar situation can be found in the Mid-Goiás Planning Region, where sugarcane crops expanded much between 2000 and 2011, increasing from 23,000 to 123,000 hectares, which pushed up either the land value or the food supply in the region [37].

The region hosts one of the largest protected areas in the Cerrado biome, the Parque Nacional das Emas, whose integrity is threatened by the land use in its surroundings. Plateaus, cliffs and gorges, forest fragments, and water sources that drain their waters to three major watersheds (Paraguay, Araguaia and Paraná rivers), as well as cultural and archaeological sites, provide the local population and visitors with the scenic beauty and the cultural wealth of the region. Monocultures devoid of the tree element—which some cattle ranches still have in their pastures—are dominated by heavy machinery and, in some cases, use burning as an agricultural practice. Thus, they represent a high risk to the integrity of rural landscapes in the SGPR.

As for the environmental dimension, by taking into consideration the importance given to water resources in both productive activities and in the supply for human consumption, as well as the high water demand in the sugarcane culture, the insertion of the conservation of abiotic resources LUF was proposed. It was considered comprehensive and included issues such as water resources and erosion and loss of soil and nutrients.

Table 2. Indicators to assess sugarcane expansion sustainability in the SGPR.

Sustainability	LUF	Indicators
SOCIAL (a)	Labor quality and supply	Hiring, firing and balance Average income Employment rate by sector Use of local labor
	Health care and quality of life	Access to basic sanitation Hospital facilities and beds/100,000 inhabitants Student attendance per school year Crime rate
	Local socio-cultural development	Number of public leisure facilities GINI index Cultural groups Number of Municipal Councils Number of high school graduates
ECONOMIC (b)	Industrial activities and construction	Urbanization rate Industrial diversification Industrial gross domestic product (GDP) Availability of public transportation Electricity consumption
	Rural production and local consumption	Agricultural GDP Agricultural diversification Area occupied by crops Consumption of local agricultural products
	Infrastructure	Electrical power generation Electric transmission network Road network diversity Electricity cogeneration
ENVIRONMENTAL (c)	Conservation of abiotic resources	Consumption of pesticides and fertilizers Use/Agroecology Zoning (ZAE) discrepancy Percentage of preserved PPA Total use/sugarcane expansion rate Soil loss Burned area/harvested area
	Conservation of biotic resources	Percentage of preserved permanent preservation area (PPA) Total use/sugarcane expansion rate Burned points/year Deforestation rate—pasture clearing Number of fragments (measures of associated landscape) Pesticide consumption
	Maintenance of ecosystem processes	Percentage of preserved PPA Carbon stock and sequestration Water body sedimentation rate Percentage of contiguous production area (landscape/permeability matrix) Fragmentation level of the remaining forests

3.3. Session III: Impact Indicators

The work groups in the Session III were the same as session II and, according to criteria such as relevance, simplicity, robustness, measurability, operability and spatial and temporal availability, they chose the indicators that would be most appropriate to represent each LUF defined in the previous session. Each work group proposed three indicators for each LUF, according to the sustainability dimension they represented (Table 2).

Fifty-three indicators were pointed out by the six work groups. The analysis and removal of redundancies and inconsistencies resulted in 43 proposed indicators: 13 social, 13 economic and 17 environmental indicators (Table 2). The experts understood that the indicators must reflect the impacts of land use changes on the municipality or region, based on a broader scale than that observed exclusively in production areas such as plants or rural properties.

Regarding the social dimension, most experts agreed to include indicators capable to evaluate the quality of issues such as job opportunities, education, public security and leisure for the people.

Similarly, not only did the suggested economic indicators measure the industrial and agricultural production values, but they also assessed the diversification of these products. They measured power generation and assessed its distribution. They also measured the availability of public transportation, as well as the capacity of the road network to meet the demand for transport. Thus, the experts selected variables to identify the relationship between social and economic factors in order to expand the potential to improve the local population's quality of life.

The environmental issue is reflected in the indicators representing the relevant impacts to sugarcane monocultures, namely the contamination of soils and water resources by pesticides and excessive nutrients (especially nitrogen) [38]; the biodiversity loss due to the sugarcane monoculture and the way it is managed by agricultural companies and landowners [39,40]; the air quality, which is strongly compromised by the smoke from post-harvest burning [41]; and the soil compaction and its subsequent incapacity to retain sediments, nutrients and water [42].

4. Conclusions

The sugarcane expansion scenarios in the SGPR indicate the dynamics of the expansion and point out some weaknesses and potentialities of ethanol production in this region. This information can be analyzed in the light of the influence of the driving forces that operate in the local industrial and agro-industrial sectors, which may favor or restrict the cultivation of sugarcane in the region. The study collected information that allows us to conclude that the edaphoclimatic conditions, the availability of areas for cultivation, and governmental policies at the federal and state levels are among the main attraction factors for the implantation, expansion and revitalization of the sugar and alcohol industry in the SGPR.

The ZAE-CANA was the main instrument of public policies selected for the construction of sugarcane expansion scenarios in SGPR. Other public policies and drivers should be considered in scenarios for the expansion of sugarcane. However, due to the socio-environmental diversity of the region, the same set of public policy instruments can result in very different social, economic and environmental impacts. The participatory methodological approach provided the basis for evaluating the sustainability impacts caused by the expansion process of sugarcane cultivation and the implementation of ethanol agribusiness in the SGPR.

Based on technical knowledge integrated with the stakeholders of the local society, it was possible to indicate the LUF and the indicators that should be considered in the FoPIA in the SGPR sugarcane expansion scenarios. The LUF defined in this study were similar to those originally proposed, demonstrating that the LUF set by the experts address the main sustainability issues of the sugarcane expansion in SGPR. Participants, however, placed a special emphasis on the health care issue of quality of life as a new LUF proposal. They also stressed the local socio-cultural development, which would include recreational, educational, religious, scientific and cultural land use as the region shelters landscapes with remarkable scenic beauty. The prioritization stage of the indicators was not carried out due to the low availability of time during the workshop so that the participating experts could have access to more information and reflect on the issue. However, they showed a clear need to select a minimum set of indicators in a participatory way associated with each LUF to evaluate and monitor the sustainability of scenarios for expansion of sugarcane cultivation geared to the industrial production of sugar and ethanol in SGPR.

Finally, the results indicate the potential of the FoPIA methodology as a tool to assess sustainability in a participatory way, bringing together stakeholders to discuss and promote guidelines to achieve sustainability. However, as the methodology was not fully applied, it was not possible to access stakeholders' preferences and feedback.

FoPIA has proven to be a powerful tool—although it is complex and demands a great amount of energy input. We highly recommend this tool; however, we must stress that studies that use it must require thorough previous planning, especially concerning the workshops and stakeholders mobilization.

Acknowledgments: We gratefully acknowledge the time and efforts of the experts who joined the workshop. We also thank the Sixth EU's Framework Programme for the financial support and the anonymous reviewers for their critical and constructive comments.

Author Contributions: Heitor Coutinho designed the FoPIA application for the sugarcane expansion in SGPR; Ana Turetta wrote the manuscript and compiled the figures and tables; Joyce Monteiro wrote the manuscript mainly the public policies section and conclusions; Selma Castro and José Pietrafesa contributed with the workshop organization, map elaboration and text suggestions. All authors contributed to the discussion.

Conflicts of Interest: The authors declare no conflict of interest.

References

1. Buckeridge, M.S.; de Souza, A.P.; Arundale, R.A.; Teixeira, K.J.; De Lucia, E. Ethanol from sugarcane in Brazil: a 'midway' strategy for increasing ethanol production while maximizing environmental benefits. *GCB Bioenergy* **2012**, *4*, 119–126. [CrossRef]

2. Martinelli, L.A.; Filoso, S. Expansion of Sugarcane Ethanol Production in Brazil: Environmental and Social Challenges. Available online: http://onlinelibrary.wiley.com/doi/10.1890/07-1813.1/epdf (accessed on 14 August 2017).

3. Walter, A.; Galdos, M.V.; Scarpare, F.V.; Scarpare, F.V.; Leal, M.R.L.V.; Seabra, J.E.A.; da Cunha, M.P.; Picoli, M.A.; de Oliveira, C.O.F. Brazilian sugarcane ethanol: Developments so far and challenges for the future: Brazilian sugarcane ethanol. *Wiley Interdiscip. Rev. Energy Environ.* **2014**, *3*, 70–92. [CrossRef]

4. Marin, R.M. Understanding sugarcane production, biofuels, and market volatility in Brazil—A research perspective. *Outlook Agric.* **2016**, *45*, 75–77. [CrossRef]

5. Goldemberg, J. Ethanol for a sustainable energy future. *Science* **2007**, *315*, 808–810. [CrossRef] [PubMed]

6. Brasil. Pretendida Contribuição Nacionalmente Determinada. Available online: http://www.itamaraty.gov. br/images/ed_desenvsust/brasil-indc-portugues.pdf (accessed on 19 June 2017).

7. Brasil. Política Nacional Sobre Mudança do Clima—PNMC e dá Outras Providências. Available online: http://www.planalto.gov.br/ccivil_03/_ato2007-2010/2009/lei/l12187.htm (accessed on 19 June 2017).

8. IBGE/SIDRA Pesquisa Agrícola Municipal. Available online: http://www.sidra.ibge.gov.br/bda/tabela/ listabl.asp?z=t&o=11&i=P&c=1181 (accessed on 5 November 2016).

9. Brasil. Aprova o Zoneamento Agroecológico da Cana-de-açúcar e Determina ao Conselho Monetário Nacional o Estabelecimento de Normas para as Operações de Financiamento ao setor Sucroalcooleiro, nos termos do Zoneamento. Available online: http://www.planalto.gov.br/ccivil_03/_Ato2007-2010/2009/ Decreto/D6961.htm (accessed on 19 June 2017).

10. Manzatto, C.V.; Assad, E.D.; Mansilla Bacca, J.F.; Zaroni, M.J.; Pereira, S.E.M. Zoneamento Agroecológico da Cana-de-Açúcar—Expandir a Produção, Preservar a Vida, Garantir o Future. Available online: https://ainfo. cnptia.embrapa.br/digital/bitstream/CNPS-2010/14408/1/ZonCana.pdf (accessed on 14 August 2017).

11. Rathmann, R.; Silveira, S.J.C.; Santos, O.I.B. Governança e configuração da cadeia produtiva do biodiesel no Rio Grande do Sul. *Rev. Ext. Rural* **2008**, *15*, 69–101.

12. Hardi, P.; Zdan, T.J. *Assessing Sustainable Development: Principles in Practice*; International Institute for Sustainable Development: Winnipeg, MB, Canada, 1997.

13. Vezzoli, C.; Manzini, E. *O Desenvolvimento de Produtos Sustentáveis: Os Requisitos Ambientais dos Produtos Industriais*; EDUSP: São Paulo, Brazil, 2002.

14. Lawn, P. *Sustainable Development Indicators in Ecological Economics*; Edward Elgar: Cheltenham, London, UK, 2006.

15. Sachs, I. Barricadas de ontem, campos de futuro. *Estud. Av.* **2010**, *24*, 25–38. [CrossRef]

16. Helming, K.; Diehl, K.; Bach, H.; Dilly, O.; König, B.; Kuhlman, T.; Pérez-Soba, M.; Sieber, S.; Tabbush, S.; Tscherning, K. Ex ante impact assessment of policies affecting land use, Part A: Analytical framework. *Ecol. Soc.* **2011**, *16*, 29. [CrossRef]

17. Morris, J.B.; Tassone, V.; de Groot, R.; Camilleri, M.; Moncada, S. A framework for participatory impact assessment: Involving stakeholders in European policy making, a case study of land use change in Malta. *Ecol. Soc.* **2011**, *16*, 12. [CrossRef]

18. Van Bellen, H.M. Indicadores de Sustentabilidade: Uma Análise Comparativa. Ph.D. Thesis, Programa de Pós-Graduação em Engenharia de Produção, Universidade Federal de Santa Catarina, Florianópolis, Brazil, 2002.

19. Veiga, J.E. *Indicadores de Sustentabilidade*; Estudos Avançados: São Paulo, Brazil, 2010.
20. EU—European Union. Directive 2009/28/EC—Renewable Energy Directive. Available online: http://eur-lex.europa.eu/legal-content/EN/ALL/?uri=CELEX:32009L0028 (accessed on 6 April 2017).
21. EU—European Union. Directive 2009/30/EC Amending Directive 98/70/EC on Fuel. Available online: http://ec.europa.eu/environment/air/transport/pdf/art7a.pdf (accessed on 6 April 2017).
22. König, H.J.; Schuler, J.; Suarma, U.; Mcneill, D.; Imbernon, J.; Damayanti, F.; Dalimunthe, S.A.; Uthes, S.; Sartohadi, J.; Helming, K. Assessing the impact of land use policy on urban-rural sustainability using the FoPIA approach in Yogyakarta, Indonesia. *Sustainability* **2010**, *2*, 1991–2009. [CrossRef]
23. König, H.; Sghaier, M.; Schuler, J.; Abdeladhim, M.; Helming, K.; Tonneau, J.P.; Ounalli, N.; Imbernon, J.M.; Wiggering, H. Participatory Impact assessment of soil and water conservation scenarios in Oum Zessar Watershed, Tunisia. *Environ. Manag.* **2012**, *50*, 153–165. [CrossRef] [PubMed]
24. König, H.J.; Zhen, L.; Helming, K.; Uthes, S.; Yang, L.; Cao, X.; Wiggering, H. Assessing the impact of the sloping land conversion programme on rural sustainability in Guyuan, Western China. *Land Degrad. Dev.* **2014**, *25*, 385–396. [CrossRef]
25. Trindade, S.P. Aptidão Agrícola, Mudanças de usos dos Solos, Conflitos e Impactos Diretos e Indiretos da Expansão da Cana-de-Açúcar na Região Sudoeste Goiano. Tese Doutorado Ufg/Programa de Pós-Graduação em Ciências Ambientais. Available online: http://repositorio.bc.ufg.br/tede/handle/tede/5047 (accessed on 29 August 2017).
26. Lima, D.L. Estrutura e Expansão da Agroindústria Canavieira no Sudoeste Goiano: Impactos No Uso Do Solo e na Estrutura Fundiária a Partir de 1990. Ph.D. Thesis, IE/UNICAMP, Campinas, Brazil, 2010.
27. Pérez-Soba, M.; Petit, S.; Jones, L.; Bertrand, N.; Briquel, V.; Omodei-Zorini, L.; Contini, C.; Helming, K.; Farrington, J.H.; Mossello, M.T. Land use functions: a multifunctionality approach to assess the impact of land use changes on land use sustainability. In *Sustainability Impact Assessment of Land Use Changes*; Helming, K., Pérez-Soba, M., Tabbush, P., Eds.; Springer: Berlin, Germany, 2008; cap. 19; pp. 375–404.
28. Kuhlman, T. Scenarios: Driving forces and policies. In *Sustainability Impact Assessment of Land Use Changes*; Helming, K., Pérez-Soba, M., Tabbush, P., Eds.; Springer: Berlin, Germany, 2008; cap. 19; pp. 131–157.
29. Paracchini, M.L.; Pacini, C.; Jones, M.L.M.; Pérez-Soba, M. An aggregation framework to link indicators associated with multifunctional land use to the stakeholder evaluation of policy options. *Ecol. Indic.* **2011**, *11*, 71–80. [CrossRef]
30. OECD—Organisation for Economic Co-Operation and Development. *Core Set of Indicators for Environmental Performance Reviews*; OECD: Paris, France, 1993. Available online: http://enrin.grida.no/htmls/armenia/soe2000/eng/oecdind.pdf (accessed on 23 June 2017).
31. OECD—Organisation for Economic Co-Operation and Development. *Environmental Indicators: Development, Measurement and Use*; OECD: Paris, France, 2003. Available online: https://www.oecd.org/env/indicators-modelling-outlooks/24993546.pdf (accessed on 23 June 2017).
32. Ficarelli, T.R.d.A.; Ribeiro, H. Queimadas nos canaviais e perspectivas dos cortadores de cana-de-açúcar em macatuba, São Paulo. *Saúde Soc.* **2010**, *19*, 48–63.
33. Vieira, J.N.d.S. A Agroenergia e os Novos Desafios para a Política Agrícola no Brasil. In *O Futuro da Indústria: Biodiesel*; Série Política Industrial, Tecnológica e de Comércio Exterior; Ministério do Desenvolvimento, Indústria e Comércio Exterior-MDIC/Instituto Euvaldo Lodi-IEL/Núcleo Central: Brasília, Brazil, 2006; pp. 37–48.
34. Rosillo-Calle, F.; Cortez, L.A.B. Towards Proalcool II: A review of the Brazilian Bioethanol Programme. *Biomass Bioenergy* **1998**, *14*, 115–124. [CrossRef]
35. Brasília, D.F. Plano Nacional de Agroenergia 2006–2011/Ministério da Agricultura. In *Pecuária e Abastecimento, Secretaria de Produção e Agroenergia*, 2nd ed.; Embrapa Informação Tecnológica: Brasília, Brazil, 2006.
36. Goiás. Aprova o Regulamento do Programa de Desenvolvimento Industrial de Goiás—PRODUZIR (Decreto No 5.265). Available online: http://www.gabinetecivil.goias.gov.br/decretos/numerados/2000/decreto_5265.htm (accessed on 19 June 2017).
37. Pietrafesa, J.P.; Sauer, S.; Santos, A.E.A. Políticas e recursos públicos na expansão dos agrocombustíveis em Goiás: ocupação de novos espaços em áreas de Cerrado. In *Transformação do Cerrado: Progresso, Consumo e Natureza*; Pietrafesa, J.P., e Silva, S.D., Eds.; Editora da PUC Goiás: Goiânia, Brazil, 2011; pp. 93–121.

38. Silva, M.A.M.; Martins, R.C. *Produção de Etanol e Impactos Sobre os Recursos Hídricos*; IBASE/BNDES: Hong Kong, China; Brasília, Brazil, 2008; pp. 50–63.

39. Carvalho, S.P.; Marin, J.O.B. Agricultura familiar e agroindústria canavieira: Impasses sociais. *Rev. Econ. Sociol. Rural* **2011**, *49*, 3. [CrossRef]

40. Pasqualetto, A.; Zito, R.K. *Impactos Ambientais da Monocultura da Cana-de-Açúcar*; UFG: Goiânia, Brazil, 2000.

41. Gonçalves, D.B. Sob as cinzas dos canaviais: o perigoso impasse das queimadas no estado de São Paulo. *Inf. Econ.* **2005**, *35*, 32–44.

42. Severiano, E.C.; Oliveira, G.C.; Dias Júnior, M.S.; Costa, K.A.P.; Castro, M.B.; Magalhães, E.N. Potencial de descompactação de um Argissolo promovido pelo capim-tifton 85. *Rev. Bras. Eng. Agríc. Ambient.* **2010**, *14*, 39–45. [CrossRef]

sustainability

MDPI

Article

Conservation Farming and Changing Climate: More Beneficial than Conventional Methods for Degraded Ugandan Soils

Drake N. Mubiru [1,*], **Jalia Namakula** [1], **James Lwasa** [1], **Godfrey A. Otim** [1], **Joselyn Kashagama** [1], **Milly Nakafeero** [1], **William Nanyeenya** [1] and **Mark S. Coyne** [2]

[1] National Agricultural Research Organization (NARO), P.O. Box 7065, Kampala, Uganda;
    jalianamakula@kari.go.ug (J.N.); jlwasa@kari.go.ug (J.L.); ogodfrey@kari.go.ug (G.A.O.);
    jkashagama@kari.go.ug (J.K.); mnakafeero@kari.go.ug (M.N.); nwilliam@kari.go.ug (W.N.)
[2] Department of Plant and Soil Sciences, University of Kentucky, 1100 S. Limestone St.,
    Lexington, KY 40546-0091, USA; mark.coyne@uky.edu
* Correspondence: dnmubiru@kari.go.ug; Tel.: +256-782-415-843

Received: 13 March 2017; Accepted: 14 June 2017; Published: 30 June 2017

Abstract: The extent of land affected by degradation in Uganda ranges from 20% in relatively flat and vegetation-covered areas to 90% in the eastern and southwestern highlands. Land degradation has adversely affected smallholder agro-ecosystems including direct damage and loss of critical ecosystem services such as agricultural land/soil and biodiversity. This study evaluated the extent of bare grounds in Nakasongola, one of the districts in the Cattle Corridor of Uganda and the yield responses of maize (*Zea mays*) and common bean (*Phaseolus vulgaris* L.) to different tillage methods in the district. Bare ground was determined by a supervised multi-band satellite image classification using the Maximum Likelihood Classifier (MLC). Field trials on maize and bean grain yield responses to tillage practices used a randomized complete block design with three replications, evaluating conventional farmer practice (CFP); permanent planting basins (PPB); and rip lines, with or without fertilizer in maize and bean rotations. Bare ground coverage in the Nakasongola District was 187 km^2 (11%) of the 1741 km^2 of arable land due to extreme cases of soil compaction. All practices, whether conventional or the newly introduced conservation farming practices in combination with fertilizer increased bean and maize grain yields, albeit with minimal statistical significance in some cases. The newly introduced conservation farming tillage practices increased the bean grain yield relative to conventional practices by 41% in PPBs and 43% in rip lines. In maize, the newly introduced conservation farming tillage practices increased the grain yield by 78% on average, relative to conventional practices. Apparently, conservation farming tillage methods proved beneficial relative to conventional methods on degraded soils, with the short-term benefit of increasing land productivity leading to better harvests and food security.

Keywords: land degradation; land management; conservation farming

---

1. Introduction

Land degradation arising from inefficient and unsustainable land management is reducing crop productivity across sub-Saharan Africa (SSA). Land degradation reportedly affects 67% of SSA, and in some countries, more than 30% of the land area is severely or very severely degraded [1]. This is the case despite most households overwhelmingly relying on land resources [1]. The impacts of land degradation, which are becoming increasingly severe and are accelerating, include low crop productivity leading to food insecurity and disruption of ecosystem functions, which reduces ecosystem performance, resilience, and stability. The combined effects of the land degradation impacts are poor human livelihoods and wellbeing.

The extent of land affected by degradation in Uganda ranges from 20% in relatively flat and vegetation-covered areas, to 90% in the eastern and southwestern highlands [2,3]. Earlier observations indicated that land/soil degradation and soil fertility are major impediments in all cropping systems in Uganda, especially where there has been agricultural intensification [4]. However, as elsewhere in SSA, much of the population depends on land for their livelihoods [5–7] and therefore suffers the repercussions of land degradation. Additionally, climate change is a major influence on the food security and livelihoods of households in Uganda, which mostly depend on rain-fed agriculture, and are increasingly at risk from perpetually low yields of major staples such as maize (*Zea mays*) [8–11] and common beans (*Phaseolus vulgaris* L.) [12]. Many households must deal with degraded, nutrient-starved soils, and the inability to access or purchase inputs such as improved seeds and fertilizer [13].

To its comparative advantage over the rest of SSA, Uganda has a diverse agricultural production system within 10 agricultural production zones (APZs) [14]. The zones are characterized by different farming systems determined by soil types, climate, topography, and socio-economic and cultural factors. Due to the different zonal characteristics, the APZs experience varying levels of land degradation and vulnerability to climate-related hazards, which include drought, floods, storms, pests, and disease [5].

Soil/land degradation stemming from deforestation, burning of grasslands and organic residues, and continuous cultivation with minimum soil fertility enhancement leads to soil erosion and organic matter and nutrient depletion [13,15,16]. Other unsustainable land-use practices, such as overgrazing, have produced compacted soil layers and often bare grounds in extreme cases [13]. Another underlying factor in the development of compacted soil layers is that hand-hoeing, which only disturbs the first 15 to 20 cm—or sometimes as little as 5 cm—of the topsoil, when done consistently and regularly, can potentially produce restrictive layers below 0–20 cm of the topsoil. Under these soil conditions, nutrient- and water-use efficiency is reportedly very low [17,18]. These soil layers act as barriers to root and water movement and soil water-holding capacity (WHC), making land susceptible to the frequency and intensity of rainfall. Soil compaction in these layers affects agricultural land in several ways, including inhibiting root and water movement, limiting water infiltration and retention (hence facilitating runoff), and making plowing difficult. As a consequence, this directly affects agricultural productivity and contributes to the yield gap between potential output vis-à-vis farmer outputs. In that regard, land degradation and a total dependence on rain-fed agriculture has increased the vulnerability of farming systems and predisposed rural households to food insecurity and poverty [13]. Furthermore, it has led to significant adverse impacts on smallholder agro-ecosystems, including direct damage and loss of critical ecosystem services such as agricultural land/soil and biodiversity.

Due to climate change, the frequency and severity of climate-related hazards have increased, severely affecting agricultural production and in many cases leading to instability in agricultural production systems [19,20]. For example, poor rains severely affect pastures and livestock in most pastoral areas of the country, resulting in the migration of thousands of people and animals in search of water and food [5]. Jennings and Magrath [21] noted that excessive rains, both in intensity and duration, lead to water logging and negatively affect crops and pastures.

Past climatic scenarios make the outlook for the future unsettling; empirical evidence shows that seven droughts were experienced between 1991 and 2000. This caused severe water shortages, which seriously affected the animal industry [5]. Other impacts included low crop yields and increased food prices, culminating in food insecurity and negative effects on the economy. An increase in the intensity and frequency of heavy rains, floods, and landslides in the highland areas in the eastern, western, and southwestern parts of the country has been documented [22]. The recent severe drought in 2016 affected thousands of people, mainly in the Karamoja and Teso sub-regions and Isingiro District of southwestern Uganda. This was followed by the outbreak of the fall armyworm (*Spodoptera frugiperda*), affecting thousands of hectares of maize planted in the early 2017 season. The effects of climate change and variability in Uganda are compounded by existing developmental challenges of high population growth rates, high and increasing poverty levels, and declining GDP growth rates. Thus, climate change can undermine and even undo significant gains in social and economic developments in the country.

Unsustainable land-use practices lead to land degradation, and reduce the ecological and social resilience of landscapes. The overall impact of degradation has been the disruption of ecosystem services, particularly provisioning services, due to habitat fragmentation that reduces complexity and diversity, and soil erosion with consequent declining fertility and productivity. The situation is further aggravated by high population growth rates, which have led to extensive land fragmentation—a problem for sustainable land management [23]. Average landholdings in Uganda range from 0.4 to 3 ha for each typical household of seven persons [24]. High population areas are also often associated with poverty, thus requiring improved management systems to increase food security. Without a doubt, Uganda needs meaningful mitigation measures for the protection, recovery, and rehabilitation of the ecosystem services. The viability, functionality, and quality of ecosystem services are essential in enhancing and supporting community health and wellbeing, prosperity, and sustainability [25].

Ecosystem-based land management practices, such as conservation farming, bestow adaptive benefits that reduce the negative impacts of extreme weather events by buffering temperature extremes, harvesting and conserving rainwater, reducing soil loss within the agro-ecosystem, improving soil physicochemical and biological conditions, and regulating pest and disease cycles. Conservation farming practices can potentially address the soil and water management constraints faced by smallholder farmers [26]. The conservation farming package entails dry-season land preparation using minimum tillage systems, crop residue retention, seeding and input applications in permanent planting stations, and nitrogen-fixing crop rotations [27].

Permanent planting basins (PPBs) and rip lines are two major components of the recently introduced conservation farming package for renovating degraded landscapes that are being extensively promoted for smallholder farming [26,28–31]. PPBs and rip lines, as used in conservation farming, are crop management methods that enhance the capture and storage of rainwater, and allow sustainable precision management of limited nutrient resources. Both methods reduce the risk of crop failure due to erratic rainfall and extended droughts. The use of these methods in combination with improved seed and crop residues to create a mulch cover that reduces evaporation losses has consistently increased average yields by 50–200%, depending on the amount of rainfall, soil type, and fertility [32]. PPBs are being targeted for households with limited or no access to oxen, while ripping is meant for smallholder farmers with oxen [26].

Maize and beans are major staple foods for much of the population, and are a major source of food security in Uganda. The annual per capita maize consumption is estimated to be 28 kg, and bean consumption, 58 kg [33]. Both crops are cash crops for some smallholder farmers. Maize is also an important animal feed. At the household level, the importance of maize and beans is centered on their dietary roles of supplying proteins, carbohydrates, minerals, and vitamins to resource-constrained rural and urban households with rampant shortages of these dietary elements. Reportedly, the dietary intake for the most resource-constrained households in Uganda comprises 70% carbohydrates, and these are mainly from maize, supplying 451 kcal/person/day and 11 g protein/person/day. Beans provide about 25% of the total calories and 45% of the protein intake in the diets of many Ugandans [34].

Unfortunately, due to the biophysical and socio-economic factors previously noted, the average maize and bean grain yields on smallholder farms, which on average are less than 1 ha, are less than 30% of their potentials [8–12]. The potential maize yield in Uganda is estimated to range from 3.8 to 8.0 t ha^{-1} [9], while that of beans is 2.0 t ha^{-1} [12]. Poor soil conditions (low soil fertility, compacted soils, and moisture stress) coupled with a low nutrient- and water-use efficiency are major contributing factors to this yield gap.

We postulate that employing ecosystem-based land management practices such as conservation farming will increase water- and nutrient-use efficiency, provide greater rooting depth, and improve WHC that would increase land productivity, leading to better grain harvests and food security. The long-term benefits would be an increased soil organic matter content, increased return on fertilizer use, and greater resilience of dry-land smallholder plots to erratic rainfall patterns from climate change.

This study: (i) assessed the extent of bare grounds in Nakasongola, one of the districts in the Cattle Corridor of Uganda; and (ii) evaluated yield responses of maize and beans to different tillage methods in the district.

2. Materials and Methods

2.1. Site Description

The Nakasongola District is in central Uganda, between 00°57′44.89″ and 10°40′42.76″ North latitude and between 310°58′03.77″ and 320°48′00.29″ East longitude. The district is in the Pastoral Rangelands agro-ecological zone (AEZ), which is one of the AEZs that comprise the Cattle Corridor of Uganda (Figure 1).

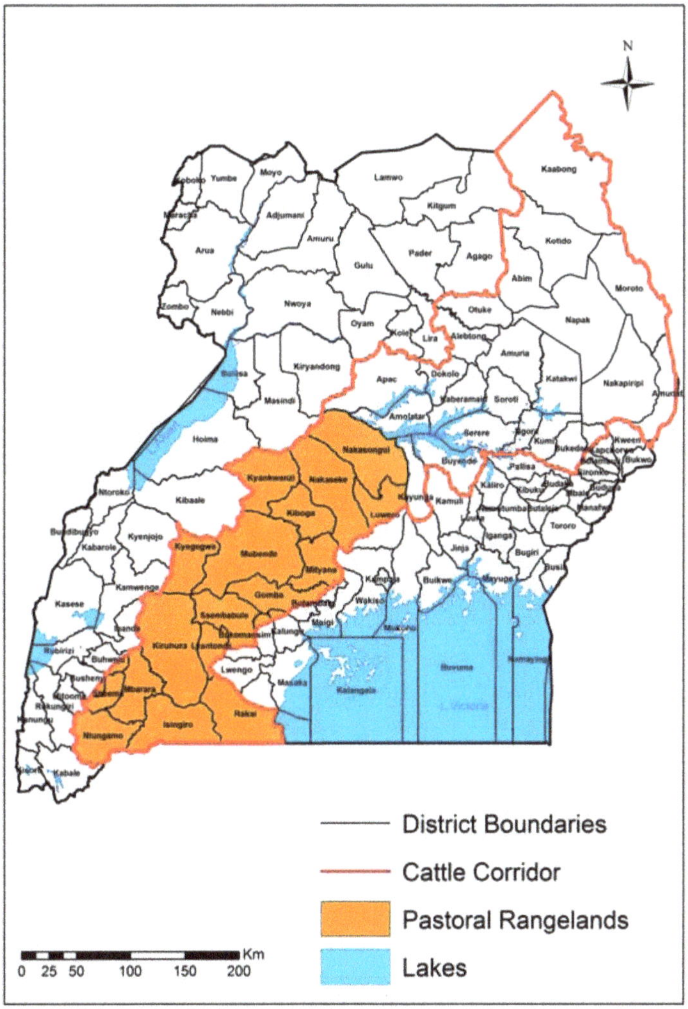

Figure 1. Uganda's Cattle Corridor (Source: Land Resources Database, NARL-Kawanda).

Constituting the country's rangelands, the Cattle Corridor has a total area coverage of 84,000 km^2, which is approximately one-third of the total 241,000 km^2 of the land area in Uganda, and is home to a population of 6.6 million people. The corridor is host to a mixed production system comprising nomadic pastoralists, agro-pastoralists, and subsistence farmers. On average, it receives 500–1000 mm of rainfall annually, which is spatially variable, from about 400 mm in some parts of the northeastern corridor, to about 1200 mm in parts of the southwestern and central corridor. The rainfall pattern is bimodal in the southwestern and central parts of the corridor, and transitions into one rainy season of about $5\frac{1}{2}$ months in the northern and northeastern areas [14]. Dry spells are frequent, and droughts of significant magnitudes occur, causing hardship to peoples' livelihoods and economy in the districts that comprise the corridor.

Specifically, the Pastoral Rangelands AEZ receives moderate rainfall, spatially varying from 915 to 1021 mm/year with a bimodal pattern [14]. The main rainy season is from March to May with a peak in April, and the secondary season is from September to December with a modest peak in November. Dry periods are from June to August and January to February. The daily average temperature ranges from 12.5 to 30 °C. Evaporation exceeds rainfall by a factor of about 6 during the dry months from June to August, while during the main rainy months (April and May), rainfall equals evaporation. Altitude in the zone spatially ranges from 129 to 1524 m ASL (above sea level), with the land characterized by rolling hills with some flat areas and moderate-to-poor soils. The farming system and socio-economic characteristics are characterized by smallholders with many communal grazing and agro-pastoral practices; low literacy levels; absentee landlords with a squatter population; and infrastructure and marketing systems that are poor to moderate [13,14].

2.2. Assessing the State of the Soil in the Nakasongola District

2.2.1. Quantification of Bare Ground Coverage

Based on the assumption that bare grounds are one of the visible indicators of extreme land degradation, the approach was to physically survey and capture, using GPS, the spatial extent of some bare grounds, and use the data to locate the same features on a satellite image captured during a fairly dry month. These points were used to develop digital signatures for searching similar features in the rest of the image, and generating coverage statistics using Geographic Information System (GIS) tools.

2.2.2. Data Sources/Analysis

A supervised multi-band satellite image classification using the Maximum Likelihood Classifier (MLC) was used [35]. A high resolution (<5 M) image from 2013, covering a greater part of the district, was used for the analysis. Bands 1, 4, and 7 of the Landsat Thematic Mapper image (p171r059_7t20011127_z36_nn10) were used.

2.2.3. Soil Physicochemical Analysis

Soil samples from depths of 0 to 20 cm were collected from geo-referenced sites in eight sub-counties comprising the Nakasongola District. The samples were dried in open air, ground to pass a 2 mm sieve, and analyzed according to Okalebo et al. [36] and Foster [37]. Texture analysis was performed by the hydrometer method [38]. Soil pH was measured with a soil/water ratio of 1:2.5. Extractable P, K, and Ca were measured in a single ammonium lactate/acetic acid extract buffered at pH 3.8 [36]. Total nitrogen (N) was determined by a micro-Kjeldahl block digestion apparatus, and soil organic matter was determined by acid-dichromate digestion. Soil samples were also collected using a double-cylinder, hammer-driven core sampler to determine the bulk density according to methods by Blake and Hartge [39].

2.2.4. Statistical Analysis

The soils' physicochemical data was subjected to Pearson's correlation to establish relationships among the parameters, using Statistix V. 2.0. Furthermore, Principal Component Analysis (PCA) was used to determine similarities among soils from different farms and sub-counties as manifested in the status of their physicochemical properties.

2.3. Sustainable Agricultural Production

Sustainable agriculture has been defined as a means of production that seeks to sustain farmers, resources, and communities by promoting farming practices and methods that are profitable, environmentally sound, and good for communities [40–44]. Sustainability rests on the principle that the present generation must meet its own needs without compromising the ability of future generations to meet their own needs [44]. In this study, we assessed how conservation farming practices could contribute to sustainable agriculture production.

Trials were conducted on 16 randomly-selected farmer fields in the first and second seasons of 2015 in two sub-counties in the Nakasongola District. The first season (A) ran from March to May with a rainfall peak in April, while the second season (B) ran from October to December with a peak in November.

2.3.1. Field Design

The experiment design was a randomized complete block with three replications. The different tillage methods under assessment were: Conventional Farmer Practices (CFPs), PPBs, and rip lines, all with or without fertilizer. CFP entailed the preparation of a seedbed followed by at least two hand-hoe weeding passes, with crop residues incorporated into the soil.

Prior to the trial's establishment, in conservation farming treatments, the fields were slashed and sprayed with glyphosate (500 mg L^{-1}) at a rate of 7.5 L ha^{-1}, two weeks after slashing. In the preceding cropping season, most fields had been used to grow maize, beans, or sweet potatoes (*Ipomoea batatas*). Due diligence was made to ensure that there was no continuous cropping of a particular crop in the same plot. The traditional crop rotations in this area are: sweet potato, bean or groundnut (*Arachis hypogea*), maize, then cassava (*Manihot esculenta*). Sweet potato is important as a first crop in the rotation because it helps to loosen, as well as increase, the soil volume, while cassava, which is tolerant to low soil nutrient levels, comes last in the rotation (Sarah Nakamya per. Comm., [45]). Due to multiple uses of crop residues, little material was laid down on the plots. In the conservation farming treatments, weeds were controlled by light weeding with a hand-hoe or by hand. A high-yielding and drought-tolerant hybrid maize variety (PH5052) and bean variety (NABE 15) were used in all treatments. The average plot size was 513 m^2.

2.3.2. Seeding Rates

Conventional Farmer Practice: Planting holes for maize were designated by planting lines and digging with a hand-hoe at a spacing of 75 cm between rows and 60 cm within rows. Each hole was seeded with two seeds, giving a total of 44,444 plants/ha. In the case of beans, spacing was 50 cm × 10 cm and each hole was seeded with one seed to give a total of 200,000 plants/ha.

Permanent Planting Basins: Basins were designated using planting lines and digging planting basins 35 cm (long) × 15 cm (wide) × 15 cm (deep), with a spacing of 75 cm between rows and 70 cm within rows from center-to-center of the PPB, before the onset of rains. Available crop residues were laid between rows to create a mulch cover. The basins were seeded with three maize seeds per basin (57,143 plants/ha) and six bean seeds per basin (114,286 plants/ha).

Rip lines: Rip lines were ripped before the onset of rains by an ox ripper set at a depth of 15 cm. Available crop residues were laid between rows to create a mulch cover. Maize was seeded at a spacing of 75 cm × 30 cm with one seed per hill (44,444 plants/ha). Beans were seeded at a spacing of 75 cm × 10 cm with two seeds per hill (266,667 plants/ha).

In the maize and bean trials, micro-doses of basal fertilizer (DAP) at a rate of 92.5 kg ha^{-1} were applied and covered with topsoil before planting the seeds. For maize, nitrogen as urea at a rate of 150 kg ha^{-1} was evenly side-dressed when the maize was at knee height, approximately at vegetative stage 9 (V9).

2.4. Statistical Analysis

Data was examined by ANOVA to determine significant ($p \leq 0.05$) treatment effects. Comparisons of means were made by LSD all-pair-wise comparisons. All analyses were done using Statistix V. 2.0.

3. Results and Discussion

3.1. Assessment of the State of the Soil in the Nakasongola District

Quantification of Bare Ground Coverage

Bare ground coverage in the Nakasongola District due to extreme cases of soil compaction was 187 km^2 (11%) of the 1741 km^2 of arable land (Figure 2 and Table 1). At present, Uganda has 7.2 million hectares of arable land under crop agriculture, which is less than 50% of the arable land estimated at 16.8 million hectares [6]. Pessimistic forecasts indicate that the available arable land for agriculture will run out in most parts of the country by around 2022. With such grim statistics, the country cannot afford to lose any arable land. It is therefore imperative that Uganda embraces sustainable land management to reverse this trend of land degradation.

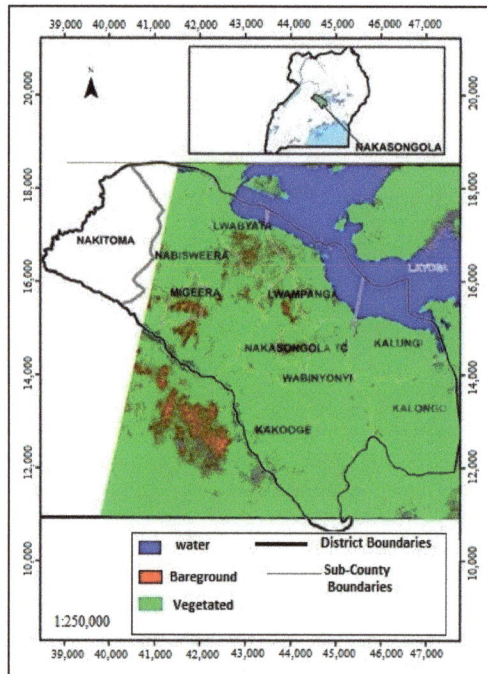

Figure 2. Spatial distribution of bare-grounds in the Nakasongola District and surrounding areas.

Table 1. Spatial distribution of different land cover classes in the Nakasongola District.

	Class	Area (km^2)	% Cover
1	Open water	233	7.9
2	Vegetated	1527	51.7
3	Bare ground	187	6.3
4	Seasonal wetland	915	31.0
5	Cloud cover	48	1.6
6	Permanent wetland	46	1.6
Total		2956	100

Pearson's correlation (Table 2) of soil physicochemical data from all sub-counties revealed that the bulk density, which was used as an indicator of soil compaction, was significantly correlated only with clay ($r = -0.54$, $p < 0.0003$) and sand ($r = 0.48$, $p < 0.002$). Therefore, clay and sand were the most important determinant parameters for bulk density or soil compaction. Observations from our study are well corroborated by several workers [46–50], who observed, from different areas and soil types, that the higher the amount of sand in the soil, the greater the bulk density, while the higher the amount of clay, the lower the bulk density.

Table 2. Pearson's correlation of soil physicochemical data from all sub-counties in the Nakasongola District.

	pH	OM [¥]	N	P	K	Ca	Mg	Sand	Silt	Clay
OM	0.27 *									
N	0.28 *	0.97 ***								
P	0.57 ***	0.19	0.20							
K	0.42 ***	−0.05	-0.03	0.30 *						
Ca	0.82 ***	0.33 **	0.31 **	0.40 ***	0.26					
Mg	0.79 ***	0.31 **	0.30 *	0.36 **	0.38 **	0.97 ***				
Sand	−0.14	−0.48 ***	−0.49 ***	−0.07	−0.13	−0.26	−0.29 *			
Silt	0.50 ***	0.53 ***	0.52 ***	0.14	0.03	0.60 ***	0.55 ***	−0.45 ***		
Clay	−0.07	0.30 *	0.32 **	0.01	0.13	0.03	0.08	−0.92 ***	0.05	
BD [†]	0.23	−0.18	−0.16	0.19	0.03	0.16	0.13	0.48 ***	0.01	−0.54 ***

[†] BD = Bulk Density; [¥] OM = Organic Matter; * significant at 0.1 level; ** significant at the 0.05 level; *** significant at the 0.01 level.

PCA was used to determine if there were similarity clusters of soils from different farms and sub-counties with respect to soil properties. All soils from the different farms and sub-counties formed one cluster, indicating that there were no exceptional differences in the soil properties among the sub-counties. Means of all soil properties (Table 3) in the topsoils and subsoils were below normal for the soils of Uganda [36]. For example, the critical value of soil pH in Ugandan soils is 5.6, while that of organic carbon is 3.0% [51,52]; this is an indication that all soils in this study were, to some extent, chemically and physically degraded. On analyzing the properties of the topsoil and subsoil, the average pH of the subsoil was slightly higher than that of the topsoil, which was unusual. However, the concentration of calcium in the subsoil was also higher than in the topsoil, which might explain this phenomenon.

Although there was no distinct differentiation for soils from the different farms and sub-counties, separately, soils from the Kalungi sub-county had the highest average bulk density (Table 4), which was significantly different ($p \leq 0.05$) from the other sub-counties, except Kalongo and Lwampanga. Soils from the Wabinyonyi and Kakoge sub-counties had the lowest ($p \leq 0.05$) average bulk densities compared to the other sub-counties. Correspondingly, the Wabinyonyi and Kakoge sub-counties also had a higher ($p \leq 0.05$) percentage of clay and a significantly lower percentage of sand than all the other sub-counties, with a few exceptions.

Table 3. Cluster means of soil properties for soil samples from sub-counties of the Nakasongola District.

Soil Layer	BD [†]	pH	OM *	N	P	K	Ca	Mg	Sand	Clay	Silt
	(g/cc)		(%)				(ppm)			(%)	
Topsoil [¥]	1.4	4.4	2.2	0.2	6.3	98.8	459	283	51	41	8
Subsoil [¥]	-	4.6	2.1	0.1	3.1	45.4	571	217	50	42	8
Critical levels		5.6	3.0	0.2	35.5	72.5	1640	87			

[†] BD = Bulk Density; * OM = Organic Matter; [¥] Topsoil = Top layer of soil collected at 0–20 cm depths; Subsoil = Soil samples collected at 20–40 cm depths.

Table 4. Soil properties well-correlated with bulk density from the different sub-counties.

Sub-County	Soil Property [1]		
	Bulk Density (g/cc)	Clay (%)	Sand (%)
Kalungi	1.58a	42bc	51ab
Kalongo	1.57ab	38c	57a
Lwampanga	1.56ab	40c	50ab
Rwabyata	1.49bc	38c	53a
Nakitoma	1.47c	37c	56a
Nabisweera	1.44c	37c	54a
Wabinyonyi	1.34d	47ab	44b
Kakooge	1.33d	50a	44b
SE	0.04	3	4

[1] Different letters within each column indicate significant differences between treatments at the $p \leq 0.05$ level, using the LSD method.

3.2. Sustainable Agricultural Production

3.2.1. Bean Grain Yield Response to Tillage Practices and Fertilizer

There were no significant seasonal differences in the bean grain yield (Table 5). There were also no significant season × tillage interactions, indicating that treatment effects on the grain yield were independent of seasonal characteristics. Since the season × tillage interactions were not significant, the yield means were averaged across the seasons (Table 5). All tillage practices, whether conventional or the newly introduced conservation farming practices in combination with a fertilizer increased bean grain yield. However, the increases were only significantly different between rip lines with and without fertilizer. On average, fertilizer use in combination with the tillage practices increased the bean grain yield from 436 kg ha^{-1} to 743 kg ha^{-1}, a 70% increase. Separately, the highest average percentage yield increase (102%) was between rip lines with and without fertilizer; this was followed by conventional practices without and with fertilizer (56%), and lastly between PPBs without and with fertilizer (53%). The average bean grain yield from conventional practices was 460 kg ha^{-1}; from PPBs, 648 kg ha^{-1}; and from rip lines, 661 kg ha^{-1}. Apparently, the newly introduced conservation farming tillage practices increased the bean grain yield relative to conventional practices by 41% in PPBs, and 43% in rip lines.

Table 5. Average bean and maize grain yields as a response to different tillage practices [†].

Tillage Practice	Bean Yield		Maize Yield	
	(kg ha^{-1})	SE	(kg ha^{-1})	SE
Conventional	359c	±138	1536b	±879
Conventional + fertilizer	560abc	±138	2481ab	±879
PPB	512abc	±138	3328ab	±918
PPB + fertilizer	784ab	±138	4963a	±918
Rip line	438bc	±148	2086b	±963
Rip line + fertilizer	884a	±148	3921ab	±963

[†] Yield means for a particular crop followed by the same letter are not significantly different according to LSD at $p = 0.05$.

3.2.2. Potential versus Actual Bean Grain Yield

The potential bean grain yield in Uganda is about 2.0 t ha^{-1} [12]. In our study, the response of bean grain yields to fertilizer and the newly introduced conservation farming tillage practices was below the yield potential, notwithstanding the remarkable increase. Other workers [53–56] have observed that yields from on-farm trials were enhanced by using improved seeds and fertilizers, but yields still remained below the genetic potential. This has been attributed to management factors that contributed to poor early-season vigor, in-season plant loss, and environmental stresses.

The tillage effects increased the bean grain yield in the newly introduced conservation farming practices relative to conventional practices. However, the yield differences between rip lines and PPBs could partly be attributed to differences in plant population; that is, 266,667 plants/ha in rip line tillage vis-à-vis 114,286 plants ha^{-1} in PPBs. In an earlier study (not published) conducted to determine the optimum seeding rates in PBBs, it was established that six bean seeds per basin, as were used in the current study, was optimal. It is plausible that increasing the seeding rate in PPBs creates competition among the plants, thus affecting productivity.

Ghaffarzadeh et al. [57] observed that the potential for stress could be increased when crops compete among themselves. Ghaffarzadeh et al. [58] further intimated that competition for resources might develop because of root growth patterns and/or different resource demands, although they acknowledged that there is limited information available about light, water, and nutrient competition in regard to plant position. Some studies suggest that spatial and temporal arrangement of crops may influence competition for water and light [59,60]. Under water-limiting conditions, production advantages could diminish [61–63].

3.2.3. Maize Grain Yield Response to Tillage Practices and Fertilizer

Unlike for beans, there were significant seasonal differences in the maize grain yield (Table 5). In the first season (2015A), the maize grain yield was 2113 kg ha^{-1} (106%) greater than in the second season (2015B). It is plausible that the yield difference was a result of water stress experienced in the 2015B season. This effect was more pronounced in maize than in beans because beans are short-term compared to maize, and it is likely that by the time drought manifested, the bean crop was already in advanced stages of development.

Although there were significant seasonal yield differences, the season × tillage interactions were not significant. As was the case with beans, this indicated that the tillage effects on the maize grain yield were independent of the seasonal characteristics. Correspondingly, the yield means were averaged across seasons (Table 5). As would be expected, there were yield responses to fertilizer applications in all tillage practices, however, the differences between particular tillage practices without and with fertilizer were minimally significant. Suffice to note also that the newly introduced conservation farming practices, on average, increased the grain yield more than the conventional practice, by 78%. In their study spanning three seasons, Mazvimavi et al. [64] realized that maize in conservation farming tillage practices out-yielded that in conventional tillage practices by 59%.

When each season was critically examined, this demonstrated the performance differences between the two conservation farming tillage practices. In season 2015A, which was deemed to have normal rainfall, rip line tillage had a higher maize grain yield compared to the PPBs. Conversely, in 2015B, which is believed to have had below-average rainfall, the PPBs had a higher maize grain yield compared to rip lines. Although it cannot be conclusively concluded from our study results, it is plausible that in years with below-average rainfall, the PPBs are better at harvesting and conserving rainwater than rip lines, and are thus the superior performer. In their study on conservation tillage for soil water management, Mupangwa et al. [26] concluded that planting basin tillage is better at controlling water losses than ripper, double, and single conventional ploughing techniques.

3.2.4. Potential versus Actual Maize Grain Yield

The potential maize yield in Uganda is estimated to range from 3.8 to 8.0 t ha^{-1} [8–11], with the open pollinated varieties (OPV) being on the lower end compared to hybrid varieties. However, according to the Food and Agriculture Organization Statistical Database (FAOSTAT), the actual maize productivity is stagnant, at a low level of between 1.5 and 2.5 t ha^{-1} [11]. The yield gap is attributed to the limited use of inputs such as improved seed and fertilizer, now coupled with soil moisture stress due to climate variability. In the current study, the newly introduced conservation farming practices apparently brought the maize grain yield within the potential yield range, although there was still room for improvement.

4. Conclusions

This study showed that 11% of the arable land in the Nakasongola District is bare ground, an extreme case of soil compaction and land degradation. Because this is not an isolated case, it is imperative that the country embraces sustainable land management and agricultural production to meet the food needs of its people and to spur economic development, while at the same time conserving the environment.

The newly introduced conservation farming tillage practices increased the bean grain yield relative to conventional practices by 41% in PPBs and 43% in rip lines. For maize, the newly introduced conservation farming tillage practices on average increased the grain yield by 78%, relative to the conventional practices. Conservation farming tillage methods, that is, PPBs and rip lines, proved to be more beneficial than conventional methods for degraded soils, with a short-term benefit of increasing land productivity, leading to better harvests and food security. The long-term benefits are expected to be an increased soil organic matter content, an increased return on fertilizer use, and a greater resilience of dryland smallholder plots to erratic rainfall patterns, occasioned by climate change. Conservation farming practices, as empirically tested in this study, facilitated the rehabilitation and recovery of degraded farmer fields, as evidenced by increased grain yields, thus fitting well within the league of sustainable agricultural production practices.

Long-term studies are needed to establish the effects of variable rainfall on the performance of planting basins vis-à-vis rip lines. Furthermore, considering the variable costs of inputs and the variability of outputs among the different tillage practices, there is a need to conduct a cost-benefit analysis to determine the cost effectiveness of each tillage practice.

Acknowledgments: The authors are grateful to the participating farmers and local government of the Nakasongola District. The research was supported by the National Agricultural Research Organization, the National Agricultural Research Laboratories–Kawanda, the International Maize and Wheat Improvement Centre (CIMMYT), and the Australian Centre for International Agricultural Research (ACIAR) through the Sustainable Intensification of Maize-Legume cropping systems for Food Security in Eastern and Southern Africa (SIMLESA) program. M.S. Coyne was supported by the USDA National Institute of Food and Agriculture Hatch project KY007090, with additional support from a Natural Resources Conservation Service Conservation Innovation Grant, and a USDA-ARS specific cooperative agreement. Funds to publish in open access were sourced from the Uganda Sustainable Land Management (SLM) project, financed by a grant from the Global Environment Facility (GEF).

Author Contributions: Drake N. Mubiru was the principal investigator and lead person in the experimentation process and preparation of the manuscript. Jalia Namakula, Godfrey Otim, Joselyn Kashagama, and Milly Nakafeero established and managed the trials, and collected the yield data. James Lwasa led the team in assessing the state of the soils in Nakasongola District using, Global Positioning System (GPS) and Geographic Information System (GIS) tools. William Nanyeenya collated all the socio-economic and biophysical information of Nakasongola District. Mark S. Coyne made significant intellectual contributions during the designing of the trials and for improving the technical content of the manuscript.

Conflicts of Interest: The authors declare no conflict of interest.

References

1. FAO. World Soils Report. No. 90. 2000. Available online: www.fao.org/soils-portal/resources/world-soil-resources-reports/en (accessed on 13 December 2016).
2. Magunda, M.K.; Tenwya, M.M. Soil and water conservation. In *Agriculture in Uganda*; Mukiibi, J.K., Ed.; Uganda National Agricultural Research Organization (NARO): Entebbe, Uganda, 2001; Volume I, pp. 145–168.

3.   Nakileza, B.; Nsubuga, E.N.B. *Rethinking Natural Resource Degradation in Semiarid Sub-Saharan Africa: A Review of Soil and Water Conservation Research and Practice in Uganda, with Particular Emphasis on the Semiarid Areas*; Overseas Development Institute: Kampala, Uganda, 1999.

4.   Nkonya, E.; Pender, J.; Jagger, P.; Sserunkuma, D.; Kaizzi, C.K.; Ssali, H. *Strategies for Sustainable Land Management and Poverty Reduction in Uganda*; Research Report 133; IFPRI: Washington, DC, USA, 2004.

5.   Government of Uganda (GOU). *Climate Change: Uganda National Adaptation Programmes of Action*; Environmental Alert, GEF, UNEP, Ministry of Water and Environment: Kampala, Uganda, 2007.

6.   National Environment Management Authority (NEMA). *State of the Environment Report for Uganda 2006/2007*; NEMA: Kampala, Uganda, 2007.

7.   Uganda Bureau of Statistics (UBOS). Statistical Abstract. 2015. Available online: http://www.ubos.org (accessed on 11 February 2016).

8.   Otunge, D.; Muchiri, N.; Wachoro, G.; Anguzu, R.; Wamboga-Mugirya, P. *Enhancing Maize Productivity in Uganda Through the WEMA Project*; A Policy Brief; National Agricultural Research Organisation of Uganda (NARO)/African Agricultural Technology Foundation (AATF): Entebbe, Uganda, 2010.

9.   Semaana, H.R. *Crop Production Handbook for Good Quality Cereals, Pulses and Tuber Crops*; Ministry of Agriculture Animal Industry and Fisheries/Sasakawa Africa Association (SAA): Entebbe, Uganda, 2010.

10. Regional Agricultural Expansion Support (RATES). *Maize Market Assessment and Baseline Study for Uganda*; Regional Agricultural Expansion Support Program: Nairobi, Kenya, 2003.

11. Okoboi, G.; Muwanga, J.; Mwebaze, T. Use of improved inputs and its effects on maize yield and profit in Uganda. *Afr. J. Food Agric. Nutr. Dev.* **2012**, *12*, 6932–6944.

12. Sebuwufu, G.; Mazur, R.; Westgate, M.; Ugen, M. Improving the Yield and Quality of Common Beans in Uganda. Available online: www.soc.iastate.edu/staff/.../CRSP (accessed on 9 September 2014).

13. World Bank. *Uganda Sustainable Land Management: Public Expenditure Review*; Report No. 45781-UG AFTAR; Sustainable Development Department Country Department 1, Uganda Africa Region: Kampala, Uganda, 2008.

14. Government of Uganda (GOU). *Increasing Incomes through Exports: A Plan for Zonal Agricultural Production, Agro-Processing and Marketing for Uganda*; MAAIF: Entebbe, Uganda, 2004.

15. Magunda, M.; Majaliwa, M. *Soil Erosion in Uganda: A Review*; Nile Basin Initiative Issue Paper; NBI: Kampala, Uganda, 2006.

16. Zake, J.; Magunda, M.; Nkwiine, C. Integrated soil management for sustainable agriculture and food security: The Uganda case. Presented at the FAO Workshop on Integrated Soil Management for Sustainable Agriculture and Food Security in Southern and Eastern Africa, Harare, Zimbabwe, 8–12 December 1997.

17. Ewel, J.J. Nutrient Use Efficiency and the Management of Degraded Lands. In *Ecology Today: An Anthology of Contemporary Ecological Research*; Gopal, B., Pathak, P.S., Saxena, K.G., Eds.; International Scientific Publications: New Delhi, India, 1988; pp. 199–215.

18. Fatondji, D.; Martius, C.; Bielders, C.L.; Vlek, P.L.G.; Bationo, A.; Gerard, B. Effect of planting technique and amendment type on pearl millet yield, nutrient uptake, and water use on degraded land in Niger. *Nutr. Cycl. Agroecosyst.* **2006**, *76*, 203. [CrossRef]

19. Mubiru, D.N.; Komutunga, E.; Agona, A.; Apok, A.; Ngara, T. Characterising agrometeorological climate risks and uncertainties: Crop production in Uganda. *S. Afr. J. Sci.* **2012**, *108*. [CrossRef]

20. Ogallo, L.A.; Boulahya, M.S.; Keane, T. Applications of seasonal to interannual climate prediction in agricultural planning and operations. *Int. J. Agric. For. Met.* **2002**, *103*, 159–166. [CrossRef]

21. Jennings, S.; Magrath, J. What Happened to the Seasons? 2009. Available online: http://www.oxfam.org.uk/resources/policy/climate_change/ (accessed on 11 August 2014).

22. National Environment Management Authority (NEMA). *National State of the Environment Report for Uganda 2014*; NEMA: Kampala, Uganda, 2016.

23. United Nations Development Programme (UNDP). *Uganda Strategic Investment Framework for Sustainable Land Management*; UNDP: Kampala, Uganda, 2014.

24. Okwi, P.O.; Hoogeveen, J.G.; Emwanu, T.; Linderhof, V.; Begumana, J. Welfare and environment in rural Uganda: Results from a small-area estimation approach. *Afr. Stat. J.* **2016**, *3*, 135–188. [CrossRef]

25. Serrao-Neumann, S.; Turetta, A.P.; Prado, R.; Choy, D.L. Improving the management of climate change impacts to support resilient regional landscapes. In Proceedings of the Conference of the Ecosystem Services Partnership Local Action for the Common Good, San Jose, Costa Rica, 7–12 September 2014.

26. Mupangwa, W.; Twomlow, S.; Walker, S. Conservation Tillage for Soil Water Management in the Semiarid Southern Zimbabwe. Available online: http://www.cgspace.cgiar.org/bitstream/handle/ (accessed on 10 December 2014).

27. Haggblade, S.; Tembo, G. Early Evidence on Conservation Farming in Zambia. Presented at the International Workshop on Reconciling Rural Poverty and Resource Conservation: Identify Relationships and Remedies, Ithaca, NY, USA, 2–3 May 2003.

28. Hove, L.; Twomlow, S. Is conservation agriculture an option for vulnerable households in Southern Africa? Presented at the Conservation Agriculture for Sustainable Land Management to Improve the Livelihood of People in Dry Areas Workshop, Damascus, Syria, 7–9 May 2007.

29. Twomlow, S.; Urolov, J.; Oldrieve, J.C.; Jenrich, B. Lessons from the Field: Zimbabwe's Conservation Agriculture Task Force. *J. SAT Agric. Res.* **2008**, *6*, 1–11.

30. Mazvimavi, K.; Twomlow, S. Socioeconomic and institutional factors influencing adoption of conservation farming by vulnerable households in Zimbabwe. *Agric. Syst.* **2009**, *101*, 20–29. [CrossRef]

31. Pedzisa, I.; Minde, I.; Twomlow, S. An evaluation of the use of participatory processes in wide-scale dissemination of research in micro dosing and conservation agriculture in Zimbabwe. *Res. Eval.* **2010**, *19*, 145–155. [CrossRef]

32. Twomlow, S. *Integrated Soil Fertility Management Case Study*; SLM Technology, Precision Conservation Agriculture: Bulawayo, Zimbabwe, 2012.

33. Soniia, D.; Sperling, L. Improving technology delivery mechanisms: Lessons from bean seed systems research in eastern and central Africa. *Agric. Hum. Values* **1999**, *16*, 381–388.

34. National Agricultural Research Organization (NARO). *Annual Report*; NARO: Entebbe, Uganda, 2000.

35. Lillesand, T.M.; Kiefer, R.W. *Remote Sensing and Image Interpretation*; John Wiley & Sons: New York, NY, USA, 1987; p. 669.

36. Okalebo, J.R.; Gathau, K.W.; Woomer, P.L. *Laboratory Methods of Soil and Plant Analysis: A Working Manual*; TSBF-CIAT and SACRED Africa: Nairobi, Kenya, 2002.

37. Foster, H.L. Rapid routine soil and plant analysis without automatic equipment: I. Routine soil analysis. *E .Afr. Agric. For. J.* **1971**, *37*, 160–170.

38. Gee, G.W.; Bauder, J.W. Particle-size analysis. In *Methods of Soil Analysis, Part 1*; SSSA Book Series 5; Klute, A., Ed.; Soil Science Society of America, American Society of Agronomy: Madison, WI, USA, 1986; pp. 383–411.

39. Blake, G.R.; Hartge, K.H. Bulk density. In *Methods of Soil Analysis, Part 1*; SSSA Book Series 5; Klute, A., Ed.; Soil Science Society of America, American Society of Agronomy: Madison, WI, USA, 1986; pp. 363–375.

40. National Sustainable Agriculture Coalition. Available online: http://sustainableagriculture.net/about-us/what-is-sustainable-ag/ (accessed on 31 May 2017).

41. Union of Concerned Scientists. Science for a Healthy Planet and Safer World. Available online: http://www.ucsusa.org/food-agriculture/advance-sustainable-agriculture/what-is-sustainable-agriculture#.WS6WkpKGPIU (accessed on 31 May 2017).

42. UWestern SARE. Sustainable Agricultural Research and Education. Available online: http://www.westernsare.org/About-Us/What-is-Sustainable-Agriculture (accessed on 31 May 2017).

43. Grace Communications Foundation. Available online: http://www.sustaianble.org/246/sustainable-agriculture-the-basics (accessed on 31 May 2017).

44. UCDAVIS Agricultural Sustainability Institute. Available online: http://asi.ucdavis.edu/programs/sarep/about/what-is-sustainable-agriculture (accessed on 31 May 2017).

45. Musiitwa, F.; Komutunga, E. Agricultural systems. In *Agriculture in Uganda*; Mukiibi, J.K., Ed.; Uganda National Agricultural Research Organization (NARO): Entebbe, Uganda, 2001; Volume I, pp. 220–230.

46. Nelson, L.B.; Muckenhirn, R.J. Field percolation rates of four Wisconsin soils having different drainage characteristics. *J. Am. Soc. Agron.* **1941**, *33*, 1028–1036. [CrossRef]

47. Fritton, D.D.; Olson, G.W. Bulk density of a fragipan soil in natural and disturbed profiles. *Soil Sci. Soc. Am. Proc.* **1972**, *36*, 686–689. [CrossRef]

48. Dawud, A.Y.; Gray, F. Establishment of the lower boundary of the sola of weakly developed soils that occur in Oklahoma. *Soil Sci. Soc. Am. J.* **1979**, *43*, 1201–1207. [CrossRef]

49. Larson, W.E.; Gupta, S.C.; Useche, R.A. Compression of agricultural soils from eight soil orders. *Soil Sci. Soc. Am. J.* **1980**, *44*, 450–457. [CrossRef]

50. Yule, D.F.; Ritchie, J.T. Soil shrinkage relationships of Texas Vertisols: I. Small cores. *Soil Sci. Soc. Am. J.* **1980**, *44*, 1285–1291. [CrossRef]

51. Ssali, H. Soil fertility. In *Agriculture in Uganda*; Mukiibi, J.K., Ed.; Uganda National Agricultural Research Organization (NARO): Entebbe, Uganda, 2001; Volume I, pp. 104–135.

52. Hazelton, P.; Murphy, B. *Interpreting Soil Test Reults: What do the Numbers Mean?* CSIRO: Collingwood, Australia, 2010.

53. Namugwanya, M.; Tenywa, J.S.; Etabbong, E.; Mubiru, D.N.; Basamba, T.A. Development of common bean (*Phaseolous Vulgaris* L.) production under low soil phosphorus and drought in sub Saharan Africa: A review. *J. Sustain. Dev.* **2014**, *7*, 128–139.

54. Goettsch, L.H.; Lenssen, A.W.; Yost, R.S.; Luvaga, E.S.; Semalulu, O.; Tenywa, M.; Mazur, R.E. Improved production systems for common bean on Phaeozem soil in south-central Uganda. *Afr. J. Agric. Sci.* **2016**, *11*, 4797–4809.

55. Kalyebara, R. *The Impact of Improved Bush Bean Varieties in Uganda*; Network on Bean Research in Africa, Occasional Publication Series 43; Chartered Institute of Architectural Technologists (CIAT): Kampala, Uganda, 2008.

56. Sibiko, K.W.; Ayuya, O.I.; Gido, E.O.; Mwangi, J.K. An analysis of economic efficiency in bean production: Evidence from eastern Uganda. *J. Econ. Sustain. Dev.* **2013**, *4*. Available online: www.iiste.org (accessed on 2 May 2017).

57. Ghaffarzadeh, M.; Garcia, F.; Cruse, R.M. Grain yield response of corn, soybean, and oat grown in strip intercropping system. *Am. J. Altern. Agric.* **1994**, *9*, 171–177. [CrossRef]

58. Ghaffarzadeh, M.; Garcia, F.; Cruse, R.M. Tillage effect on soil water content and corn yield in a strip intercropping system. *Agron. J.* **1997**, *89*, 893–899. [CrossRef]

59. Shaw, R.H.; Felch, R.E.; Duncan, E.R. *Soil Moisture Available for Crop Growth*; Special Report; Iowa State University: Ames, IA, USA, 1972.

60. Hulugalle, N.R.; Willatt, S.T. Seasonal variation in the water uptake and leaf water potential of intercropped and monocropped chillies. *Exp. Agric.* **1987**, *23*, 273–282. [CrossRef]

61. Francis, C.A.; Jones, A.; Crookston, K.; Wittler, K.; Goodman, S. Strip cropping corn and grain legumes: A review. *Am. J. Altern. Agric.* **1986**, *1*, 59–164.

62. Hulugalle, N.R.; Lal, R. Soil water balance in intercropped maize and cowpea in a tropical hydromorphic soil in western Nigeria. *Agron. J.* **1986**, *77*, 86–90. [CrossRef]

63. Fortin, M.C.; Culley, J.; Edwards, M. Soil water, plant growth, and yield of strip-intercropped corn. *J. Prod. Agric.* **1994**, *7*, 63–69. [CrossRef]

64. Mazvimavi, K.; Ndlovu, P.V.; Nyathi, P.; Minde, I.J. Conservation agriculture practices and adoption by smallholder farmers in Zimbabwe. Presented at the 3rd African Association of Agricultural Economists (AAAE) and 48th Agricultural Economists Association of South Africa (AEASA) Conference, Cape Town, South Africa, 19–23 September 2010.

sustainability

MDPI

Review

Comparison of Organic and Integrated Nutrient Management Strategies for Reducing Soil N₂O Emissions

Rebecca F. Graham [1],*, Sam E. Wortman [2] and Cameron M. Pittelkow [1]

[1] Department of Crop Sciences, University of Illinois at Urbana-Champaign, Champaign, IL 61801, USA; cmpitt@illinois.edu

[2] Department of Agronomy and Horticulture, University of Nebraska-Lincoln, Lincoln, NE 68583, USA; swortman@unl.edu

* Correspondence: rfgraha2@illinois.edu; Tel.: +1-217-333-3420

Academic Editors: Suren N. Kulshreshtha and Elaine Wheaton

Received: 25 January 2017; Accepted: 22 March 2017; Published: 28 March 2017

Abstract: To prevent nutrient limitations to crop growth, nitrogen is often applied in agricultural systems in the form of organic inputs (e.g., crop residues, manure, compost, etc.) or inorganic fertilizer. Inorganic nitrogen fertilizer has large environmental and economic costs, particularly for low-input smallholder farming systems. The concept of combining organic, inorganic, and biological nutrient sources through Integrated Nutrient Management (INM) is increasingly promoted as a means of improving nutrient use efficiency by matching soil nutrient availability with crop demand. While the majority of previous research on INM has focused on soil quality and yield, potential climate change impacts have rarely been assessed. In particular, it remains unclear whether INM increases or decreases soil nitrous oxide (N₂O) emissions compared to organic nitrogen inputs, which may represent an overlooked environmental tradeoff. The objectives of this review were to (i) summarize the mechanisms influencing N₂O emissions in response to organic and inorganic nitrogen (N) fertilizer sources, (ii) synthesize findings from the limited number of field experiments that have directly compared N₂O emissions for organic N inputs vs. INM treatments, (iii) develop a hypothesis for conditions under which INM reduces N₂O emissions and (iv) identify key knowledge gaps to address in future research. In general, INM treatments having low carbon to nitrogen ratio C:N (<8) tended to reduce emissions compared to organic amendments alone, while INM treatments with higher C:N resulted in no change or increased N₂O emissions.

Keywords: integrated nutrient management (INM); nitrous oxide emissions; organic nitrogen inputs; inorganic nitrogen fertilizer

1. Introduction

Soil nitrous oxide (N₂O) emissions are one of the largest sustainability concerns facing agriculture. Atmospheric N₂O concentrations have been increasing since the mid 19th century when humans started applying nitrogen (N) fertilizer to cultivated land [1]. Nitrous oxide is naturally produced through the denitrification and nitrification in the nitrogen (N) cycle, there is a clear link between increased N application rates and increased N₂O emissions [2,3]. Agricultural soil management accounts for approximately 75% of anthropogenic N₂O emissions in the United States [4]. Nitrous oxide is a key greenhouse gas contributing to global climate change, with a global warming potential nearly 300 times that of carbon dioxide (CO₂) over a 100-year period and an atmospheric lifespan of about 120 years [5]. Moreover, N₂O contributes to roughly 6% of the overall radiative forcing in the atmosphere and is considered to be the single most important ozone-depleting substance in our atmosphere [6]. Due to these harmful environmental impacts, and the fact that emissions are largely anthropogenic in nature, it is critical to identify options for reducing N_2O emissions from agriculture. Although it results in increased N_2O emissions, additional N is often applied either in the form of inorganic N fertilizers or organic amendments (e.g., crop residues, manure, compost etc.) to prevent N limitations to crop growth. Soil organic matter is an important, yet sometimes overlooked, source of nutrients in agricultural systems. The current challenge of large amounts of reactive N losses from agricultural systems is largely driven by inorganic fertilizer N inputs, while low-input cropping systems that primarily rely on organic amendments and soil organic matter are thought to be more sustainable. Organic amendments can promote soil health by building organic matter content, increasing aeration, and enhancing microbial abundance and diversity [7,8]. Organic amendments also provide plant-available nutrients and often act as a slow-release fertilizer throughout the growing season. However, it is difficult to achieve the soil health and nutrient-provisioning benefits of organic N sources. High-quality organic amendments with a low carbon to nitrogen ratio (C:N) decompose quickly and contribute less to stable organic matter in the soil, whereas amendments with a high C:N (low quality) decompose slowly and may not supply sufficient N to meet crop demand [9], potentially resulting in lower yields. By contrast, inorganic N fertilizers easily dissolve in soil solution and are quickly available for plant uptake upon application. Given their contrasting properties, the integrated use of both organic amendments and inorganic fertilizers may contribute to improved soil quality without sacrificing crop nutrition or yield.

Integrated nutrient management (INM) is the concept of using a combination of organic, inorganic, and biological amendments to increase nitrogen use efficiency (NUE) and reduce nutrient loss by synchronizing crop demand with nutrient availability in soil [10]. There are three main principles that govern INM: (1) use all possible sources of nutrients to optimize their input; (2) match soil nutrient supply with crop demand spatially and temporally; and (3) reduce N losses while improving crop yield [11]. Integrated nutrient management is a broad concept and particular versions of this approach have gained popularity in different regions. For example, Integrated Soil Fertility Management (ISFM) has a long history of research and application in smallholder farming systems of Africa. To meet current food security challenges, ISFM is now viewed as an important framework for boosting crop productivity while improving soil quality by several international initiatives [12]. A meta-analysis of the effects of applying both inorganic fertilizer and organic amendments together found that combining the two increased maize yield between 60% and 114% compared to either N source alone [13]. A two-year experiment in hybrid rice systems using INM reported higher yield, NUE, and soil organic carbon compared to both organic and inorganic amendments alone [14]. Javaria and Kahn analyzed several studies with tomatoes and found that INM improved yield, overall crop quality, and soil fertility [15]. Importantly, the principles of INM are broadly applicable and the concept of combining organic with inorganic N sources can either be integrated into low-input systems to increase soil nutrient supply or high-input systems to potentially reduce N fertilizer requirements. Given this adaptability and the numerous cropping system benefits that it provides, INM will likely expand as a practice to address nutrient losses from agriculture in the future.

Global NUE ranges from approximately 20–65% [16]. Nutrient imbalances are common in agricultural systems with N fertilizer often being over- or under-applied [17], indicating significant room for increased efficiency. A meta-analysis of experiments in sub-Saharan Africa found that combining inorganic and organic amendments significantly improved nutrient efficiency [12]. Although much research has been dedicated to the effects of INM on crop productivity, NUE, and aspects of soil quality, it remains unclear how INM impacts N_2O emissions. Importantly, increasing NUE by combining organic and inorganic fertilizers may reduce N_2O emissions, which would represent an additional positive cropping system benefit. However, available studies show that N_2O emissions from INM can both increase or decrease depending on cropping system context and management [18,19]. It is generally thought that organic N inputs would lead to lower N_2O emissions compared to INM because organic N must be mineralized before becoming susceptible to losses. Yet it is also possible that increases in labile carbon (C) and microbial activity associated with organic inputs may increase nitrification and denitrification processes, contributing to higher overall N_2O emissions. Reducing soil N_2O emissions without negatively impacting yields is a critical sustainability challenge facing global agriculture, particularly because N_2O emissions can represent more than half of the total C footprint of crop production systems. To gain a holistic understanding of the environmental performance of INM, the potential for increased N_2O emissions represents an environmental tradeoff that needs to be considered.

To our knowledge, no attempt has been made to synthesize current evidence regarding the potential climate-change impacts of N_2O emissions from INM as compared to systems entirely dependent on organic amendments. The purpose of this paper is to: (1) review the mechanisms influencing how organic and inorganic N amendments affect N_2O emissions; (2) synthesize findings from the limited number of field experiments that have directly compared INM and organic N input systems; (3) develop a hypothesis for conditions under which INM reduces N_2O emissions; and (4) identify key knowledge gaps to address in future research.

2. Effect of Organic vs. Inorganic N Sources on N_2O Emissions

2.1. Summary of Research

There is no scientific consensus on whether N_2O emissions are lower in systems using inorganic N fertilizers or organic N amendments. In fact, a recent meta-analysis found that there was no significant difference in N_2O emissions between the two soil management approaches [20]. A comprehensive analysis focused on maize production in the Midwestern U.S. found higher N_2O emissions from soils receiving manure than soils receiving inorganic N fertilizer [21]. However, these authors noted that N_2O emissions may have been greater in manure-amended experiments because the N application rate in some studies were greater for manured fields than the inorganic N application rate. By contrast, other reviews have reported lower N_2O emissions from soils managed with organic inputs compared to conventional systems using inorganic N inputs, particularly when assessed on an areal basis [22–24]. Lower emissions with organic inputs in these comparisons could be due to a number of confounding factors, including significantly lower N inputs in the organic systems [25]. There is a direct relationship between N_2O emissions and N addition [26], therefore a reduction in N inputs in organic systems is likely to reduce N_2O emissions [27]. However, it should be considered that a reduction in N rates may also contribute to lower crop yields, which may have implications for global food production [28,29]. Despite the demonstrated relationship between N inputs and N_2O emissions, many other environmental and crop management factors can influence the complex process of N_2O loss from soil [30].

2.2. Mechanisms Controlling N_2O Emissions

Nitrous oxide is produced by the activity of soil microorganisms through two phases of the nitrogen cycle: nitrification and denitrification [31,32]. A number of variable and interacting factors within the soil system control these phases of the N cycle. Soil texture, freeze/thaw cycles, precipitation events and temperature all significantly effect N_2O emissions but cannot be easily controlled through management [33]. Factors affecting N_2O emissions that can be more easily altered by crop management practices include: soil organic C content, nitrate and ammonium concentrations in soil solution, N application rate, type, and technique, soil oxygen status, microbial abundance and activity, soil pH, soil drainage and moisture, and crop species [3]. Application of inorganic versus organic N fertilizer will influence many of the above factors [34], and a number of potential interactions are expected to occur among factors, which will ultimately determine the relative change in N_2O emissions (e.g., changes in soil moisture will influence microbial activity and subsequently inorganic N concentrations). Important factors that are differentially affected by organic N amendments and inorganic N fertilizer are briefly discussed below: soil organic carbon, soil structure and moisture, soil pH, soil N status. For a more general review of microbial processes regulating N_2O emissions, the reader is referred to the following papers [32,34–37].

2.2.1. Soil Organic Carbon

Carbon availability is a key component of the denitrification process. Early research showed that denitrification is greatly influenced by C availability, with denitrification rates of (nitrate) NO_3-N remaining low when C was unavailable despite high N concentrations, but increasing rapidly in response to C addition [38]. Depending on the source of organic N amendments, addition of organic material to soils may increase N_2O emissions by providing the necessary C substrates for driving microbial nitrification and denitrification processes [34,39]. Similarly, N_2O emissions tend to increase with the C:N of soil, due in part to the potential for reduced plant N uptake and increased microbial consumption of inorganic N during soil organic matter decomposition [20]. Manure application increases total organic C and total N pools in soil at levels proportional to the application rate [40]. When comparing soil with a history of manure application to a non-manured soil, it was found that N_2O emissions were nearly 25 times greater from manured soil [41]. In contrast to organic amendments, inorganic N fertilizers do not provide additional C substrate, but this will not necessarily lead to lower emissions. Moreover, addition of inorganic N fertilizer can have a priming effect on soil microbial communities which facilitates more rapid decomposition of soil organic matter [42], potentially also increasing N_2O emissions.

2.2.2. Soil Aggregation, Drainage, and Moisture

The relationship between water-filled pore space (WFPS), drainage, and N_2O emissions is not completely understood. Generally, N_2O emissions are greatest following a significant increase in soil water content after a rainfall or irrigation event [22], likely due to a flush of microbial activity from soil wetting and drying events [43]. Both denitrification and nitrification processes contributing to N_2O emissions are stimulated at high WFPS, with nitrification playing a larger role as soils dry down [32]. Soils with restricted drainage, even if they are not completely water-saturated, are particularly prone to greater N_2O emissions [2]. For example, fine-textured soils that typically are associated with greater soil water content tend to have higher N_2O emissions [20]. Thus, an important opportunity for decreasing emissions is to increase soil aeration, potentially through soil amendments or changes in soil structure. Increased aggregate stability can create larger soil pores between aggregates in fine-textured soils, and greater pore sizes may increase oxygen (O_2) content, which has been shown to decrease N_2O emissions [32]. Accordingly, soils managed with organic amendments tend to have greater aggregate stability compared to those managed with inorganic fertilizer [25], therefore the addition of organic amendments may reduce N_2O emissions, especially in fine-textured soils.

At the same time, it must be considered that organic amendments can increase soil water holding capacity, particularly for coarse-textured soils. In addition, because O_2 concentrations in soil pores are determined by water content as well as microbial activity, elevated microbial respiration in response to higher C availability in organic inputs may decrease O_2 content and increase N_2O emissions [34]. The extent to which these physical and biological processes may interact to increase or decrease soil water content will largely depend on initial soil texture. Soils with lower water holding capacity may be more likely to experience increases in soil water content following addition of organic N inputs, whereas soils with initially higher water holding capacity may benefit more greatly from increased pore size, in turn leading to increased O_2 concentrations and decreased N_2O emissions.

2.2.3. Soil pH

Denitrification-associated N_2O emissions are generally greater in acidic soils (pH < 6) which may either be due to increased activity of relevant soil microbes in these conditions or inhibition of enzymes necessary for complete denitrification, including nitrous oxide reductase [20,37,44]. As such, the ratio of N_2O:N_2 production is generally higher in soils that are acidic to neutral compared with alkaline soils [2]. Soils receiving primarily organic inputs tend to have slightly higher pH than those managed with inorganic fertilizer due to the acidifying potential of inorganic N fertilizer [9,20,25]. Considered in isolation of other interacting effects discussed above (e.g., soil moisture and C:N), maintenance or increase of soil pH following repeated organic amendment may help to mitigate N_2O emissions from soil.

2.2.4. Soil N Availability

Increases in N inputs tend to increase soil inorganic N concentrations and N_2O emissions [45], through stimulation of both nitrification and denitrification microbial processes [32,46]. Addition of inorganic N fertilizer may cause large amounts of NO_3 to accumulate in soil because an available C source is needed to provide energy for microbial activity and C and N transformations [38]. Moreover, the ratio of $N_2O:N_2$ produced via denitrification increases with increasing soil NO_3 concentrations [38], suggesting higher N_2O emissions in soils with inorganic N application. By contrast, soils receiving primarily organic N sources tend to have lower levels of available NO_3 [47]. The majority of C and N added through organic amendments is stored in organic matter pools, which are less susceptible to N losses [48]. However, even if soil inorganic N concentrations do not increase with organic amendments, N losses are not always lower. For example, recent research indicates that when organic N sources contain a relatively high concentration of inorganic N (i.e., more than 0.3% dry weight), the percentage of applied N emitted as N_2O is generally greater than the expected ranged predicted by the Intergovernmental Panel on Climate Change default emission factors and may be higher than inorganic N addition [49].

3. INM and N_2O Emissions

3.1. Summary of Field Research

Research is limited on the effects of INM on N_2O emissions. We performed a literature search in Web of Science, Google Scholar and Scopus using the following search terms: nitrogen, organic amendments, inorganic fertilizer, integrated nutrient management, integrated soil fertility management, combined inorganic and organic amendments, nitrous oxide and N_2O emissions. Publications included in the review had to represent field experiments comparing N_2O emissions over the course of a growing season in four treatments: (1) unfertilized control plot; (2) organic-only; (3) inorganic-only, and a combination of organic and inorganic N inputs, which was considered to represent an INM approach for the purposes of this paper. A summary of experimental details and the overall effects of INM compared to organic N inputs on N_2O emissions for each study ($n = 6$) are presented in Table 1.

Meng et al. [50] compared N_2O emissions from maize (*Zea mays* L.) and winter wheat (*Triticum aestivum* L.) treated with composted manure, inorganic N fertilizer and a combined INM approach with equal amounts of N from both sources in China during 2002–2003 (treatments had been established in 1989). They found no difference in N_2O emissions among all treatments over the two growing seasons but still recommended combining inorganic and organic fertilizers to improve overall soil fertility. Emissions were positively correlated with WFPS and soil temperature, and the authors concluded that these factors, along with pH, likely accounted for the lack of difference between treatments. Yields in each treatment were not reported.

Researchers in Zimbabwe [18] measured N_2O emissions from seasonal wetland soils in rape (*Brassica napus*) production supplemented with a combination of ammonium nitrate and organic manure at varying levels. While higher N rates increased N_2O emissions, they found that INM reduced emissions compared to sole organic or inorganic treatments despite total N inputs for INM being slightly higher (125 for INM compared to 97.5 and 120 kg N ha^{-1} for organic and inorganic treatments, respectively). These authors noted that the INM treatment with the highest rate of manure led to increased N_2O emissions compared to lower rates from manure alone, which was due to increased N rate more than N source. Soil moisture was a poor predictor of N_2O emissions in this study, likely due to the consistently high moisture levels caused by irrigation. The INM treatment increased yield significantly compared to both inorganic and organic treatments.

Table 1. Description of field experiments reviewed in this paper including study location, crop, total nitrogen (N) rate, and mean nitrous oxide (N_2O) emissions from the selected organic amendment, inorganic N fertilizer, and Integrated Nutrient Management (INM) treatment comparisons, fertilizer-induced emission factor (EF) and carbon to nitrogen ratio (C:N) of the organic amendment treatment. The effect of INM on N_2O emissions and crop yield is summarized when reported.

Study	Country	Crop Rotation	Total N Rate (kg N ha⁻¹)	Mean N_2O (mg N m⁻²)	EF * (%)	C:N of Org †	N_2O Trend	INM Yield
Meng et al. (2005) [50]	China	Maize, winter wheat	Org: 150 MC #; Inorg ‡: 150 urea; INM: 75 kg MC + 75 urea (150 total INM)	Control: 15; Org †: 85.6; INM: 81.8; Inorg ‡: 76.7	Org: 0.471; INM: 0.445; Inorg: 0.411	7.75	No significant differences	Not reported
Nyamadzawo et al. (2014) [18]	Zimbabwe	Rape	Org: 97.5 M §; Inorg: 120 NH₄NO₃; INM: 65 M + 60 NH₄NO₃ (125 total INM)	Control: 250; Org: 1930; INM: 770; Inorg: 1200	Org: 17.231; INM: 4.160; Inorg: 7.917	Not reported	INM < Org, Inorg	Org, Inorg < INM
Cai et al. (2013) [51]	China	Maize, winter wheat	Org: 150 MC ‖; Inorg: 150 urea; INM:75 MC + 75 urea (150 total INM)	Control: 22; Org: 166; INM: 118; Inorg: 181	Org: 0.960; INM: 0.640; Inorg: 1.060	8	No significant differences	Not reported
Ding et al. (2013) [19]	China	Maize, winter wheat	Org: 150 MC; Inorg: 150 urea; INM: 75 MC + 75 kg urea (150 total INM)	Control: 20.4; Org: 117.8; INM: 132.5; Inorg: 142.7	Org: 0.812; INM: 0.934; Inorg: 1.019	8	INM, Org < Inorg	Not reported
Nyamadzawo et al. (2014) [52]	Zimbabwe	Maize, winter wheat	Org: 120 M; Inorg: 120 NH₄NO₃; INM: 60 M + 60 NH₄NO₃ (120 total INM)	Control: 32; Org: 27; INM: 35; Inorg: 41	Org: −0.042; INM: 0.025; Inorg: 0.075	Not reported	Org < INM < Inorg	Org < INM, Inorg
Sarkodie-Addo et al. (2003) [53]	United Kingdom	Maize	Org: Inc ¶ rye; Inorg: 250 NH₄NO₃; INM: Inc rye + 250 NH₄NO₃; Org: Inc winter wheat; Inorg: 200 NH₄NO₃; INM: Inc winter wheat + 200 NH₄NO₃	*Rye* Control: 6.07; Org: 5.27; Inorg: 9.15; INM: 7.72; *Winter Wheat* Control: 2.65; Org: 2.32; INM: 15.4; Inorg: 8.87	N/A	=Rye: 13Wheat: 18	Org, Inorg < INM Rye < winter wheat	Org < INM < Inorg

* Emission Factor (EF) calculated as = ($N_2O_{trt} − N_{ctl}$)/N_2O rate$_{trt}$ × 100; † Organic-only treatment; # Manure compost; ‡ Inorganic-only treatment; § Cattle manure; ¶ Incorporated.

Two field studies in China compared N_2O emissions from inorganic NPK fertilizer, compost, and INM with 50% of N from each source in both wheat and maize. Both studies found that combining inorganic NPK and compost significantly reduced N_2O emissions compared to compost or NPK fertilizer alone at a total N rate of 150 kg N ha^{-1} [19,51]. Cai et al. found that WFPS was significantly correlated with N_2O emissions [51], whereas Ding et al. did not observe this relationship [19]. Both researchers suggested that applying composted organic amendments with C:N lower than 20 would reduce N_2O emissions due to a lower amount of N released during decomposition into the soil [19,51]. Soil pH was also negatively correlated with N_2O emissions in Ding et al. [19], which supports the hypothesis that N_2O emissions may be mitigated in soils with organic N inputs in part due to the effect of organic amendments on soil chemistry. Despite significant differences in N_2O emissions, crop yield did not differ among N input strategies in Cai et al. [51]. Ding et al. [19] did not report yield.

Nyamadzawo et al. [52] measured the effects of INM on N_2O emissions during one growing season for winter wheat and one for maize in Zimbabwe using combinations and rates of cattle manure and ammonium nitrate. The INM treatment increased N_2O emissions compared to the cattle manure treatments, but decreased emissions compared to the inorganic fertilizer treatments. The lower N_2O emissions from the organic amendment treatments were likely due to the slow decomposition of C and N in the manure and the slow release of mineralized N. These authors noted that if the low-quality manure were the only source of N, there would be yield penalties. There was no correlation between N rate and N_2O emissions, which emphasizes the importance of N source in this experiment. Yields from the INM treatments were not different from the inorganic N treatments, but were greater than the manure treatments. These results clearly demonstrate a tradeoff between yield and N_2O emissions and the potential to balance the two with an INM approach.

Sarkodie-Addo et al. [53] compared winter rye and wheat as green manures with added ammonium nitrate and measured subsequent N_2O emissions in one growing season in the U.K. Incorporating green manure with added inorganic N fertilizer significantly increased N_2O emissions compared to inorganic N addition and green manure alone. Sarkodie-Addo et al. [53] suggest that elevated N_2O emissions were due to the supply of C from the incorporated rye and wheat residue, which combined with the added inorganic N provided energy for denitrifying microbes. Crop yields in the INM treatments were greater than in the green manure treatments, but not the inorganic N fertilizer treatments.

3.2. Potential for Minimizing N_2O Emissions with INM

We found only six field experiments that fit our criteria for inclusion in this review. Despite the limited data, several key factors appear to be controlling N_2O emissions in INM compared to organic amendment systems. Water-filled pore space and C:N of the organic amendment are likely factors explaining the variability in N_2O emissions from INM systems. Nitrous oxide emissions tend to peak following rainfall events as the soil dries during the transition from high soil moisture to low [22,50]. However, when soil moisture remains high (i.e., with irrigation), the correlation between N_2O emissions and WFPS disappears [18]. Therefore, in the absence of irrigation, a substantial proportion of N_2O emissions will be driven by changes in WFPS and cannot be mitigated via management. Integrated treatments received about half the amount of organic material as the organic-only treatments, which would not be likely to result in short-term changes in soil porosity or soil water holding capacity as discussed above. However, to our knowledge no studies have measured N_2O emissions for more than three years to determine potential long-term effects between organic and INM systems.

For the studies included in this review, C:N of organic amendments in the INM treatments had the greatest potential to control N_2O emissions. N_2O emissions from INM systems using organic amendments with C:N near 8 (composted manures and yard waste compost) were lower or equivalent to the organic amendment treatments in most cases [19,50,51]. Conversely, when the C:N of organic amendments were around 8 which included treatments of fresh cattle manure and freshly incorporated green manure, N_2O emissions from INM increased compared to the organic amendment treatments [52,53]. These results suggest that high C substrate availability relative to the amount of added N is an important contributor to N_2O emissions in INM systems [54]. Differences in C:N of amendments in these studies were typically due to composting, which has been shown to reduce the C:N of organic material [55,56], and limit N_2O emissions when compared to non-composted organic inputs [55]. This preliminary review suggests combining compost with inorganic N fertilizer holds promise for further reducing N_2O emissions compared to either inorganic or organic N inputs alone.

While field studies on N_2O emissions from INM are rare, a number of investigations have assessed the effects of organic amendment C:N on N_2O emissions in lab incubations. As noted above, microbial communities that carry out critical steps of N mineralization, nitrification, and denitrification depend on a supply of available C to function. Considering a context without inorganic N addition, organic amendments with high C:N (>20) tend to result in microbial immobilization of inorganic N, in turn reducing N available for denitrification [57]. Alternatively, amendments with low C:N are more rapidly mineralized by soil microbes [57], releasing C and N which can promote increased microbial activity and increase N_2O emissions. However, microbial activity and resulting N_2O emissions from INM treatments not only depend on C:N, but also the amount of inorganic N in soil from added fertilizer. Application of inorganic N fertilizer with an organic amendment containing large amounts of labile C may further enhance denitrification and therefore increase N_2O emissions [58]. In fine-textured soils, N_2O emissions from both soil fertilized with only inorganic N fertilizer and composted amendments applied in combination with inorganic N application are similar [59]. Moreover, a synthesis of N_2O emission factors for organic N addition recently placed compost and compost plus inorganic N fertilizer application in low- and medium-risk categories, respectively, as compared to the high-risk group containing animal slurries and biosolids [49].

When interpreting results from experiments measuring N_2O emissions using a combination of inorganic N and organic amendments with varying C:N it is critical to consider both the total N rate and type of amendment. Huang et al. [60] found decreasing N_2O emissions in response to increasing C:N of different organic residues in a laboratory experiment, and observed that this relationship became stronger when combined with inorganic N addition. However, in this study, the treatment with the highest N_2O emissions (and lowest C:N) also had the greatest total N rate, which likely accounts for the high emissions. Similarly, results from another laboratory experiment measuring N_2O in response to increasing C:N of crop residue amendments and varying levels of inorganic fertilizer suggest that reductions in N_2O emissions are more likely to occur when lower C:N organic amendments are applied alone or when higher C:N organic amendments are applied in combination with inorganic fertilizer [55]. Differences from our review of field experiments and these laboratory studies can be accounted for by the fact that crop residues were used as organic amendments to provide organic material, and not necessarily an N source, while manure or compost were used as an organic N source in the reviewed field experiments. Moreover, a recent global meta-analysis of fertilizer emission factors for organic amendments concluded that C:N only partially explains the response of N_2O emissions to organic amendments, particularly for C:N lower than 25 where other environmental and management factors appear to become increasingly important [49]. In general, research on the effects of amendment C:N at this lower range of values, including the studies reviewed here, is inconsistent and further field investigations are needed.

A third factor possibly explaining trends for N_2O emissions from INM systems is the total N rate when using a combination of organic and inorganic fertilizers. For this review the majority of studies had similar N rates between all treatments, but one study used an additive N approach (i.e., full inorganic N rate plus full organic N amendment rate) for their INM treatment. Increasing N rate generally increases N_2O emissions [26,45]. Not surprisingly, using an additive approach to N application for an INM treatment resulted in higher N_2O emissions compared to the other treatments consisting of either inorganic or organic N inputs alone which had lower total N rates [53]. Studies using a substitutive approach to assess the effects of INM (i.e., half inorganic N rate plus half organic N rate) generally found no differences or decreased N_2O emissions compared to inorganic or organic

N inputs alone [18,19,50,51]. To account for potential differences in N rate for INM treatments across studies, we also calculated fertilizer-induced emission factors (EF) (Table 1). When EFs were averaged across studies, INM resulted in a value of 1.2% compared to 3.9% for organic-only treatments and 2.1% for inorganic-only treatments. While there was a large amount of variability for these means, based on this calculation, INM stands out as a potential management strategy to reduce N_2O emissions. To assess the effect of combining inorganic and organic treatments on N_2O emissions more accurately, it is recommended that consistent total N rates, or consistent levels of different N rates, should be applied for INM, organic, and inorganic treatments in future experiments.

It is important to note that combining inorganic fertilizer and organic amendments is not a guaranteed strategy for reducing N_2O emissions from agricultural soils. However, this review suggests that INM systems employing amendments with medium to low C:N (<8) and a substitutive approach to total N application rates (proportional reduction of N rate from each N source) have the greatest potential to mitigate N_2O emissions. Previous lab experiments provide support for this hypothesis by indicating that integrating inorganic N fertilizer with organic amendments having low to medium C:N may help avoid two important processes contributing to N_2O emissions, namely stimulation of soil microbial communities through addition of excessive carbon substrates (high C:N) and rapid inorganic N mineralization (low C:N).

4. Knowledge Gaps and Additional Considerations

4.1. Knowledge Gaps for Field Research

The challenge of sustainably increasing global food production through appropriate nutrient management practices can also directly be related to soil N_2O emissions. Due to a limited number of available field experiments, definitive conclusions about amendment properties driving N_2O emissions in INM systems cannot be drawn here. To further understand the effect of C:N on N_2O emissions, future INM field experiments should include amendments with a range of C:N paired with inorganic N fertilizer at a constant total N rate (i.e., substitutive N input approach), similar to the laboratory research conducted by Frimpong and Baggs [54]. Moreover, different types of organic amendments (e.g., manure, compost and green manure from different species) with the same C:N should be paired with constant total N rates to explore the effects of other amendment properties aside from C:N on N_2O emissions. While most studies reviewed here used equal portions of organic and inorganic N sources in their INM treatment, there is limited knowledge about the ideal of organic amendment versus inorganic N fertilizer necessary to mitigate N_2O emissions while optimizing crop yield with an INM approach. Our review indicates that additional factors contributing to N_2O emissions in INM systems requiring further study include the duration and method of composting amendments; the method, timing, and location of amendment and fertilizer applications (including whether N sources were incorporated into soil); and local climate and soil characteristics.

4.2. Yield-Scaled Emissions and INM

Recent reports have highlighted the importance of considering crop yield response when measuring the environmental impact of N_2O mitigation strategies for agriculture. For example, the concept of yield-scaled emissions (expressed as mass of N_2O emissions per unit yield) has gained recognition as a practical assessment tool [61]. Importantly, if fertilizer N input strategies aimed at N_2O mitigation decrease yield in addition to reducing N_2O emissions, they may not be considered as an effective strategy for addressing both food security and environmental goals. With a focus on yield, research demonstrates that management of crop nutrients through INM can increase crop productivity compared to fields managed with inorganic N fertilizer alone [11]. Similarly, based on the few studies in this review that reported yield, INM holds the potential to increase yields compared to organic amendments alone [18,52,53]. A general consideration is that promising yield-scaled N_2O mitigation strategies either decrease emissions while maintaining yield, or maintain emissions while increasing yield. On a yield-scaled basis, soils receiving only organic amendments have been found to emit more N_2O than soils receiving inorganic N fertilizer [24,25,62]. However, this comparison is confounded by other factors that often limit yield in systems primarily managed with organic N inputs (e.g., pests [28,63]), which may not accurately represent N_2O emission potential based on soil nutrient management alone. Two studies measuring yield in this review [18,52] also calculated yield-scaled emissions and found that INM treatments generally reduced yield-scaled emissions compared to organic amendment and inorganic N fertilizer treatments. Future field experiments should evaluate yield-scaled emissions when assessing the potential environmental and agronomic benefits of INM, as this metric is likely to become increasingly important due to the ongoing challenge of achieving global food security.

4.3. Net Global Warming Potential

This review focused on direct soil N_2O emissions, but the net global warming potential (GWP) of cropping system greenhouse gas (GHG) emissions is determined by changes in soil N_2O, CO_2, and CH_4 emissions as well as changes in soil C [64]. Addition of organic amendments can be expected to increase soil CO_2 emissions, while also contributing to short- and long-term pools of soil C [65]. Increases in soil C are particularly important because they can offset higher soil CO_2 emissions due to respiration of organic C added to the field through organic amendments, as well direct N_2O emissions and embodied CO_2 costs related to organic or inorganic N inputs [34,66]. At the same time, when assessing the combined addition of inorganic N fertilizer with organic amendments, cycling of soil C and N is tightly coupled and N fertilizer addition has been shown to influence microbial activity and the breakdown of C and N substrates compared to organic amendments alone [65]. Future work is thus needed under field conditions to determine whether INM increases or decreases the buildup of soil C relative to addition of organic amendments alone, while also simultaneously monitoring for potential changes in N_2O fluxes. Finally, there are upstream environmental impacts beyond the field boundary related to inorganic and organic N sources that need to be considered. Large amounts of energy are consumed to produce and transport N fertilizer and this is associated with significant embodied CO_2 costs, which can be large enough to negate changes in soil C discussed above [67]. Likewise, organic N amendments derived from manure or compost have CO_2, N_2O, and CH_4 emissions associated with their processing and transport [34]. Therefore, to fully determine the net impact of INM practices on GWP, changes in soil C and the embodied GHG costs associated with inorganic N fertilizer and organic amendments need to be quantified and weighed against any increase or decrease in soil N_2O emissions discussed in this review.

5. Conclusions

Increasing nutrient use efficiency and reducing nutrient loss in agricultural systems while simultaneously improving crop yields is a critical sustainability challenge facing humanity. The concept of using all available sources of N inputs (organic, inorganic or biological) has been gaining momentum under the umbrella term Integrated Nutrient Management (INM) to help address this challenge. While INM is considered a sustainable approach offering a number of potential cropping system benefits, there is limited research on the effects of this management strategy on air quality and climate change, particularly N_2O emissions. Soil N_2O emissions result from microbial nitrification and denitrification processes which are affected by a number of soil properties including moisture content, texture, pH, source of organic amendments and the C and N contents of amendments. In this paper, available field studies were reviewed to identify promising INM strategies for reducing N_2O emissions compared to organic N inputs. Despite considerable variability in results and the complexity of potential mechanisms controlling N_2O emissions, INM treatments having low C:N (<8) tended to reduce emissions compared to organic amendments alone, while INM treatments with higher C:N resulted in no change or increased N_2O emissions. To further understand the effect of C:N on N_2O emissions, future INM field experiments should include amendments with a range of C:N paired with inorganic

N fertilizer at a constant total N rate (i.e., substitutive N input approach). Moreover, different types of organic amendments (e.g., manure, compost and green manure from different species) with the same C:N should be paired with constant total N rates, or levels of N rates, to explore the effects of other amendment properties aside from C:N on N_2O emissions.

Author Contributions: C.P. and R.G. conceived and designed the review, R.G. performed the review and drafted the manuscript, and all authors contributed to writing the paper.

Conflicts of Interest: The authors declare no conflict of interest.

References

1. Galloway, J.; Aber, J.D.; Erisman, J.W.; Seitzinger, S.P.; Howarth, R.W.; Cowling, E.B.; Cosby, B.J. The Nitrogen Cascade. *Bioscience* **2003**, *53*, 341–356. [CrossRef]
2. Bouwman, A.F.; Boumans, L.J.M.; Batjes, N.H. Emissions of N_2O and NO from fertilized fields: Summary of available measurement data. *Glob. Biogeochem. Cycles* **2002**, *16*. [CrossRef]
3. Snyder, C.S.; Bruulsema, T.W.; Jensen, T.L.; Fixen, P.E. Review of greenhouse gas emissions from crop production systems and fertilizer management effects. *Agric. Ecosyst. Environ.* **2009**, *133*, 247–266. [CrossRef]
4. Davidson, E.A.; Kanter, D. Inventories and scenarios of nitrous oxide emissions. *Environ. Res. Lett.* **2014**, *9*, 105012. [CrossRef]
5. Ravishankara, A.R.; Daniel, J.S.; Portmann, R.W. Nitrous Oxide (N_2O): The Dominant Ozone-Depleting Substance Emitted in the 21st Century. *Science* **2009**, *326*, 123–125. [CrossRef] [PubMed]
6. U.S. Environmental Protection Agency (US EPA). *Draft Inventory of US Greenhouse Gas Emissions and Sinks: 1990–2010*; U.S. Environmental Protection Agency: Washington, DC, USA, 2012.
7. Birkhofer, K.; Bezemer, T.M.; Bloem, J.; Bonkowski, M.; Christensen, S.; Dubois, D.; Ekelund, F.; Fließbach, A.; Gunst, L.; Hedlund, K.; et al. Long-term organic farming fosters below and aboveground biota: Implications for soil quality, biological control and productivity. *Soil Biol. Biochem.* **2008**, *40*, 2297–2308. [CrossRef]
8. Wortman, S.E.; Drijber, R.A.; Francis, C.A.; Lindquist, J.L. Arable weeds, cover crops, and tillage drive soil microbial community composition in organic cropping systems. *Appl. Soil Ecol.* **2013**, *72*, 232–241. [CrossRef]
9. Wortman, S.E.; Galusha, T.D.; Mason, S.C.; Francis, C.A. Soil fertility and crop yields in long-term organic and conventional cropping systems in Eastern Nebraska. *Renew. Agric. Food Syst.* **2012**, *27*, 200–216. [CrossRef]
10. Janssen, B.H. Integrated nutrient management: The use of organic and mineral fertilizers. In *The Role of Plant Nutrients for Sustainable Food Crop Production in Sub-Saharan Africa*; van Reuler, H., Prins, W.H., Eds.; VKP: Leidschendam, The Netherlands, 1993; pp. 89–105.
11. Wu, W.; Ma, B. Integrated nutrient management (INM) for sustaining crop productivity and reducing environmental impact: A review. *Sci. Total Environ.* **2015**, *512–513*, 415–427. [CrossRef] [PubMed]
12. Vanlauwe, B.; Kihara, J.; Chivenge, P.; Pypers, P.; Coe, R.; Six, J. Agronomic use efficiency of N fertilizer in maize-based systems in sub-Saharan Africa within the context of integrated soil fertility management. *Plant Soil* **2011**, *339*, 35–50. [CrossRef]
13. Chivenge, P.; Vanlauwe, B.; Six, J. Does the combined application of organic and mineral nutrient sources influence maize productivity? A meta-analysis. *Plant Soil* **2011**, *342*, 1–30. [CrossRef]
14. Mondal, S.; Mallikarjun, M.; Ghosh, M.; Ghosh, D.C.; Timsina, J. Influence of integrated nutrient management (INM) on nutrient use efficiency, soil fertility and productivity of hybrid rice. *Arch. Agron. Soil Sci.* **2016**, *62*, 1521–1529. [CrossRef]
15. Javaria, S.; Khan, M.Q. Impact of integrated nutrient management on tomato yield quality and soil environment. *J. Plant Nutr.* **2010**, *34*, 140–149. [CrossRef]
16. Roberts, T.L. Improving nutrient use efficiency. *Turk. J. Agric. For.* **2008**, *32*, 177–182.
17. Vitousek, P.M.; Naylor, R.; Crews, T.; David, M.B.; Drinkwater, L.E.; Holland, E.; Johnes, P.J.; Katzenberger, J.; Martinelli, L.A.; Matson, P.A.; et al. Nutrient imbalances in agricultural development. *Science* **2009**, *324*, 1519–1520. [CrossRef] [PubMed]
18. Nyamadzawo, G.; Wuta, M.; Nyamangara, J.; Smith, J.L.; Rees, R.M. Nitrous oxide and methane emissions from cultivated seasonal wetland (dambo) soils with inorganic, organic and integrated nutrient management. *Nutr. Cycl. Agroecosyst.* **2014**, *100*, 161–175. [CrossRef]

19. Ding, W.; Luo, J.; Li, J.; Yu, H.; Fan, J.; Liu, D. Effect of long-term compost and inorganic fertilizer application on background N2O and fertilizer-induced N_2O emissions from an intensively cultivated soil. *Sci. Total Environ.* **2013**, *465*, 115–124. [CrossRef] [PubMed]

20. Abalos, D.; Jeffery, S.; Drury, C.F.; Wagner-Riddle, C. Improving fertilizer management in the U.S. and Canada for N_2O mitigation: Understanding potential positive and negative side-effects on corn yields. *Agric. Ecosyst. Environ.* **2016**, *221*, 214–221. [CrossRef]

21. Decock, C. Mitigating Nitrous Oxide Emissions from Corn Cropping Systems in the Midwestern U.S.: Potential and Data Gaps. *Environ. Sci. Technol.* **2014**, *48*, 4247–4256. [CrossRef] [PubMed]

22. Aguilera, E.; Lassaletta, L.; Sanz-Cobena, A.; Garnier, J.; Vallejo, A. The potential of organic fertilizers and water management to reduce N_2O emissions in Mediterranean climate cropping systems. A review. *Agric. Ecosyst. Environ.* **2013**, *164*, 32–52. [CrossRef]

23. Tuomisto, H.L.; Hodge, I.D.; Riordan, P.; Macdonald, D.W. Does organic farming reduce environmental impacts?—A meta-analysis of European research. *J. Environ. Manag.* **2012**, *112*, 309–320. [CrossRef] [PubMed]

24. Skinner, C.; Gattinger, A.; Muller, A.; Mäder, P.; Fließbach, A.; Stolze, M.; Ruser, R.; Niggli, U. Greenhouse gas fluxes from agricultural soils under organic and non-organic management—A global meta-analysis. *Sci. Total Environ.* **2014**, *468–469*, 553–563. [CrossRef] [PubMed]

25. Mader, P. Soil Fertility and Biodiversity in Organic Farming. *Science* **2002**, *296*, 1694–1697. [CrossRef] [PubMed]

26. Shcherbak, I.; Millar, N.; Robertson, G.P. Global metaanalysis of the nonlinear response of soil nitrous oxide (N_2O) emissions to fertilizer nitrogen. *Proc. Natl. Acad. Sci. USA* **2014**, *111*, 9199–9204. [CrossRef] [PubMed]

27. Millar, N.; Robertson, G.P.; Grace, P.R.; Gehl, R.J.; Hoben, J.P. Nitrogen fertilizer management for nitrous oxide (N_2O) mitigation in intensive corn (Maize) production: An emissions reduction protocol for US Midwest agriculture. *Mitig. Adapt. Strateg. Glob. Chang.* **2010**, *15*, 185–204. [CrossRef]

28. Seufert, V.; Ramankutty, N.; Foley, J.A. Comparing the yields of organic and conventional agriculture. *Nature* **2012**, *485*, 229–232. [CrossRef] [PubMed]

29. Snyder, C.; Davidson, E.; Smith, P.; Venterea, R. Agriculture: Sustainable crop and animal production to help mitigate nitrous oxide emissions. *Curr. Opin. Environ. Sustain.* **2014**, *9–10*, 46–54. [CrossRef]

30. Venterea, R.T.; Coulter, J.A.; Dolan, M.S. Evaluation of Intensive "4R" Strategies for Decreasing Nitrous Oxide Emissions and Nitrogen Surplus in Rainfed Corn. *J. Environ. Qual.* **2016**, *45*, 1186. [CrossRef] [PubMed]

31. Bouwman, A.F. Land use related sources of greenhouse gases: Present emissions and possible future trends. *Land Use Policy* **1990**, *7*, 154–164. [CrossRef]

32. Hu, H.-W.; Chen, D.; He, J.-Z. Microbial regulation of terrestrial nitrous oxide formation: Understanding the biological pathways for prediction of emission rates. *FEMS Microbiol. Rev.* **2015**, *39*, 729–749. [CrossRef] [PubMed]

33. Hatfield, J.L.; Chatterjee, A.; Clay, D. Soil and Nitrogen Management to Reduce Nitrous Oxide Emissions. In *ACSESS Publications*; American Society of Agronomy, Crop Science Society of America, and Soil Science Society of America, Inc.: Madison, WI, USA, 2016.

34. Thangarajan, R.; Bolan, N.S.; Tian, G.; Naidu, R.; Kunhikrishnan, A. Role of organic amendment application on greenhouse gas emission from soil. *Sci. Total Environ.* **2013**, *465*, 72–96. [CrossRef] [PubMed]

35. Baggs, E.M. Soil microbial sources of nitrous oxide: Recent advances in knowledge, emerging challenges and future direction. *Curr. Opin. Environ. Sustain.* **2011**, *3*, 321–327. [CrossRef]

36. Butterbach-Bahl, K.; Baggs, E.M.; Dannenmann, M.; Kiese, R.; Zechmeister-Boltenstern, S. Nitrous oxide emissions from soils: How well do we understand the processes and their controls? *Philos. Trans. R. Soc. B Biol. Sci.* **2013**, *368*, 20130122. [CrossRef] [PubMed]

37. Thomson, A.J.; Giannopoulos, G.; Pretty, J.; Baggs, E.M.; Richardson, D.J. Biological sources and sinks of nitrous oxide and strategies to mitigate emissions. *Philos. Trans. R. Soc. B Biol. Sci.* **2012**, *367*, 1157–1168. [CrossRef] [PubMed]

38. Weier, K.L.; Doran, J.W.; Power, J.F.; Walters, D.T. Denitrification and the Dinitrogen/Nitrous Oxide Ratio as Affected by Soil Water, Available Carbon, and Nitrate. *Soil Sci. Soc. Am. J.* **1993**, *57*, 66–72. [CrossRef]

39. Jäger, N.; Stange, C.F.; Ludwig, B.; Flessa, H. Emission rates of N_2O and CO_2 from soils with different organic matter content from three long-term fertilization experiments—A laboratory study. *Biol. Fertil. Soils* **2011**, *47*, 483–494. [CrossRef]

40. Bhogal, A.; Nicholson, F.A.; Young, I.; Sturrock, C.; Whitmore, A.P.; Chambers, B.J. Effects of recent and accumulated livestock manure carbon additions on soil fertility and quality. *Eur. J. Soil Sci.* **2011**, *62*, 174–181. [CrossRef]

41. Graham, C.J.; van Es, H.M.; Melkonian, J.J. Nitrous oxide emissions are greater in silt loam soils with a legacy of manure application than without. *Biol. Fertil. Soils* **2013**, *49*, 1123–1129. [CrossRef]

42. Fontaine, S.; Mariotti, A.; Abbadie, L. The priming effect of organic matter: A question of microbial competition? *Soil Biol. Biochem.* **2003**, *35*, 837–843. [CrossRef]

43. Mikha, M.M.; Rice, C.W.; Milliken, G.A. Carbon and nitrogen mineralization as affected by drying and wetting cycles. *Soil Biol. Biochem.* **2005**, *37*, 339–347. [CrossRef]

44. Cheng, Y.; Zhang, J.-B.; Wang, J.; Cai, Z.-C.; Wang, S.-Q. Soil pH is a good predictor of the dominating N_2O production processes under aerobic conditions. *J. Plant Nutr. Soil Sci.* **2015**, *178*, 370–373. [CrossRef]

45. McSwiney, C.P.; Robertson, G.P. Nonlinear response of N_2O flux to incremental fertilizer addition in a continuous maize (*Zea mays* L.) cropping system. *Glob. Chang. Biol.* **2005**, *11*, 1712–1719. [CrossRef]

46. Mulvaney, R.L.; Khan, S.A.; Mulvaney, C.S. Nitrogen fertilizers promote denitrification. *Biol. Fertil. Soils* **1997**, *24*, 211–220. [CrossRef]

47. Doran, J.W.; Fraser, D.G.; Culik, M.N.; Liebhardt, W.C. Influence of alternative and conventional agricultural management on soil microbial processes and nitrogen availability. *Am. J. Altern. Agric.* **1987**, *2*, 99. [CrossRef]

48. Poudel, D.D.; Horwath, W.R.; Mitchell, J.P.; Temple, S.R. Impacts of cropping systems on soil nitrogen storage and loss. *Agric. Syst.* **2001**, *68*, 253–268. [CrossRef]

49. Charles, A.; Rochette, P.; Whalen, J.K.; Angers, D.A.; Chantigny, M.H.; Bertrand, N. Global nitrous oxide emission factors from agricultural soils after addition of organic amendments: A meta-analysis. *Agric. Ecosyst. Environ.* **2017**, *236*, 88–98. [CrossRef]

50. Meng, L.; Ding, W.; Cai, Z. Long-term application of organic manure and nitrogen fertilizer on N_2O emissions, soil quality and crop production in a sandy loam soil. *Soil Biol. Biochem.* **2005**, *37*, 2037–2045. [CrossRef]

51. Cai, Y.; Ding, W.; Luo, J. Nitrous oxide emissions from Chinese maize-wheat rotation systems: A 3-year field measurement. *Atmos. Environ.* **2013**, *65*, 112–122. [CrossRef]

52. Nyamadzawo, G.; Shi, Y.; Chirinda, N.; Olesen, J.E.; Mapanda, F.; Wuta, M.; Wu, W.; Meng, F.; Oelofse, M.; de Neergaard, A.; et al. Combining organic and inorganic nitrogen fertilisation reduces N_2O emissions from cereal crops: A comparative analysis of China and Zimbabwe. *Mitig. Adapt. Strateg. Glob. Chang.* **2014**, *22*, 233–245. [CrossRef]

53. Sarkodie-Addo, J.; Lee, H.G.; Braggs, E.M. Nitrous oxide emissions after application of inorganic fertilizer and incorporation of green manure residues. *Soil Use Manag.* **2003**, *19*, 331–339. [CrossRef]

54. Frimpong, K.A.; Baggs, E.M. Do combined applications of crop residues and inorganic fertilizer lower emission of N2O from soil?: N_2O from combined residue and fertilizer application. *Soil Use Manag.* **2010**, *26*, 412–424. [CrossRef]

55. Dittert, K.; Lampe, C.; Gasche, R.; Butterbach-Bahl, K.; Wachendorf, M.; Papen, H.; Sattelmacher, B.; Taube, F. Short-term effects of single or combined application of mineral N fertilizer and cattle slurry on the fluxes of radiatively active trace gases from grassland soil. *Soil Biol. Biochem.* **2005**, *37*, 1665–1674. [CrossRef]

56. Tang, J.-C.; Maie, N.; Tada, Y.; Katayama, A. Characterization of the maturing process of cattle manure compost. *Process Biochem.* **2006**, *41*, 380–389. [CrossRef]

57. Hoge, A.; Robinson, D.; Fitter, A. Are microorganisms more effective than plants at competing for nitrogen? *Trends Plant Sci.* **2000**, *5*, 304–308. [CrossRef]

58. Senbayram, M.; Chen, R.; Budai, A.; Bakken, L.; Dittert, K. N_2O emission and the $N_2O/(N_2O + N_2)$ product ratio of denitrification as controlled by available carbon substrates and nitrate concentrations. *Agric. Ecosyst. Environ.* **2012**, *147*, 4–12. [CrossRef]

59. Zhu-Barker, X.; Doane, T.A.; Horwath, W.R. Role of green waste compost in the production of N_2O from agricultural soils. *Soil Biol. Biochem.* **2015**, *83*, 57–65. [CrossRef]

60. Huang, Y.; Zou, J.; Zheng, X.; Wang, Y.; Xu, X. Nitrous oxide emissions as influenced by amendment of plant residues with different C:N ratios. *Soil Biol. Biochem.* **2004**, *36*, 973–981. [CrossRef]

61. Van Groenigen, J.W.; Velthof, G.L.; Oenema, O.; Van Groenigen, K.J.; Van Kessel, C. Towards an agronomic assessment of N_2O emissions: A case study for arable crops. *Eur. J. Soil Sci.* **2010**, *61*, 903–913. [CrossRef]

62. Mondelaers, K.; Aertsens, J.; Van Huylenbroeck, G. A meta-analysis of the differences in environmental impacts between organic and conventional farming. *Br. Food J.* **2009**, *111*, 1098–1119. [CrossRef]

63. De Ponti, T.; Rijk, B.; van Ittersum, M.K. The crop yield gap between organic and conventional agriculture. *Agric. Syst.* **2012**, *108*, 1–9. [CrossRef]

64. Robertson, G.P. Greenhouse Gases in Intensive Agriculture: Contributions of Individual Gases to the Radiative Forcing of the Atmosphere. *Science* **2000**, *289*, 1922–1925. [CrossRef] [PubMed]

65. Liang, L.L.; Eberwein, J.R.; Allsman, L.A.; Grantz, D.A.; Jenerette, G.D. Regulation of CO_2 and N_2O fluxes by coupled carbon and nitrogen availability. *Environ. Res. Lett.* **2015**, *10*, 34008. [CrossRef]

66. Gan, Y.; Liang, C.; Chai, Q.; Lemke, R.L.; Campbell, C.A.; Zentner, R.P. Improving farming practices reduces the carbon footprint of spring wheat production. *Nat. Commun.* **2014**, *5*, 5012. [CrossRef] [PubMed]

67. Schlesinger, W.H. On fertilizer-induced soil carbon sequestration in China's croplands. *Glob. Chang. Biol.* **2010**, *16*, 849–850. [CrossRef]

sustainability

MDPI

Review

Diamondback Moth, *Plutella xylostella* (L.) in Southern Africa: Research Trends, Challenges and Insights on Sustainable Management Options

Honest Machekano [1], Brighton M. Mvumi [2] and Casper Nyamukondiwa [1,*]

[1] Department of Biological and Biotechnological Sciences, Botswana International University of Science and Technology, P. Bag 16, Palapye, Gaborone 0267, Botswana; honest.machekano@studentmail.biust.ac.bw

[2] Department of Soil Science and Agricultural Engineering, Faculty of Agriculture, University of Zimbabwe, P.O. Box MP 167, Mt. Pleasant, 00263 Harare, Zimbabwe; mvumibm@agric.uz.ac.zw

* Correspondence: nyamukondiwac@biust.ac.bw; Tel.: +267-49-31528; Fax: +267-49-00102

Academic Editor: Suren N. Kulshreshtha
Received: 5 November 2016; Accepted: 30 December 2016; Published: 3 February 2017

Abstract: The diamondback moth (DBM), *Plutella xylostella*, is a global economic pest of brassicas whose pest status has been exacerbated by climate change and variability. Southern African small-scale farmers are battling to cope with increasing pressure from the pest due to limited exposure to sustainable control options. The current paper critically analysed literature with a climate change and sustainability lens. The results show that research in Southern Africa (SA) remains largely constrained despite the region's long acquaintance with the insect pest. Dependency on broad-spectrum insecticides, the absence of insecticide resistance management strategies, climate change, little research attention, poor regional research collaboration and coordination, and lack of clear policy support frameworks, are the core limitations to effective DBM management. Advances in Integrated Pest Management (IPM) technologies and climate-smart agriculture (CSA) techniques for sustainable pest management have not benefitted small-scale horticultural farmers despite the farmers' high vulnerability to crop losses due to pest attack. IPM adoption was mainly limited by lack of locally-developed packages, lack of stakeholders' concept appreciation, limited alternatives to chemical control, knowledge paucity on biocontrol, climate mismatch between biocontrol agents' origin and release sites, and poor research expertise and funding. We discuss these challenges in light of climate change and variability impacts on small-scale farmers in SA and recommend climate-smart, holistic, and sustainable homegrown IPM options propelled through IPM-Farmer Field School approaches for widespread and sustainable adoption.

Keywords: small-scale farmers; pest management; brassicas; farmer-extension-researcher networking; insecticide misuse

1. Introduction

Brassica vegetables, like cabbage (*Brassica oleracea* var. *capitata*) and cauliflower (*B. oleracea* var. *botrytis*), and open leaf kales, like rape (*Brassica napus*) and covo (*Brassica carinata*), are the popular staple relish and most widely grown leafy vegetables in the tropical and subtropical regions of Southern Africa (SA), cutting across a wide range of cultures and agro-ecologies [1–5]. These vegetables are grown throughout the year [6] and form the fastest growing agricultural subsector that contributes significantly to national and regional incomes [6,7]. With the persistent droughts, extreme temperatures, and flooding challenges faced in field crop production due to climate change [8,9], irrigable small vegetable plots remain comparatively reliable as an attractive source of food and income for rural households, who make up over 80% of the farming community in SA [6,10] and whose farming systems are more vulnerable to effects of climate change [9,11]. On the other hand, African urban areas face high food demand because of rapid rural to urban migration, which has grown from 53 million to 400 million between 1960 and 2010, with a potential to increase to 600 million by 2030 [11]. As a result, high unemployment and low per capita income in the highly populated urban areas have created an ever increasing demand for food [11]; hence the need for horticultural expansion in rural, urban, and peri-urban agriculture (UPA) to meet fresh vegetable food demand, supplement incomes, and meet nutritional needs [10–12].

Despite doubling as a household income generating enterprise, brassicas also serve as an important inexpensive source of vitamins and minerals [7,10]. Due to the simplicity with which they can be grown, numerous small-scale farmers make a living out of brassica production, relying on the proximal urban markets [10,12–14]. Similarly farmers distant from the city typically rely on alternative markets [14]. However South Africa still exports brassicas (especially cabbage) to some of its regional neighbours, including Zambia (0.2%), Mozambique (3.3%), Angola (3.4%), Namibia (5%), Swaziland (6.5%), Botswana (31.4%), and Lesotho (46.3%) [15], thus lending credence to the theory of high demand against a production deficit in the region. The global demand for organically produced vegetables [16] has also significantly opened new lucrative markets for these African economies with the potential to substantially increase their Gross Domestic Product (GDP) if the required quality standards are met. This, however, is challenged by the scourge of the diamondback moth (DBM), *Plutella xylostella* (L.) (Lepidoptera: Plutellidae), a cosmopolitan insect pest of brassicas [17,18].

The DBM is the major, ubiquitous, and year-round insect pest hindering the economic production of brassica crops in SA [17–19]. Small-scale farmers are facing difficulty coping with DBM damage-induced losses and management challenges [3,19–24]. The economic importance of DBM is derived from its exceptional pest status that originates from its genetic diversity, high and year-round abundance, high reproductive potential, high genetic elasticity, cosmopolitan distribution, multivoltinity, and continuous suppression of the pest's natural enemies by synthetic pesticides [5,18,25] and possible survival failures by efficient natural enemies in the pest's new invasion areas [26]. Global losses of leafy vegetables attributed to damage and control costs of DBM alone were estimated to be around US$ 4–5 billion [27]. Partitioning crop losses in SA under small-scale farmer conditions has not been explored in detail. However, in Kenya, an estimated 31% loss has been reported [28]. If uncontrolled, losses of up to 100% are possible [5,29], as has been reported in Botswana ([30]; and from personal observation during fieldwork in 2014 and 2015. There is little knowledge on the actual loss data of brassicas due to DBM in SA countries. However, cases of abandoning brassicas and changing production timing (i.e., concentrating only on winter production) as a means of infestation avoidance have been widely recorded [5,23,25].

Temperature is a critical climatic factor, which influences insect biological activities such as survival, reproduction, growth, development, geographical distribution, and fitness [31–33]. An increase in temperature reduces the time taken to acquire the number of degree-days required to complete the *P. xylostella* life cycle, thus decreasing its generation time and increasing the number of generations per year [27,32–34]. An increase in average temperature with global change may imply reduced overwintering time or a total absence of diapause for some economic insects [32,33], with consequent implications on pest management and food security. In addition, global warming has the potential to impair the potential of *P. xylostella* biological control if an increase in temperature disrupts the life cycle synchronisation of the host and its parasitoids [34]. Recent modelling data has predicted a decrease in ecological niches for some insects with climate change [35], and, similarly, invertebrate biocontrol agents are not an exception. Previous work reported broad lethal temperature limits for adult DBMs [36–39]: the minimum body temperature that 0% of the moths could survive, known as lower lethal temperature (LLT_0), was −16.5 °C. The maximum body temperature that 0% of the moths could survive, known as upper lethal temperature (ULT_0), was 42.6 °C. The minimum body temperature for 25% moth survival (LLT_{25}) was −15.2 °C, while the maximum body temperature for 25% moth survival (ULT_{25}) was 41.8 °C [37,38]. However thermal tolerance for its major parasitoids has not been fully studied [34,36,38]. Unless the thermal tolerance of the major parasitoids matches that of the host DBM and evidence is presented that these traits may have coevolved, parasitoid efficacy in the face of climate change may be compromised [26]. Without coevolution of thermal tolerance, DBM challenge may likely intensify due to conducive climatic conditions [37–41] that may stimulate increased pest activity (feeding, breeding, and migration) [26,38]. Therefore, without efficient control mechanisms, the DBM problem could continue to increase despite the intensive pesticide use, which to-date may have been short-term, ineffective, unsustainable and expensive [18].

In this paper, we review the status of DBM management in the context of practice in SA. Specifically, we examine the past and current DBM pest status, management practices by farmers, DBM research, and development linkages among member countries (or the lack thereof) in SA with special reference to small-scale farmers who are the most affected. We also analyse the perspectives of researchers, farmers, and agricultural extension agents regarding DBM management and identify challenges and principal areas that require cooperation. We propose research on sustainable climate-smart agriculture and the selection of compatible integrated pest management (IPM) components that provide effective management of the DBM under small-scale farmers in SA, in the context of current and projected climate change scenarios.

2. Horticulture and DBM in Southern Africa

Due to socio-economic challenges and high unemployment rates in SA [11], horticulture is fast transforming into an intensive production and high-income-generating enterprise [39,40]. However, despite large expansion in land committed to horticulture in the region, returns per unit land area are still minimal [41], mainly due to pest related losses and, in some cases, high production costs. In addition, small-scale brassica farming systems are dominated by low scale cultivation of non-rotated monocrops with heavy dependence on family labour and locally available inputs [4,13]. Due to this perception, the management of the DBM (and other pests) is an in-built farming practice based on prophylactic pesticide use with the intention to 'eliminate' rather than to 'manage' the pest; and therefore economic threshold levels based on insect pest monitoring and scouting are not observed [5,13,40].

Consumer perception is another driver to intensive pesticide use. Urban consumers are biased to aesthetically damage-free vegetables and their demand for such produce cannot be ignored as a driver to intensive insecticide use by the farmers [14,42]. For small-scale farmers, the market is typified by vegetable vendors under make-shift stalls in urban and peri-urban roadsides. These vendors are an important market link between small-scale farmers and the urban market as they not only determine the market price for different levels of pest damage but are also directly linked to consumers [2]. This vendor market, just like urban supermarkets, has the capacity to influence price and quality; it triggers the excessive use of pesticides as farmers compete to produce and supply shiny, damage-free, 'quality' brassicas to satisfy the 'market standards'. Research in SA, however, has not contextualized these and other market forces in the light of acceptable damage levels on leafy vegetables, especially with reference to DBM attack. Reports indicate that market rejection and strong legislative frameworks influence the chemical application behaviour of farmers [43], forcing them to change chemical use patterns as fear of market loss supersede concern for public health [44].

DBM damage substantially hinders production and marketing of brassicas in SA [17,18,25]. Farmers' perceptions and practices on the management of this pest in the region are not yet fully understood [39]. Research to date has been survey-based [5,20,22,40] and generalised on both insect pests and diseases for all horticultural crops. This approach generalised and limited the information that could be generated regarding a specific pest. One of the main features of climate change, amongst others, is the rise in global mean temperatures and prolonged hot weather conditions [9]. In SA, temperature is projected to increase by 1–3 °C by 2050 [45–47] and its effects are likely to be more pronounced in the drier tropics than the humid subtropics [8,9]. In laboratory experiments, DBM showed activity over a broad temperature range, measured as LLTs and ULTs [36,38]. This may mean that, under the currently projected climate change in SA, DBM pest status is likely to increase, exacerbating already failing management practices [18,25,48]. Field population peaks, determined by both pheromone trap catches and crop infestation scouting, were observed in the warmer austral summer [19,22,49]. Regardless of the population source, high temperatures were shown to hasten development and thus shorten life cycles in *P. xylostella* [36,37]. However, temperature may differentially affect organisms, such that different insect pests (hosts) and their associated natural enemies may develop at different rates and thus affect host-prey/parasitoids synchronisation [26,33,50,51]. Extreme temperatures eliminate natural enemies that are susceptible to very high/low temperatures, whereas divergence from thermal preferences also disrupts the temporal and spatial synchronization of host/parasitoid phenologies, resulting in a high risk of challenging pest (host) outbreaks [34,35]. An increase in atmospheric carbon dioxide levels associated with global climate change may also reduce the efficacy of biological control agents against DBM by precluding or reducing the production of plants' secondary metabolites, which are necessary for the recruitment of natural enemies as part of the plants' natural defence mechanisms [52,53]. This and the misalignment of host-natural enemy life cycles may affect the natural enemy's efficacy and thus jeopardise the future of biological control programs [26]. There is a scarcity of published literature on climate-related coevolution of DBM and its natural enemies for optimising the efficacy of biological control. Nevertheless, some researchers have recommended that IPM programmes aimed at improving efficacy under global climate change should develop resilient agro-ecosystems, which incorporate populations' evolutionary potential and buffers against climate change effects [26,51,54].

3. Why is Southern Africa Hard Hit by the DBM Scourge?

3.1. Vulnerability to Effects of Climate Change

Sub-Saharan Africa will continue to be the area most hard-hit by climate change effects, due to increased mean temperature and increased rainfall variability [9]. With a record of 0.5 °C regional temperature increase, [9,35,45,46] predicts a projected increase in temperature of 3–4 °C by 2080, reduced rainfall, and increased degree days, aridity index, and evapotranspiration gradients. These factors will increase stress on already debilitating horticultural ecosystems, especially pest management, through changed pest dynamics, spatio-temporal distribution and increased pressure. Insecticide resistance associated with high temperatures has been recorded in different species [55], including variations in *P. xylostella* susceptibility to some organophosphates [55,56]. Therefore, under current climatic projections in SA [9,38,57], it is highly unlikely that DBM populations will decline due to the physiological stress associated with high or low temperature scenarios [36]. Sub-Saharan Africa's majority of rural small-scale farmers remains at the core of food production, but their production ecosystems are the most prone to climate change effects [9,45,57,58]. Using a prediction model [45], between 8% and 22% field crop losses have already been reported in sub Saharan Africa.

3.2. Farmers' Behaviour and Insecticide Use

Details and comprehensive data on farmers' behaviour relating to pesticide usage on DBM in SA are lacking [18]. However, survey baseline results show that between 75% and 100% of farmers in SA totally rely on chemical insecticides (Table 1). By global standards, these farmers use the greatest variety of chemicals, highest application rates, and the highest application frequency [5,43]. Frequency of application ranges from once every three weeks to three times a week [5,20,43,59].

At any given time, brassica farmers possess at least two to six different insecticides [42] and up to five different insecticides have been mixed in a single sprayer tank without technical recommendation or manufacturer instructions [60]. This might result in unknown phytotoxicity and unwanted (and seldom known) chemical reactions into compounds, which are possibly more hazardous and persistent in the environment [61]. Such hazardous compounds, even when geographically concentrated in pattern, could create significant exposure to the environment and the public through non-occupational exposure, where individuals not directly involved with chemical use get exposed to the chemical hazards through a contaminated environment [61,62]. Magauzi et al. [63] and Macharia et al. [28] detailed pesticide-related illnesses in Zimbabwe and Kenya respectively, and it has been reported that various symptoms related to pesticide poisoning have significantly increased as most small-scale farmers misuse chemicals and do not use personal protective equipment (PPE) [64]. Moreover farmers tend to ignore or take for granted certain levels of illnesses from synthetic chemicals, which they feel do not warrant medical attention, as an expected normal part of farmwork [60]. Consequently, there is scant information on the details of health effects and costs related to pesticide exposure, as most cases go unreported [28,60,61]. However, Magauzi et al. [63] reported high organophosphate levels in young horticultural farmworkers' blood and also recorded 24.1% abnormal cholinesterase activity in 50% of the sprayers (occupational exposure) and 49% of workers entering previously sprayed fields (non-occupational exposure) in Zimbabwe. Khoza et al. [64] reported similar results with both organophosphates and organochlorines and further reported chronic illnesses that were often misdiagnosed and mistreated in health centres; possibly due to rampant pesticide incorrect use [65]. Similar results were also recently reported in Kenya [28], as supported by reports of high proportion of small-scale horticulture farmers using insecticides in Africa (Table 1).

Table 1. Proportion of small-scale horticultural farmers using synthetic insecticides in Southern Africa and other parts of Africa.

Country	Farmers Using Pesticides (%)	Reference
Southern Africa		
Mozambique	100	[5]
Botswana	98	[20]
Zimbabwe	No data	
Zambia	75	[40]
Malawi	75	[40]
Other selected African countries		
Tanzania	98	[60]
Cameron	90	[66]
Ghana	85	[67]
Kenya	No data	

Occupational exposure is exacerbated by inefficient chemical use by small-scale farmers [42,43,60]. This ranges from using inappropriate chemicals, incorrect dosages, and wrong application timing and targeting, to non-calibrated or poorly maintained and defective (often leaky) application equipment [1,42,68]. Mvumi et al. [65] also reported the first three problems on synthetic grain protectants in Zimbabwe. Leakages were observed to lead to about 29 mL of dermal exposure per person per hour [60], depending on leakage rates, which might be currently higher due to cheap and faulty spraying equipment from non-reputable manufacturers flooding the market. Other forms of inefficient use resulting in exposure include the choice of extremely hazardous chemicals (Class 1a and 1b by WHO standards) [1,20,40,42,61], the use of banned chemicals [42,60,61], applying chemicals using twig/leaf bunches or home-made grass brushes/brooms, making homemade 'insecticide cocktails', and tongue-testing to assess concentrations [43]. Due to economic challenges, farmers sometimes often procure pesticides from unlicensed and unscrupulous dealers, thus increasing the risk of exposure and the chances of fraud and adulteration [60,65]. Reports from Zimbabwe indicate a failure to adhere to safety withdrawal periods, presumably due to market pressure; inefficient chemical use (only 35%–50% of sprayed chemical reach the target organism) [67,68]; application of the wrong pesticides (e.g., fungicides on insects); and abuse associated with the need to clear last seasons' expired pesticides [13,67]. This uncontrolled misuse and overuse of insecticide was reported to have significantly contributed to the increased resistance and suppression of potential biological control agents [17,18,25,30].

3.3. Lack of Insecticide Resistance Research

The DBM has shown resistance to 91 active ingredients of agricultural chemicals worldwide, including 12 strains of *Bacillus thuringiensis* (*Bt*), between 1953 and 2014 [48,69–71]. Compared to other parts of the world, DBM insecticide resistance in SA is relatively low (see Figure 1). Farmers tend to rely on their personal observation of insecticide efficacy failures to detect resistance. Following resistance 'detection', farmers usually continue using the same active ingredients at higher frequency, higher dosages, or in cocktails with other 'powerful' chemicals, which exacerbates the situation [5,42,43,60]. Despite the widespread use of hard chemicals to combat DBM in SA, we have not found any published comprehensive study on DBM resistance to commercially registered pesticides in this region (Figure 1). Management options and extension recommendations have been based on reports from the relatively advanced economies (China, Brazil, India, Australia, Nicaragua, Pakistan and USA) (Figure 1). However, resistance is highly geographical and highly correlated to insect strain as regards chemical exposure history, hence 'foreign' recommendations may not be directly applicable to the spatially heterogenous nature of the SA small-scale farming communities.

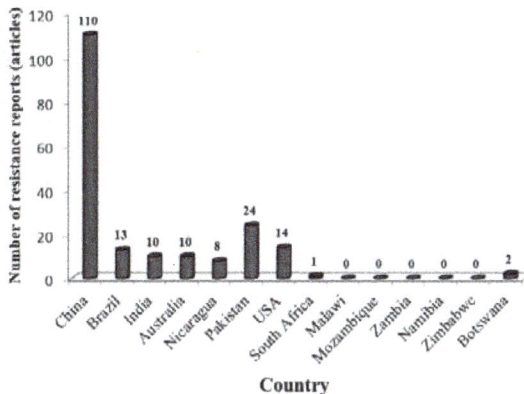

Figure 1. Selected country published reports on diamondback moth (DBM) insecticide resistance [48].

Consequently, farmers lack information on DBM resistance status in their respective localities to aid their pest control planning. This forces them to make their own, often ill-informed, decisions, mainly influenced by chemical manufacturers' advertisements, agro-dealers, and sometimes pesticide vendors ([40,43]; personal observation, 2014). Due to a lack of active pesticide control policies, farmers practice independent chemical choices and application (personal observation, 2014) without adequate consultation, resulting in 'dangerous' experimentations and haphazard chemical use with no regional or area-wide territorial regulations to aid Integrated Resistance Management [42,60]. In some cases, farmers smuggle 'effective' chemicals with noble modes of action from other countries into their home countries, where the chemicals have not yet been registered. Uncontrolled and inefficient use of these new pesticides results in early resistance development [54], which renders the modern pesticides ineffective by the time they are officially registered in the farmers' countries e.g., Hunter 500EC (Chlorfenapyr (pyrrole) 240 g/L) in Botswana (personal observation, 2014).

3.4. Low Research Attention

DBM research in SA is dominated by the public sector [2,24,25] where the agricultural ministries are custodians of agriculture and related work. In SA, countries with active DBM research are limited (see Figure 2). Conventionally research findings are delivered to the farmers through extension departments. This system is increasingly becoming inefficient due to declining public sector resources, the lack of farmer empowerment, and a lack of specialist staff in the sector [2,4]. The majority of SA research and development grants are funded externally [4,13], often coming with specific thematic areas that restrict researchers' flexibility. This may be a setback, as it limits scientists on tackling locally critical issues affecting small-scale farmers. This, coupled with low per capita funding and low capacity–building, exacerbated by 'brain-drain' to developed countries, limits research achievements [4,13]. Only a few SA countries can afford to keep specialist staff in the public sector, resulting in the disproportional distribution of research among SA member countries (Figure 2). South Africa and Kenya seem better off than the other countries, probably because of the presence of the Agricultural Research Council (ARC) and International Centre for Insect Physiology and Ecology (ICIPE), respectively, where DBM genetic, ecological, and IPM studies have mostly been conducted [23,28,71–76]. In South Africa, the ARC, in collaboration with industry and academic institutions, has conducted numerous studies on the DBM (Figure 3) on aspects including population dynamics, ecology, parasitism and predation, tritrophic interactions, and resistance breeding to *Bt* brassicas [21,71,72,76,77].

Figure 2. Countries in the Southern African region where DBM research has been conducted. This region appears in the high eco-climatic index of the world, where DBM persistence is year round and high [18]. * Namibia and Angola have very limited accessible research information on DBM, hence they were omitted from the map.

In contrast, the other SA countries have limited research on DBM, with Zimbabwe being the only country contributing just over 10% of DBM research. Most of the DBM research is conducted by incapacitated horticultural research institutions that are often poorly funded [24]. The bulk of the research was survey-based, covering general farmer practices, identification, and spatio-temporal distribution of the DBM natural enemies (Figure 4). These surveys brought about vast knowledge on DBM predation and parasitism rates in the region [5,19,22,77]. Crop systems approached through intercropping with mustard, *Brassica juncea* (L) (Czern); onion, *Allium cepa* (L.); and/or garlic, *Allium sativum* (L.) (also making 22% (Figure 4)), have been over studied and duplicated in many SA counties, due to a lack of research coordination and information sharing [12,78,79].

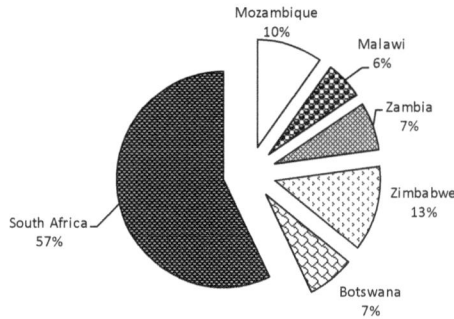

Figure 3. Proportion of publications on DBM in Southern African countries (1995–2015). (The data is based on physical counts of published papers and conference proceedings from respective countries).

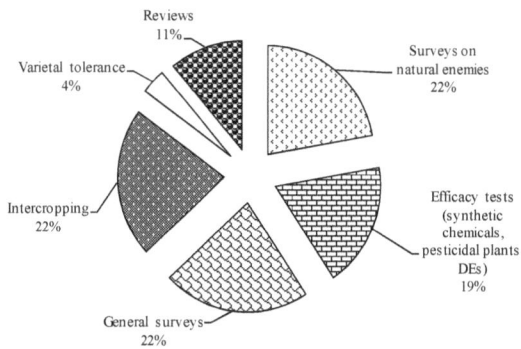

Figure 4. Proportion of published research articles on DBM in Southern Africa by theme (1995–2015). This is based on the physical checking of research themes for each of the publications in Figure 3).

Although SA has a long history of brassica production and an equally long acquaintance with DBM [25], research on its management seems to have started only about a decade ago with no data on the preceding years. Only recently, a synthetic pyrethroid (Cypermethrin), a *Bt* product (*B. thuringiensis* (*var. kurstaki*)), was tested for efficacy against DBM in Southern Africa, specifically Botswana [71,72]. Though this may be an important first step towards generating knowledge on DBM response to insecticides in the region, it needs to be expanded through testing area-specific populations for detailed territorial resistance profiling in all horticultural hotspots to aid planning on area-wide resistance management. Area-specific resistance assays may be critical in determining the susceptibility of DBM strains to current and future insecticides in different high production areas.

3.5. Lack of Regional Coordination in DBM Research

In 1984, the Southern African Development Community (SADC) (known as SADCC then) mooted and commissioned the Southern African Centre for Cooperation in Agricultural Research (SACCAR) for coordinated agricultural research in SA, which was partly funded by the Asian Vegetable Research and Development Centre (AVRDC) in the 1990s [80]. The AVRDC objectives in SACCAR were to coordinate vegetable research between and within SA member countries, develop novel vegetable postharvest preservation techniques, and, most importantly, develop an IPM program for the control of DBM in cruciferous vegetables for small-scale farmers [80]. Apart from coordinating regional research, SACCAR aimed to align agricultural research policies and priorities, identify constraints, promote cooperative research projects, strengthen national vegetable research centres, and encourage regional sharing and utilisation of scientific and technical information [80]. With headquarters in Botswana, SACCAR had sub-regional offices at reputable research institutions in Tanzania, Zimbabwe, Zambia, and Malawi in the early 1990s. However, as individual funding contributions from member states dwindled, independent donor organisations stepped in, diverting the organisation from its core mandates. To date, the organisation's activities are less visible on agriculture compared to the past, with high visibility on general economic constraints, labour-related issues, and the socio-economic welfares of selected member states. Thus, regional coordination and alignment of agricultural policies for concerted insect pest control efforts remain limited. However, there is hope in the recently formed Centre for Coordination of Agricultural Research and Development in Southern Africa (CCARDESA) (under SADC), which is targeting productivity and competitiveness of small-scale farmers across the region. The results of its activities are yet to be assessed.

The AVRDC, which is entirely committed to vegetable research, significantly sponsored regionally-coordinated research and capacitated national vegetable research centres in East Africa [2,4], but full expansion to SA was hampered by funding constrains [25,74]. Its major thrust was resistance breeding, farmer training, pest management, and general promotion of new technologies in SA that had proven successful in Asian nations [2,4]. To date, the results of its activities in the horticultural farming community in the region are certainly unclear, as is the case with SACCAR. The Asian Vegetable Network (AVNET), formed in 1989, successfully coordinated vegetable breeding and pest and disease control in Asia through the formation of strong dedicated sub-networks [81]. The advent of ICIPE in Kenya was an example of coordinated regional research in insect science, particularly in DBM crucifers [23]. Through this institution, Eastern Africa managed to conduct coordinated research aimed at DBM IPM [23,74,75]. ICIPE achieved DBM control in brassicas through the development and dissemination of biocontrol based IPM, using *Diadegma semiclausum* (Hellen) (Hymenoptera: Ichneumonidae) with complementary emphasis on a cropping systems approach [23]. Success was also achieved through a multidisciplinary approach, expert contributions, research funding, and supportive policies (national and regional) that enabled the granting of permissions to importations and releases of biocontrol agents [23]. Though this work did not effectively extend to SA, due to funding challenges (see discussion in [25]), the same model could be adopted in the DBM hard-hit SA region. Following the successful models of AVRDC and AVNET in Asia and ICIPE in East Africa, research networking may be a key mechanism for effective research aimed at achieving common goals for participating countries [23,81]. Southern Africa member states (Figure 2) can collaborate in the same manner for regionally consented efforts targeted at holistic DBM management. This networking is important to enable area-wide (regional) DBM management, as the pest's migration patterns and dispersal behaviour makes individual (farmer or country) methods ineffective [18,25]. A good example of this sub-regional collaboration is the recently-ended project aimed at combating the Asian fruit fly, *Bactrocera dorsalis* (Hendel), in Botswana, Namibia, Zambia, and Zimbabwe (BONAZAZI) under the technical assistance of FAO.

4. Possible Novel DBM Control Options

Climate-smart technologies aimed at maximizing production while promoting adaptation and mitigating the effects of changing environments are required [58]. IPM is a huge component of climate-smart agriculture, which, since the 1990s, has been generally agreed as the only sustainable and effective method of containing or managing economic pests, including the DBM [7,17,18,25,30,74]. Since synthetic chemicals offer short-time relief, several other management strategies have been investigated on a wide range of brassica agroecosystems, but IPM remains the most viable option [18,25,82]. IPM is that method of pest management that utilises all available and compatible techniques of pest management to reduce pest populations and maintain them below the crop economic injury levels [80,82]. The concept is aimed at eliminating the reliance on a single method of pest management in order to achieve better control and reduce or prevent development of pest resistance to a particular method [82]. This includes, but is not limited to, seasonal cropping (synchronised cropping calendar to minimize host plant availability), crop rotation, intercropping with non-host plants, enabling conducive environments for biological control agents, legislative plant host control (dead periods), the use of resistant varieties, and the judicious and minimal use (e.g., spot application) of environmentally benign insecticides (see [82]), which are applied only when absolutely necessary. In this system, insecticides from different chemical groups may also be rotated following legislation-enforced programs implemented and monitored by plant protection departments. Without legislative enforcement, synthetic insecticides continue to be used without due diligence despite widespread IPM awareness worldwide [43]. This is a practice that has caused deleterious consequences on DBM natural enemies including the reduction of their abundance and reduced efficacy in IPM systems [19,22,43,73].

Southern Africa is rich in natural enemies for DBM biological control [5,19,22,73]; therefore, the ecological consequence of widespread insecticide use, especially on these biological control agents, is a major concern [83,84]. Hence, a form of IPM aimed at reducing pesticide use and the promotion of selective soft insecticides (e.g., Pirimicarb, Pymetrozine and Spinosad (see Figure 5)) as its central tenets is the most crucial step in reducing the pesticide burden on the environment [84]. As explained earlier, biological control agents are currently dwindling due to intensive broad-spectrum chemical pesticide use and there is a danger that some of the natural enemy species may be completely lost unrecorded [83,84]. Therefore, unless the overreliance and unrestrained use of synthetic insecticides is significantly reduced, IPM and biological control measures in SA will continue to be hampered.

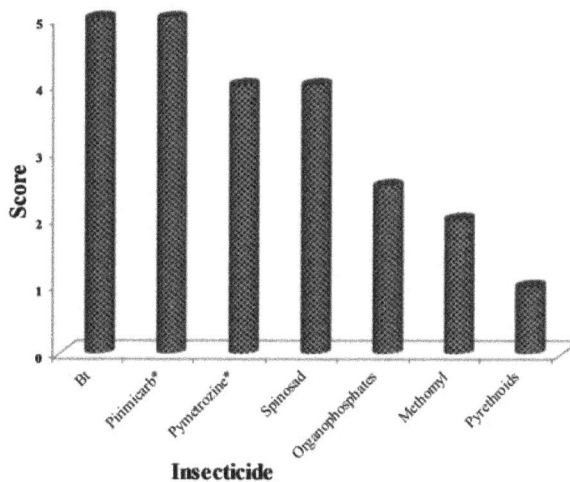

Figure 5. Common soft insecticides with high efficacy on the DBM and a low effect on its natural enemies (*Trichogramma* sp., *Cotesia* sp., spiders, lacewings and damsel bugs [84]). (Score: 5 = lowest effect on natural enemies, 1 = highest effect). *Insecticides not readily available on the market in Southern Africa (SA).

According to Walsh [84], *Bt*-based insecticides and Pirimicarb are the softest pesticides on DMB natural enemies, followed by Pymetrozine and Spinosad (Figure 5). Organophosphates, methomyl (a carbamate), and synthetic pyrethroids have high negative effects, particularly on *Trichograma* sp. and *Cotesia* sp., the most abundant and efficacious parasitoid species in SA [73,84,85]. Ironically survey results from SA, particularly Botswana [20], Zimbabwe [22], Malawi, Zambia [40], and Mozambique [5], show that synthetic pyrethroids, organophosphates, and carbamates are among the most commonly used insecticides in vegetable production. However, since genetically modified (*Bt*) brassicas were not accepted in SA due to social and environmental concerns [25], one of the sustainable management options is the rotation of *Bt*-based insecticides.

DBM is highly host-specific [86] (except in one observation of its survival on sugar-snap, *Pisum sativum* var. *macrocarpon* and snow peas, *P. sativum* var. *saccharatum* in Kenya [87]). Generally, moths do not oviposit on non-host plants; their host acceptance and oviposition is associated with a complicated integrated suite of chemical and physical cues [86]. Therefore, where soft insecticides are utilised, crop systems approached through the modification of agro-ecosystems and cropping practices can also be manipulated to confuse the adults' host finding techniques [88]. Research has shown partial DBM repellence success of cabbage intercrops with alliums through confusion in the chemical cues [12,78,79]. In such intercrops, natural enemies were shown to disperse and parasitise DBM at similar rates as in monocrops [88], evidence of compatibility between natural enemies and a cropping systems approach. There is potential in further improving this concept into a 'push-pull' cropping system technology by selecting appropriate repellent and attractant crops. Parasitoids are known to have originated from intricate mechanisms and are more efficient in heterogeneous than homogeneous landscapes. Push-pull intercropping that simultaneously improves habitat heterogeneity, conserves biodiversity by reducing hard chemical use, and improves refugia and nectar sources would improve parasitoid survival and efficiency [89]. This concept integrates climate-smart technologies as it utilizes ecosystem services for improved crop yields and quality.

Mass rearing and augmentative release of *Cotesia vestalis* (Haliday) (Hymenoptera: Braconidae) can be used to complement the conservation of existing faunal guilds [89] through the use of softer insecticides, as previously explained. Among the diverse range of DBM parasitoids, *C. vestalis* is the most widely distributed in SA [5,22,73,89], with the highest parasitism rates [5,21,22,73] and the only one tolerating the hot and arid tropical climates [5,17] typical of SA. The use of DBM entomopathogens naturally occurring in SA environments is also a novel possibility; for example, using *Metarhizium anisopliae* (Metchnikoff) [10] and a variety of other fungal microbes (as discussed in [25]).

5. Constraints to IPM Implementation and Adoption of Novel Sustainable Control Methods

5.1. Poor Understanding of the IPM Concept and Information Flow among DBM Management Actors

Currently, despite IPM being common, there is no evident decrease in pesticide usage even in areas where the concept is favourably viewed [83]. Farmers tend to adopt IPM based more on personal commitment level or influence by peers, rather than on recommendations from agricultural extension officers or researchers [52,82,83]. In Malaysia, [90] observed very little change in farming systems over a decade, particularly the use of synthetic insecticides despite widespread IPM campaigns. This can be attributed in part to lack of documented systematic IPM methodology or commercially prepared IPM packages with step-by-step instructions on how to use them [59,91]. Intensive research for a locally developed IPM system with simplified methodology, and inexpensive and accessible materials is therefore essential.

The major constraints to IPM adoption include a lack of awareness and knowledge [2,40], both of which are driven by the weak links and poor networking among the key players (Figure 6). Each player in the production and marketing chain has a crucial role to play; researchers develop the technology, extension officers transfer the technology to the end user (farmers), policy-makers create an enabling environment, and the agrochemical industry supplies the inputs (Figure 6). Vendors, supermarket chains, and horticultural export agents are key actors on the market and should be considered as part of the chain. Journalists, high profile multi-media agricultural reporters, and national broadcasters need to understand the principles of IPM for positive reporting to avoid misrepresentation of facts. The conceptual framework (Figure 6) shows that currently strong links (solid arrows) only exist between policy-makers and the agrochemical industry; researchers and funding agencies; and the agrochemical industry and media, all of which affect the farmers. Policy-makers have weak links with researchers, farmers, and the markets. The media also has weak links with researchers and extension agents, while having strong links with the agrochemical industry, explaining why horticultural programs on national broadcasters are currently dominated by product advertisements rather than IPM knowledge packages. It is therefore hypothesized that improving direct links between policy-makers and the markets, as well as the farmers, through pesticide residue limit assessment and enforcement, coupled with the development of knowledge packages that can also be passed through media and extension (Figure 6), would improve IPM adoption and reduce reliance on chemical pesticide usage. This would also improve consumer and worker safety against pesticide exposure. Knowledge packages may include case studies of successful local IPM programmes in vernacular languages to enable farmers to appreciate and fully understand the techniques, the principles, and the benefits of the IPM technology.

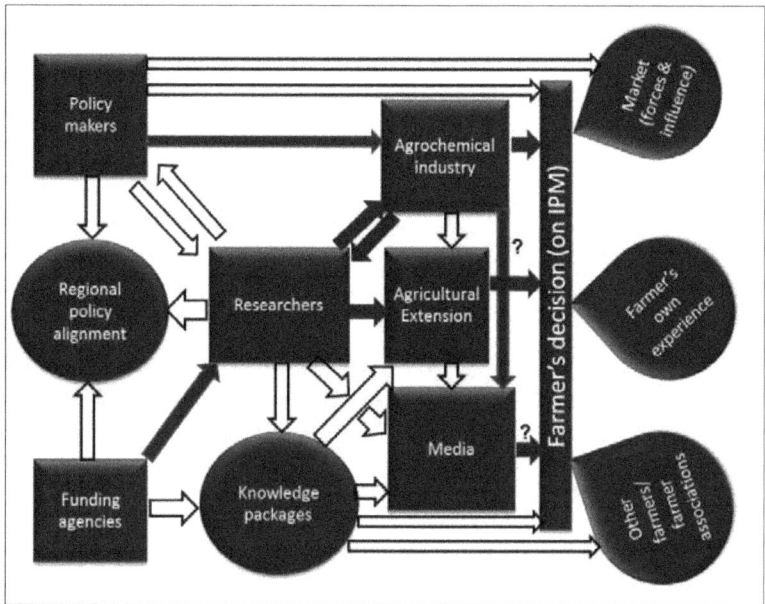

Figure 6. Perceived conceptual framework of links and information flow in DBM research in Southern Africa. The currently existing framework (solid arrows), the proposed framework (blank arrows), new suggested structures (circles), and links with both positive and negative influence on farmers' decision making (?) (Authors' own construction).

The introduction of IPM technology requires initial intensive training of the extension agents so that they cascade accurate and up-to-date information to farmers. IPM is complex process as it involves multiple components [82] and researchers often overestimate and equate their understanding of the concept with those of the extension agents, who also overestimate farmers' understanding [82]. In addition, donation of free agrochemicals by governments or donors and disproportional advertisements by the agrochemical industry, or a combination of such, does not only impede farmers' freedom of insect pest control options but also reduces their flexibility in decision-making [68,82]. Domination of synthetic pesticide research, manufacturing, and advertisements by the agrochemical industry, often in collaboration with academic researchers, coupled with lack of funding for research on non-chemical options, has further driven most agricultural extension agents and, subsequently, farmers to believe that the use of chemical pesticides is modern in agriculture [24,42,43,82]. This then overshadows the advances made in non-chemical pest control research, making farmers consider synthetic chemicals rather than non-chemical options as modern and first line of defence in DBM control [82].

Non-chemical control options, or a combination of such (in an IPM programme), are still largely considered as primitive due to a lack of understanding [40]. In most SA countries, extension work is dominated by the distribution of farming inputs (mainly fertilizers, seed, chemicals, etc.), with synthetic pesticides often being part of the package to the farmers [82]. This is also exacerbated by the farmers' high concern for access to inputs and the priority placed on these inputs [20] rather than the desire for knowledge or the use of non-chemical pest control methods [92]. This leaves little room for delivery of IPM knowledge packages through various training channels without input incentives. Requisite knowledge delivery to farmers is thus not valued as it should be, though it is key to understanding the concepts behind technologies enabling farmers to assess their risk and value for money invested, in order to make informed and independent adoption decisions [40,90,91]. Researchers and extension agents alike underestimate the amount of knowledge and information needed to convince small-scale farmers to adopt new technologies [91]. The latter's knowledge has not been able to keep pace with rapid agricultural technological changes, especially the dynamic DBM pest severity and management needs that continue to evolve in brassica production agro-ecosystems [42,91]. Increased knowledge has been proven to correlate with better pest management behaviour [43,75]. An understanding of the science behind building this knowledge in farmers is lacking among most extension agents in SA [86,91]. Knowledge is a dynamic system of cognition and is a sum of what has been learned, experienced, and perceived [91]. It involves observation, fact, and interpretative theory requiring intensive farmer participation [86,91,93]. As researchers and extension agents are more often providing information than knowledge, farmers' behaviour is unlikely to change under current scenarios [91,93]. Currently information is presented to farmers in a broad-spectrum format [91], but this has resulted in low uptake of technologies, as evidenced by low adoption. Information presented as such is often perceived by farmers as external rhetoric, associated with extension staff messages outside their farming systems [91,93]. This is so because most small-scale farmers in SA are risk-averse and unwilling to partake in voluntary schemes without immediate tangible incentives to which they are traditionally accustomed [82,91]. Until this mindset is changed through imparting knowledge and skills rather than information, for example through participatory IPM, the Farmer Field School (FFS) approach, or participatory action research/learning, co-learning and co-innovation approaches; adoption may remain a challenge.

There is a need to improve farmers' environmental knowledge base first, before the principles and practices of IPM can be emphasised [91,93]. For sustainability, the IPM packages need to be developed from local resources to avoid the constraints associated with external inputs and reduce strain on natural resources. For example, through development of participatory IPM in FFS, farmers may need to be trained in tritrophic interactions (plant-pest-natural enemy), pesticide toxicity, and its ecological consequences using farmer-tailored IPM curricula and approaches [90]. To foster positive attitudes towards IPM and improve its eventual implementation and adoption, there is need for awareness campaigns along the whole chain of stakeholders, alongside regular farmer trainings. As part of the reinforcement, it may also be beneficial if governments could feed eco-toxicological data into national pesticide registration policies to improve the adoption of IPM through the enforced use of softer and safer insecticides [43,62].

5.2. Weak Links between the Players in the Agroindustry

Parastatals, non-governmental organisations (NGOs), public national and international research institutions, independent researchers, private companies, and universities are not linked in a synergistic coordinated network, resulting in individual researchers and/or institutions independently presenting different technologies to farmers [2], sometimes with conflicting messages being conveyed. Sometimes host farmers may entertain a couple of researchers whose objectives are contradictory (personal observation, 2014), creating confusion and lack of trust among the farmers and extension agents alike [2]. The activities of the private sector, particularly the agrochemical industry, are scarce in literature. However, they are key to the procurement and distribution of chemicals and have a strong direct link with the farmers, which can be harnessed to propel other pest management options. Hence, there is need for collaboration of all stakeholders doing similar research and development work to fine-tune the broad-spectrum recommendations to specific relevant practices that enhance the fusion of the *emic* (inner perspective of the farmer) and the *etic* (outer perspective of the research/extension) [91,93]. Unfortunately, such platforms are rare.

5.3. Lack of Locally-Developed Well-Packaged IPM Practices and Procedures

The introduction of IPM should touch on various technical and social interventions [82]. The technical aspects mainly involve the techniques that farmers need to use to implement IPM in their brassica production systems. The development of step-by-step IPM methodology for cabbages in Asia through AVRDC and AVNET was key in the implementation and success of IPM in that region [82]. However, this has not been the case for SA. Direct adoption of Asian methodology may not necessarily apply in Africa due to different biophysical conditions, farmer practices, and socio-economic perceptions and circumstances [91]. Consequently, a SA IPM methodology tailored to specific local needs must be developed using participatory approaches to get farmers' buy-in. Furthermore, IPM monitoring tools to determine DBM economic threshold levels need to be scientifically investigated [29]. Local scientists and institutions have not developed IPM programs with regulatory and territorial chemical use boundaries for area-wide IRM, hence they still 'encourage' the use of any new chemicals [42,44]. These technical aspects also need to be locally refined and packaged within the small-scale farmer's contextual framework before a full IPM package can be presented and adapted for dissemination.

5.4. Lack of Policy Support

In Asia, the success of IPM programs was partly attributed to the crafting of enabling policies [80]. These included country agreements and harmonised policies to enable collaborative research, information-sharing, and the importation of natural enemies for key regional horticultural pests [79]. However, to the best of our knowledge, such enabling policies may still be lacking in SA. Global politics, as regards chemical use controls, is such that toxic pesticides are first banned in developed nations with effective regulatory and legislative policies. As regulations tighten in these countries, chemical manufacturing is reduced and the burden is passed on to developing countries by relocating factories and establishing subsidiaries in poor countries with governments that do not have effective regulatory controls [59,86,94]. In SA, this results in the uncontrolled use of extremely hazardous compounds, even years after they have been banned [5,20,61,94]. Some of the banned pesticides include DDT (only limited to mosquito control), Chlordane, Monochrotophos, Dieldrin, and Arsenic [94–96]. Therefore, this calls for strong technical and legislative capacitation of SA governments on issues of pesticide harmonized regulation and financial resources needed to develop and implement such legislations [42,81,82,94]. A classic example was Zimbabwe's successful development and implementation of a within-season pesticide rotational scheme and a closed season for the cotton bollworm *Helicoverpa armigera* (Hübner), achieved after a few years of strong legislative enforcement [24]. In SA, brassica farmers independently decide on the type of pesticide to buy, where to buy it, and when to apply it without any enforced regulation or legislation to consider. Though some general chemical regulatory frameworks may exist on paper, implementation is still a challenge in the region. Since brassicas are produced all-year-round, this promotes all-year-round unrestrained insecticide use on fresh vegetables that are supplied for public consumption, most of which are sometimes eaten raw. We therefore recommend a strong policy regulatory framework to control, minimize, and synchronize chemical use across all major horticultural production areas and markets. Mechanisms to implement and monitor the policy may also need to be clearly laid out right from the outset.

In some developing middle-income countries (e.g., Malaysia), threshold levels of pesticide residues permissible in crop products are well-laid out and monitored at different levels of the market value chain [88]. The lack of such policies in SA and the subsequent lack of regulatory frameworks account for the high pesticide residues in fresh products [2,42]. This is exacerbated by the cosmetic urban consumers' unconscious demand for damage-free brassicas [2,39,42]. The chronic nature of accumulated pesticide effect in humans makes the danger 'invisible' [63,64]. Though implementing residue-monitoring systems through the whole production and supply chain may prove logistically and financially infeasible for SA governments, the development of policy, legislation, and relevant monitoring tools may allow government officials to implement checkpoint systems across the vegetable production and supply chains.

Due to DBM notoriety and economic importance, we suggest that it may be necessary for SA governments to declare it a pest of regional economic importance, warranting policy recognition and consented governments' intervention, as is the case for tsetse flies, *Glossina morsitans* (Wiedemann); invasive fruit flies *B. dorsalis*; the larger grain borer, *Prostephanus truncatus* (Horn); migratory insect pests like the African armyworm, *Spodoptera exempta* (Walker); and African migratory locusts *Locusta migratoria migratorioides* (Fairmaire & Reiche). We recommend regional policy synchronisation, collaborative research, public awareness, farmer training, and IPM through Farmer Field School (IPM-FFS) initiatives in the management of *P. xylostella* synonymous with efforts applied to these other economic pests.

5.5. Taxonomic Confusion and Insufficient Adaptation of Biocontrol Agents to Release-Sites Climate and Bio-Ecological Conditions

The introduction of efficacious natural enemies has been marred by parasitoid taxonomic confusion and misidentification [18,97,98]. We have not found any reports of SA field-sourced parasitoid populations reared for mass release in DBM biological control programs in the region. *Diadegma semiclausum* and *C. vestalis* are currently the most common and efficacious DBM parasitoids in Africa [17,18,97,98]. *Diadegma semiclausum*, used for east African biocontrol programs, was once misidentified and exported as *D. mollipla* [99]. Due to misidentifications, release populations for the DBM control were imported from Taiwan, regardless of its local abundance in the horticultural hot-spots of Kenyan Eastern Highlands [74,98]. *Cotesia vestalis* is the most abundant and most efficient DBM parasitoid in SA [5,19,22,74,89], but some literature still refer to it as *Cotesia plutellae* (Kurdjumov) [18]. However, currently, *C. vestalis* populations from different climates are lumped together and considered as one species, despite observed biological differences [98]. Thus, molecular methods that can reliably separate biologically distinct but morphologically identical populations may be useful tools that can reliably confirm species' identity and hence improve the success of future biological control programmes [18,99].

Climate mismatching between parasitoid source areas and target release sites has led to the failure of most foreign reared but African released natural enemies [26,85]. This has now been exacerbated by unpredictably variable weather, increasing temperatures and fluctuating humidity caused by global climate change [9]. For DBM control, climate mismatching has previously been reported as a major setback for most biological control attempts [85]. Under the circumstances, climate matching between source area and target release site becomes an integral component of biological control programs based on parasitoid mass releases. This can only be achieved by a careful study of the thermal biology of the target parasitoid species. Mass introductions may then be targeted through acclimation, to suit areas of release [26,99,100]. Indeed, previous studies have recommended that thermal acclimation can significantly improve the fitness of laboratory reared insects upon introduction to wild conditions [100–103], and this approach has even been recommended for field releases using Sterile Insect Technique (SIT) [26]. It has been documented that biological control using predators and parasitoids should aim at developing resilient agro-ecosystems which maintain species' evolutionary potential to improve efficacy. This may be done through direct improvement in natural enemy genetic diversity and processes that encourage continuous in situ evolutionary adaptation [54].

5.6. Limited Alternative Control Options

In SA, the use of *Bt* transgenic brassicas have so far only been done in South Africa [96]. However, due to socio-political and controversial environment-related risks, it is yet to be commercialised in other SA countries [86]. Field and market observations from 2014 to 2016 showed that *Bt*-based insecticides are slowly filtering into the regional market. For example, pioneer *B. thuringiensis* (*var. kurstaki*) bioassays in Botswana showed 85.7%–94.6% reduction in DBM damage [71,72], but market availability still remains a challenge. However, this efficacy may also be short-lived, due to resistance development [69] as the insecticide is applied without a technical insecticide rotating scheme [68].

SIT has been successfully used for DBM management in Myanmmar [104], but its implementation in SA requires huge capital investment and substantial financial backup in addition to specialised human resources [105]. Furthermore, SIT is only effective in an area-wide approach, which may be challenging due to scattered distribution of small-scale subsistence farmers in SA.

Similarly, genetic engineering, through the release of insects carrying a male-selecting transgene, has equally managed to suppress DBM populations through the prevention of female progeny survival [106]. The same technique has been used successfully to control the fruit flies, *B. oleae* [107] and *Ceratitis capitata* (Wiedemann) [108], and the mosquito, *Aedes aegypti* (L.) [109]. However, this has not yet been considered in SA, probably due to the controversy surrounding genetically modified organisms. Nevertheless, it is an option worth considering in future DBM management programmes.

Strategies for developing varietal resistance in brassicas against DBM have not yet been fully exploited [82]. Modification of biochemical and morphological plant characteristics has also been unsuccessful [25,83]. Thus, despite its potential as an alternative non-chemical DBM control method, resistant variety development is still a huge challenge to biochemists and plant breeders [82] in SA. We have not found any research identifying chemical compounds or genes that are necessary to manipulate and cause brassicas to be completely non-preferred hosts for DBM [82].

6. Future Prospects and Research Needs

Future prospects in the sustainable management of DBM in SA lie in two principles, as outlined by [110].

(1) 'Do no harm'—the use of biologically- and environmentally-safe pest control methods with no or selective soft and safe insecticide use.
(2) 'Do good'—Improving farmer knowledge, consumer, agrochemical industry, and policy-maker awareness; policy reforms and regional policy harmonization; strengthening regulatory frameworks; and national and regional institutional capacitation.

Based on these principles, future research needs to identify and develop IPM-compatible components for the sustainable management of the DBM applicable to small-scale farmer circumstances in SA. Complementary to this, baseline information on the spatio-temporal population dynamics of DBM in relation to climatic parameters is needed. This may assist the area-specific determination of current population and pest management trends and how they correlate with environmental factors. This knowledge is important for the identification of gaps where development of new or improvement in existing IPM interventions is needed. Modelling the population and climate data will also assist in the development of predictive models that can be used in early warning systems to prepare farmers for possible outbreaks.

Farmer and extension staff capacity development can be achieved through participatory research using IPM in the FFS approach, on-farm farmer-managed, and researcher-managed trials. This will not only connect scientific findings with farmer's traditional knowledge and experience but will develop sustainable farmer-to-farmer knowledge-sharing platforms [58]. This also promotes co-learning, co-innovation and ownership of findings amongst all stakeholders which are essential ingredients for adoption. Success in technology adoption in various areas of agriculture was achieved in Asia through FFSs [44,91]. Farmer behavioural change may be possible through training, mass media awareness, legislative enforcement, and market condemnation of plant products exceeding set thresholds of pesticide residues [42,43,90]. Therefore, regular pesticide residue analysis may provide convincing evidence for governments to enforce regulations on chemical use on fresh vegetables. Where the regulations do not exist, they should be developed.

The future abundance and efficacy of *C. vestalis* under climate change remains uncertain. In addition, the synchrony and co-evolutionary adaptation between the host and the parasitoid also remains unpredictable [26]. Therefore, comparative abiotic stress tolerance studies of both the host and the parasitoid will provide insights into the needs for improvement of environmental fitness and efficiency of the potential parasitoids as an integral component of future IPM designs. Climate change was observed to impact insects negatively on the timing of life history traits, geographical shifts in species ranges, and the alteration of ecosystem interactions [55]. In addition, there are high-predicted rates of extinction in some species [111]. Therefore, comparative abiotic stress tolerance studies will not only enable the determination of whether it is the host or the parasitoid that is at high risk of extinction due to the impact of climate change, but will also be necessary to improve parasitoid field fitness for future release programs. Insect thermal biology and the ability to predict the impact of climate change on insect species are some of the most noble research findings of our time, yet adequate utilisation of this knowledge to improve pest management is still lacking, especially in Africa [35,55]. In the case of DBM, IPM systems need to have climate-resilient parasitoids capable of absorbing the 'shock' associated with the environmental changes due to global warming as a critical component of a broader climate-resilient IPM-FFS pest management systems approach [26,55].

Non-chemical control of *P. xylostella* may also be achieved through the manipulation of insect–host interactions. This may be achieved through brassica varietal resistance breeding and/or modification of the habitat by careful intercropping with attractant and repellent crops. As DBM larvae is generally monophagous, a varietal resistance option is promising if given full attention [6,82]. Research has also shown that *P. xylostella* moths do not oviposit on non-hosts. This means habitat management through agro-ecosystem manipulation may be an effective strategy to incorporate in IPM systems [81]. In light of the current knowledge, mere intercropping without careful selection of the repellent and attractant crops to enhance a 'push-pull' effect has not been very effective [12,78,79]. The 'push-pull' technology has been used for the successful management of cereal stem borers in eastern Africa [112,113]. This technology may be expanded, improved, and applied to economic pests such as the DBM in SA. For sustainability and cost-effectiveness, this technology may need to be geographically flexible in repellent crop selection to enable farmers to choose repellent crops naturally occurring and readily available in their localities. However, initial investment may be needed to conduct farmer participatory field research in the initial selection of potential candidate repellent and attractant crops.

7. Conclusions

SA is facing a serious DBM challenge and efforts towards its management are characterised by a variety of constraints. These vary from farmers' behaviour regarding insecticide choice and its use and/or misuse, a lack of health and environmental consciousness, a lack of locally-developed alternative control methods, a lack of regulatory enforcements, weak policy frameworks, and low research attention that is neither regionally-coordinated nor aligned for the achievement of common goals, all exacerbated by climate change and variability. The future of sustainable DBM control lies in IPM-FFS holistic approaches that include territorial IRM, cropping systems approaches (push-pull intercrops), soft and selective insecticides, area-wide pest management, biological control, the use of entomopathogens, and varietal resistance breeding developed in an IPM package. This should be supported by farmer and extension staff training as the founding principle of the approach to enhance in-depth knowledge and understanding of the IPM concepts, principles and procedures in a changing climate. There is also a need for exploring institutional or structural transformations to facilitate effective information flow and collaboration, sustainable uptake of IPM packages for improved crop protection systems, especially with respect to DBM and overall sustainable development.

Acknowledgments: The authors would like to acknowledge financial and technical support from Botswana International University of Science and Technology (BIUST) and technical support from the University of Zimbabwe (UZ). Many thanks to several anonymous referees, and the Eco-Physiological Entomology Research Team (BIUST) for the comments on an earlier version of the manuscript.

Author Contributions: All authors contributed equally to the manuscript. All authors have read and approved the final manuscript.

Conflicts of Interest: The authors declare no conflict of interest.

References

1. Sibanda, T.; Dobson, H.M.; Cooper, J.F.; Manyangarirwa, W.; Chiimba, W. Pest management challenges for small-holder vegetable farmers in Zimbabwe. *Crop Prot.* **2000**, *19*, 807–815. [CrossRef]
2. Saka, A.R.; Mtukuso, A.P.; Mbale, B.J.; Phiri, I.M.G. The role of research-extension-farmer linkages in vegetable production and development in Malawi. In *Vegetable Research and Development in Malawi. Review and Planning Workshop Proceedings, Lilongwe, Malawi, 23–24 September 2003*; Chadha, M.L., Oluoch, M.O., Saka, A.R., Mtukuso, A.P., Daudi, A., Eds.; World Vegetable Center (AVRDC): Shanhua, Taiwan, 2003.
3. Munthali, D.C. Evaluation of cabbage varieties to cabbage aphid. *Afr. Entomol.* **2009**, *17*, 1–7. [CrossRef]
4. Food and Agriculture Organization (FAO). *Evolving a Plant Breeding and Seed System in Sub-Saharan Africa in an Era of Donor Dependence*; FAO Plant Production and Protection Paper 210; FAO: Rome, Italy, 2011.
5. Canico, A.; Santos, L.; Massing, R. Development and adult longevity of diamondback moth and its parasitoids *Cotesia plutellae* and *Diadegma semiclausum* in uncontrolled conditions. *Afr. Crop Sci. Conf. Proc.* **2013**, *11*, 257–262.
6. Khonje, A.A. Research trends in horticultural crops in Malawi. *J. Crop Weed* **2013**, *9*, 13–25.
7. Ekesi, S.; Chabi-Olaye, A.; Subramanian, S.; Borgeimeister, C. Horticultural pest management and African Economy: Successes, Challenges and Opportunities in a changing global environment. *Acta Hortic.* **2009**, *911*, 165–183. [CrossRef]
8. Stathers, T.; Lamboll, R.; Mvumi, B.M. Postharvest agriculture in changing climate: Its importance to African smallholder farmers. *Food Secur.* **2013**, *5*, 361–392. [CrossRef]
9. Intergovernmental Panel on Climate Change (IPCC). *Climate Change 2014: Synthesis Report*; Fifth Assessment Report (AR5); Contribution of Working Groups I, II and III to the Fifth Assessment Report of the Intergovernmental Panel on Climate Change; IPCC: Geneva, Switzerland, 2014.
10. Maniania, N.K.; Takasu, K. Development of microbial control agents at the International Centre of Insect Physiology and Ecology. *Bull. Inst. Trop. Agric.* **2006**, *29*, 1–9.
11. Food and Agriculture Organization (FAO). *Growing Greener Cities in Africa*; First Status Report on Urban and Peri-Urban Horticulture in Africa; Food and Agriculture Organisation of the United Nations: Rome, Italy, 2009.
12. Katsaruware, R.D.; Dubiwa, M. Onion (*Allium cepa*) and garlic (*Allium sativum*) as pest control intercrops in cabbage based intercrop system in Zimbabwe. *J. Agric. Vet. Sci.* **2014**, *7*, 13–17.
13. Abate, T.; van Huis, A.; Ampofo, J.K.O. Pest Management Strategies in traditional agriculture: An African Perspective. *Ann. Rev. Entomol.* **2000**, *45*, 631–659. [CrossRef] [PubMed]
14. Momanyi, D.; Lagat, K.J.; Ayuya, O.I. Determinants of smallholder African indigenous leafy vegetables farmers' market participation behaviour in Nyamira County, Kenya. *J. Econ. Sustain.* **2015**, *16*, 212–217.
15. Department of Agriculture, Forestry and Fisheries. *A Profile of the South African Cabbage Market Value Chain*; Department of Agriculture, Forestry and Fisheries: Arcadia, South Africa, 2014.
16. Lubinga, M.; Ogundeji, A.; Jordaan, H. East African community trade potential and performance with European Union: A perspective of selected fruit and vegetable commodities. *ESJ* **2014**, *1*, 430–443.
17. Talekar, N.S.; Shelton, A.M. Biology, ecology and management of the diamondback moth. *Ann. Rev. Entomol.* **1993**, *38*, 275–301. [CrossRef]
18. Furlong, M.J.; Wright, D.J.; Dosdall, L.M. Diamondback moth ecology and management: Problems, progress and prospects. *Ann. Rev. Entomol.* **2013**, *58*, 517–554. [CrossRef] [PubMed]

19. Sithole, R. Life History Parameters of *Diadegma mollipla* (Holmgren), Competition with *Diadegma semiclausum* Hellen (Hymenoptera: Ichneumonidae) and Spatial and Temporal Distribution of the Host, *Plutella xylostella* (L.) and Its Indigenous Parasitoids in Zimbabwe. Ph.D. Thesis, University of Zimbabwe, Harare, Zimbabwe, 2005.

20. Obopile, M.; Munthali, D.C.; Matilo, B. Farmers' knowledge, perceptions and management of vegetable pests and diseases in Botswana. *Crop Prot.* **2008**, *27*, 1220–1224. [CrossRef]

21. Nofemela, R.; Kfir, R. The pest status of Diamondback moth (Lepidoptera: Plutellidae) in South Africa: The role of parasitoids in suppressing the pest populations. In Proceedings of the Fifth International Workshops on the Management of Diamondback Moth and other Crucifer Pests, Beijing, China, 21–24 October 2008.

22. Manyangarirwa, W.; Zehnder, G.W.; McCutcheon, G.S.; Smith, J.P.; Adler, P.H.; Mphuru, A.N. Parasitoids of the diamondback moth on brassicas in Zimbabwe. *Afr. Crop Sci. Conf. Proc.* **2009**, *9*, 565–570.

23. Nyambo, B.; Sevgan, S.; Chabi-Olaye, A.; Ekesi, S. Management of alien invasive insect pest species and diseases of fruits of vegetables: Experiences from East Africa. *Acta Hort.* **2009**, *911*. [CrossRef]

24. Tibugari, H.; Mandumbu, R.; Jowah, P.; Karavina, C. Farmer knowledge, attitude and practice on cotton (*Gossypium hirsutum* L.) pest resistance and management practices in Zimbabwe. *Arch. Phytopathol. Pflanzenschutz.* **2012**, *45*, 2395–2405. [CrossRef]

25. Gryzwacz, D.; Rosbach, A.; Rauf, A.; Russel, D.A.; Srivansan, R.; Shelton, A.M. Current control methods for diamondback moth and other brassica insect pests and the prospexcts for improved management with lepidopteran resistant *Bt* vegetable brassicas in Asia and Africa. *J. Crop Prot.* **2010**, *29*, 68–79. [CrossRef]

26. Chidawanyika, F.; Mudavanhu, P.; Nyamukondiwa, C. Biologically based methods for pest management under changing climates: Challenges and future directions. *Insects* **2012**, *3*, 1171–1189. [CrossRef] [PubMed]

27. Zalucki, M.P.; Shabbir, A.; Silva, R.; Adamson, D.; Shu-shen, L.; Furlong, M.J. Estimating the Economic Cost of One of the World's Major Insect Pests, *Plutella xylostella* (Lepidoptera: Plutellidae): Just How Long Is a Piece of String? *J. Econ. Entomol.* **2012**, *105*, 1115–1129. [CrossRef] [PubMed]

28. Macharia, I.; Mithofer, D.; Waibel, H. Health effects of pesticide use among vegetable farmers in Kenya. In Proceedings of the 4th International Conference of the African Association of Agricultural Economists, Hammamet, Tunisia, 22–25 September 2013.

29. Ayalew, G. Comparison of yield losses on cabbage from diamondback moth, *Plutella xylostella* L. (Lepidoptera: Plutellidae) using two insecticides. *Crop Prot.* **2006**, *25*, 915–919. [CrossRef]

30. Safraz, M.; Keddie, A.B.; Dosdall, L.M. Biological control of the diamondback moth, *Plutella xylostella*. A Review. *Biocontrol Sci. Technol.* **2005**, *15*, 763–789. [CrossRef]

31. Marchioro, C.A.; Foerster, L.A. Development and survival of diamondback moth, *Plutella xylostella* L. (Lepidoptera: Plutellidae) as a function of temperature: Effect on the number of generations in tropical and sub-tropical regions. *J. Neotrop. Entomol.* **2011**, *40*, 533–541.

32. Chown, S.L.; Nicolson, S. *Insect Physiological Ecology: Mechanisms and Patterns*; Oxford University Press: Oxford, UK, 2004.

33. Denlinger, D.L.; Lee, R.E., Jr. *Low Temperature Biology of Insects*; Cambridge University Press: Cambridge, UK, 2010.

34. Bahar, M.H.; Hegedus, D.; Soroka, J.; Coutu, C.; Bekkaoui, D. Survival and *Hsp70* gene expression in *Plutella xylostella* and its larval parasitoid *Diadegma insulare* varied between slowly ramping and abrupt extreme temperature regimes. *PLoS ONE* **2013**, *8*, e73901. [CrossRef] [PubMed]

35. Hill, M.P.; Bertelsmeier, C.; Clusella-Trullas, S.; Garnas, J.R.; Robertson, M.P.; Terblanche, J.S. Predicted decrease in global climate suitability masks regional complexity of invasive fruit fly species response to climate change. *Biol. Invasions* **2016**, *18*, 1105–1119. [CrossRef]

36. Andrew, N.R.; Hill, S.J.; Binns, M.; Bahar, M.H.; Ridley, E.V.; Jung, M.P.; Fyfe, C.; Yates, M.; Khusro, M. Assessing insect responses to climate change: What are we testing for? Where should we be heading? *Peer J.* **2013**, *1*, e11. [CrossRef] [PubMed]

37. Shirai, Y. Temperature tolerance of diamondback moth, *Plutella xylostella* (Lepidoptera: Yponomeutidae) in tropical and temperate regions of Asia. *Bull. Entomol. Res.* **2000**, *90*, 357–364. [CrossRef] [PubMed]

38. Nguyen, C.; Bahar, M.H.; Baker, G.; Andrew, N.R. Thermal tolerance limits of DBM in ramping and plunging assays. *PLoS ONE* **2014**, *9*, e87535. [CrossRef] [PubMed]

39. Kuntashula, E.; Silesh, G.; Mafongoya, P.L.; Bond, J. Farmer participatory evaluation of the potential for organic vegetable production in the wetlands of Zambia. *Outlook Agric.* **2006**, *35*, 299–305. [CrossRef]

40. Nyirenda, S.P.; Sileshi, G.W.; Belmain, S.R.; Kamanula, J.F.; Mvumi, B.M.; Sola, P.; Nyirenda, G.K.C.; Stevenson, P.C. Farmers' ethno-ecological knowledge of vegetable pests and pesticidal plant use in Northern Malawi and Eastern Zambia. *Afr. J. Agric. Res.* **2011**, *6*, 1525–1537.

41. Madisa, M.E.; Obopile, M.; Assefa, Y. Analysis of horticultural production trends in Botswana. *J. Plant Stud.* **2012**, *1*. [CrossRef]

42. Wandaat, E.Y.; Kugbe, J.X. Pesticide misuse in rural-urban agriculture: A case study of vegetable production in Tano South of Ghana. *AJAFS* **2015**, *3*, 343–360.

43. Williamson, S.; Ball, A.; Pretty, J. Trends in pesticide use and drivers for safer pest management in four African countries. *Crop Prot.* **2008**, *27*, 1327–1334. [CrossRef]

44. Timprasert, S.; Datta, A.; Ranamukhaarachchi, S.L. Factors determining adoption of integrated pest management by vegetable growers in Nakhon Ratchasima Province, Thailand. *Crop Prot.* **2014**, *62*, 32–39. [CrossRef]

45. Lobell, D.B.; Burke, M.B.; Tebaldi, C.; Mastrandrea, M.D.; Falcon, W.P.; Naylor, R.L. Prioritizing climate change adaptation needs for food security in 2030. *Science* **2008**, *319*, 607–610. [CrossRef] [PubMed]

46. Jack, C. *Climate projections for United Republic of Tanzania*; Climate Systems Analysis Group (CSAG)—University of Cape Town: Cape Town, South Africa, 2010.

47. Steynour, A.; Jack, C.; Taylor, A. *Information on Zimbabwe's Climate and How It Is Changing*; Climate Systems Analysis Group—University of Cape Town: Cape Town, South Africa, 2012.

48. Arthropod Pesticide Resistance Database (IRAC). Michigan State University. 2015. Available online: http://www.pesticideresistance.com/display.php?page=speciesarId=571 (accessed on 22 August 2015).

49. Mosiane, S.M.; Kfir, R.; Villet, M.H. Seasonal phenology of the diamondback moth, *Plutella xylostella* L. (Lepidoptera: Plutellidae) and its parasitoids on canola *Brassica napus* (L.) in Gauteng Province, South Africa. *Afr. Entomol.* **2010**, *11*, 277–285.

50. Dosdall, L.M.; Zalucki, M.P.; Tansey, J.A.; Furlong, M.J. Developmental responses of the diamondback moth parasitoid *Diadegma semiclausum* (Hellén) (Hymenoptera: Ichneumonidae) to temperature and host plant species. *Bull. Entomol. Res.* **2012**, *102*, 373–384. [CrossRef] [PubMed]

51. Sgrò, C.M.; Lowe, A.J.; Hoffmann, A.A. Building evolutionary resilience for conserving biodiversity under climate change. *Evol. Appl.* **2010**, *4*, 326–337. [CrossRef] [PubMed]

52. Kopper, B.J.; Lindroth, R. Effects of elevated carbon dioxide and ozone on the phytochemistry of aspen performance of an herbivore. *Oecologia* **2003**, *134*, 95–103. [CrossRef] [PubMed]

53. Sanders, N.T.; Belote, R.T.; Weltzin, K.F. Multitrophic effects of elevated carbon dioxide on understory plant and arthropod communities. *Environ. Entomol.* **2004**, *33*, 1609–1616. [CrossRef]

54. Gill, H.K.; Garg, H. Pesticide: Environmental impacts and management strategies. In *Pesticides—Toxic Effects*; Solenski, S., Larramenday, M.L., Eds.; Intech: Rijeka, Croatia, 2014; pp. 187–230.

55. Polson, K.A.; Brogdon, W.G.; Rawlins, S.C.; Chadee, D.D. Impact of environmental temperatures on resistance to organophosphate insecticides in Aedes aegypti from Trinidad. *Rev. Panam. Salud Publ.* **2012**, *32*, 1–8. [CrossRef]

56. Liu, F.; Miyata, T.; Wu, Z.J.; Li, C.W.; Wu, G.; Zhao, S.X.; Xie, L.H. Effects of temperature and fitness costs, insecticide susceptibility and heat shock protein in insecticide resistant and susceptible *Plutella xylostella*. *Pestic. Biochem. Physiol.* **2008**, *91*, 45–52. [CrossRef]

57. Metzger, M.J.; Bunce, R.G.H.; Trabucco, A.; Sayre, R.; Jangman, R.H.G.; Zomer, R.J. A high resolution bioclimate map for the world: A unifying framework for global biodiversity research and monitoring. *Glob. Ecol. Biogeogr.* **2013**, *22*, 630–638. [CrossRef]

58. Food and Agriculture Organization (FAO). *Identifying Opportunities for Climate-Smart Agriculture Investment in Africa*; Economics & Policy Innovations for Climate-Smart Agriculture, FAO: Rome, Italy, 2012.

59. Harvey, C.D. Integrated Pest Management in temperate horticulture: Seeing the wood for trees. *CAB Rev.* **2015**, *10*, 028. [CrossRef]

60. Ngowi, A.V.; Maeda, D.W.; Partanen, T.J. Knowledge, Attitudes and Practices (KAP) among agricultural extension workers concerning the reduction of the adverse impact in agricultural areas in Tanzania. *Crop Prot.* **2007**, *26*, 1617–1624. [CrossRef] [PubMed]

61. Baliga, S.S.; Repetto, R. *Pesticides and the Immune System: The Public Health Risks*; World Resources Institute: Washington, DC, USA, 1996.

62. Tsimbiri, P.F.; Moturi, W.N.; Sawe, J.; Henley, P.; Bend, J.R. Health impact of pesticides on residents and horticultural workers in the Lake Naivasha Region, Kenya. *Occup. Dis. Environ. Med.* **2015**, *3*, 24–34. [CrossRef]

63. Magauzi, R.; Mabaera, B.; Rusakaniko, S.; Chimusoro, A.; Ndlovu, N.; Tshimanga, M.; Shambira, G.; Chadambuka, A.; Gombe, N. Health effects of agrochemicals among farm workers in commercial farms of Kwekwe district, Zimbabwe. *Pan Afr. Med. J.* **2011**, *9*, 26. [CrossRef] [PubMed]

64. Khoza, S.; Nhachi, C.F.B.; Chikumo, O.; Murambiwa, W.; Ndudzo, A.; Bwakura, E.; Mhonda, M. Organophosphate and organochlorine poisoning in selected horticultural farms in Zimbabwe. *JASSA* **2003**, *9*, 7–15.

65. Mvumi, B.M.; Giga, D.P.; Chiuswa, D.V. The maize (*Zea mays* L.) post-production practices of smallholder farmers in Zimbabwe: Findings from surveys. *JASSA* **1995**, *1*, 115–130. [CrossRef]

66. Horna, D.; Falk-Zepeda, J.; Timpo, S.E. *Insecticide Use on Vegetables in Ghana. Would GM Seeds Benefit Farmers*; International Food Policy Research Institute (IFPRI), Environment and Production Technology Division: Accra, Ghana, 2008.

67. Abang, A.; Kouame, C.M.; Abang, M.M.; Hanna, R.; Kuate, A.F. Vegetable growers perception of pesticide use practices, cost and health effects in the tropical region of Cameroon. *Int. J. Agron. Plant Prod.* **2013**, *4*, 873–883.

68. Mudimu, G.D.; Waibel, H.; Fleischer, S. *Pesticide Policies in Zimbabwe: Status and Implications for Change*; Pesticide Policy Project; Special Issue Publication Series 1; Institute of Horticultural Economics: Hannover, Germany, 1999.

69. Tabashnik, B.E.; Malvar, T.; Liu, Y.B.; Finson, N.; Borthakur, D.; Shin, B.S.; Parck, S.H.; Masson, L.; Maard, R.A.; Bosch, D. Cross resistance of the diamondback moth indicates altered interactions with domain II of *Bacillus thuringiensis* Toxins. *J. Appl. Environ. Microbiol.* **1996**, *62*, 2839–2844.

70. Xia, Y.; Lu, Y.; Shen, J.; Gao, X.; Qiu, H.; Li, J. Resistance monitoring for eight insecticides in *Plutella xylostella* in central China. *Crop Prot.* **2014**, *63*, 131–137. [CrossRef]

71. Legwaila, M.M.; Munthali, D.C.; Kwerepe, B.C.; Obopile, M. Effectiveness of cypermethrin against diamondback moth (*Plutella xylostella* L.) eggs and larvae in cabbage under Botswana conditions. *Afr. J. Agric. Res.* **2014**, *9*, 3704–3710.

72. Legwaila, M.M.; Munthali, D.C.; Kwerepe, B.C.; Obopile, M. Efficacy of *Bacillus thuringiensis* (var. kurstaki) against diamondback moth *Plutella xylostella* (L.) Eggs and larvae on cabbage under semi controlled greenhouse conditions. *Int. J. Trop. Insect Sci.* **2015**, *7*, 39–45.

73. Kfir, R. Effect of parasitoid elimination on populations of diamondback moth in cabbage. In *The Management of Diamondback Moth and Other Crucifer Pests, Proceedings of the 4th International Workshop, Melbourne, Australia, 26–29 November 2001*; Endersby, N.M., Ridland, P.M., Eds.; The Regional Institute Ltd.: Gosford, Australia, 2004.

74. Löhr, B. Toward biocontrol based IPM for the diamondback moth in Eastern and Southern Africa. In *The Management of Diamondback Moth and Other Crucifer Pests, Proceedings of the 4th International Workshop. Melbourne, Australia, 26–29 November 2001*; Endersby, N.M., Ridland, P.M., Eds.; The Regional Institute Ltd.: Gosford, Australia, 2004; pp. 197–206.

75. Nyambo, B.; Löhr, B. The role and significance of farmer participation in biological control based IPM for brassica crops in East Africa. In Proceedings of the Second International Symposium on Biological Control of Arthropods, Davos, Switzerland, 12–16 September 2005.

76. Tonnang, N.E.Z.; Nodorezov, L.V.; Owino, O.; Ochanda, H.; Löhr, B. Evaluation of Discrete host-parasitoid models for diamondback moth and *Diadegma semiclausum* field time population densities. *Ecol. Model.* **2009**, *220*, 1735–1744. [CrossRef]

77. Bopape, M.J. The Management of Diamondback Moth, *Plutella xylostella* (L.) (Lepidoptera: Plutellidae), Population Density on Cabbage Using Chemical and Biological Control Methods. Master's Thesis, University of South Africa, Pretoria, South Africa, 2013.

78. Luchen, S.W.S. Effects of Intercropping Cabbage with Alliums and Tomato on the Incidences of Diamondback Moth *Plutella xylostella* (L.). Master's Thesis, University of Zambia, Lusaka, Zambia, 2001.

79. Karavina, C.; Mandumbu, R.; Zivenge, E.; Munetsi, T. Use of garlic Allium sativum as a repellent crop to control diamondback moth (*Plutella xylostella*) in cabbage (*Brassica oleracia* var. *capitata*). *J. Agric. Res.* **2014**, *52*, 615–622.

80. Opena, R.T.; Kyomo, M.L. Vegetable Research and Development in SADCC Countries. Available online: http://trove.nla.gov.au/work/7530097?selectedversion=NBD7890552 (accessed on 5 January 2017).

81. Green, S.K.; Shanmugasundaram, S. *AVRDC's International Networks to Deal with the Tomato Leaf Curl Disease: The Needs for Developing Countries*; Springer: Dordrecht, The Netherlands, 2007; pp. 417–439.

82. Maredia, M.K.; Dakomo, D.; Mota-Sanchez, D. *Integrated Pest Management in the Global Arena*; Commonwealth Agricultural Bureau International (CABI): Wallingford, UK, 2003.

83. Devine, G.J.; Furlong, M.J. Insecticide use: Contexts and Ecological Consequences. *Agric. Hum. Values* **2007**, *24*, 281–306. [CrossRef]

84. Walsh, B. *Impact of Insecticides on Natural Enemies in Brassica Vegetables*; Horticulture Australia Ltd.: Sydney, Australia, 2005.

85. Kfir, R. Biological control of the diamondback moth *Plutella xylostella* in Africa. In *Biological Control in IPM Systems in Africa*; Neuenschwander, P., Borgemeister, J., Langewald, J., Eds.; CABI Publishing: Wallingford, UK; Cambridge, MA, USA, 2003; pp. 363–376.

86. Safraz, M.; Dosdall, L.M.; Keddie, B.A. Diamondback moth host plant interactions: Implications for pest management. *Crop Prot.* **2006**, *25*, 625–639. [CrossRef]

87. Rossbach, A.; Löhr, B.; Vidal, S. Host shift to peas in the diamondback moth *Plutella xylostella* (Lepidoptera: Plutellidae) and response of its parasitoid *Diadegma mollipla* (Hymenoptera: Ichneumonidae). *Bull. Entomol. Res.* **2006**, *96*, 413–419. [CrossRef] [PubMed]

88. Silva-Torres, C.S.A.; Torres, J.B.; Barros, R. Can cruciferous agro-ecosystems grown under variable conditions influence biological control of *Plutella xylostella* (L.) (Lepidoptera: Plutellidae). *Biocontrol Sci. Technol.* **2011**, *21*, 625–641. [CrossRef]

89. Sohati, H.P. Establishment of *Cotesia vestalis* (Haliday) and *Diadromus collaris* (Grav.) Parasitoids of the Diamondback Moth *Plutella xylostella* (L.) and Assessment of the Effectiveness of *Cotesia vestalis* as a Biological Control Agent in Zambia. Ph.D. Thesis, University of Zambia, Lusaka, Zambia, 2012.

90. Mazlan, N.; Mumford, J. Insecticide use in cabbage pest management in Cameron highlands, Malaysia. *Crop Prot.* **2004**, *24*, 31–39. [CrossRef]

91. Price, L.L. Demystifying farmers' entomological and pest management knowledge: A methodology for assessing the impacts on knowledge from IPM-FFS and NES Interventions. *Agric. Hum. Values* **2001**, *18*, 153–176. [CrossRef]

92. Stadlinger, N.; Mmochi, A.J.; Dobo, S.; Glyllback, E.; Kumblad, L. Pesticide use among smallholder rice farmers in Tanzania. *Environ. Dev. Sustain.* **2011**, *13*, 641–656. [CrossRef]

93. Williamson, S. Understanding natural enemies: A review of training and information in the practical use of biological control. *Biocontrol News Inf.* **1998**, *19*, 117–126.

94. Hough, P. *The Global Politics of Pesticides: Forging Consensus from Conflicting Interests*; Earthscan Publications Ltd.: London, UK, 1998.

95. Association of Veterinary and Crop Associations of South Africa (AVCASA). *Hands off Banned Pesticides*; Media Statement; AVCASA: Midrand, South Africa, 2008.

96. Quinn, L.P.; de Vos, B.J.; Fernandes-Whaley, M.; Roos, C.; Bouwman, H.; Kylin, H.; Pieters, R.; van den Berg, J. Pesticide Use in South Africa: One of the Largest Importers of Pesticides in Africa. 2012. Available online: http://www.intechopen.com/books/pesticides-in-the-modern-world-pesticides-use-and-management/pesticide-use-in-south-africa-one-of-the-largest-importers-ofpesticides-in-africa (accessed on 5 January 2017).

97. Löhr, B.; Kfir, R. Diamond back in Africa: A review with emphasis on Biological Control. In *Improving Biocontrol of Plutella xylostella*; Kirk, A.A., Bordat, D., Eds.; Agricultural Research for Development: Montpellier, France, 2004.

98. Momanyi, C.; Löhr, B.; Gitonga, L. Biological Impact of the Exotic Parasitoid *Diadegma semiclausum* (Hellen) of diamondback moth *Plutella xylostella* (L.) in Kenya. *Biol. Control* **2006**, *38*, 254–263. [CrossRef]

99. Löhr, B.; Kfir, R. Diamondback moth *Plutella xylostella* (L.) in Africa: A review with emphasis on biological control. In Proceedings of the International Symposium, Montpellier, France, 21–24 October 2002.

100. Chidawanyika, F.; Terblanche, J.S. Costs and benefits of thermal acclimation for codling moth, *Cydia pomonella* (Lepidoptera: Tortricidae): Implications for pest control and the sterile insect release programme. *Evol. Appl.* **2011**, *4*, 534–544. [CrossRef] [PubMed]

101. Sørensen, J.; Addison, M.; Terblanche, J.S. Mass rearing of insects for pest management: Challenges, synergies and advances from evolutionary physiology. *Crop Prot.* **2012**, *38*, 87–94. [CrossRef]
102. Overgaard, J.; Sørensen, J.G. Rapid thermal adaptation during field temperature variations in *Drosophila melanogaster*. *Cryobiology* **2008**, *56*, 159–162. [CrossRef] [PubMed]
103. Kristensen, T.N.; Hoffmann, A.A.; Overgaard, J.; Sørensen, J.G.; Hallas, R. Costs and benefits of cold acclimation in field released *Drosophila*. *Proc. Natl. Acad. Sci. USA* **2008**, *105*, 216–221. [CrossRef] [PubMed]
104. Htun, P.W.; Myint, M. Radiation induced sterility for biological control of diamondback moth *Plutella xylostella* (L.). *Int. J. Adv. Sci. Eng. Technol.* **2014**, *4*, 285–291.
105. Dyck, V.A.; Hendrichs, J.; Robinson, A.S. *Sterile Insect Technique: Principles and Practice in Area-Wide Integrated Pest Management*; Dyck, V.A., Hendrichs, J., Robinson, A.S., Eds.; Springer: Dordrecht, The Netherlands, 2005.
106. Harvey-Samuel, T.; Morrison, N.I.; Walker, A.I.; Marubbi, T.; Yao, J.; Collins, H.L.; Gorman, K.; Davies, T.G.E.; Alphey, N.; Warner, S.; et al. Pest control and resistance management through release of insects carrying a male sterile transgene. *BMC Biol.* **2015**, *13*, 49–64. [CrossRef] [PubMed]
107. Ant, T.; Koukidou, M.; Rempoulakis, P.; Gong, H.F.; Economopoulos, A.; Vontas, J.; Alphey, L. Control of the olive fruit fly using genetics-enhanced sterile insect technique. *BMC Biol.* **2012**, *10*, 51. [CrossRef] [PubMed]
108. Leftwich, P.T.; Koukidou, M.; Rempoulakis, P.; Gong, H.F.; Zacharopoulou, A.; Fu, G.; Chapman, T.; Economopoulos, A.; Vontas, J.; Alphey, L. Genetic elimination of field-cage populations of Mediterranean fruit flies. *Proc. R. Soc. Biol. Sci.* **2014**, *281*, 1792. [CrossRef] [PubMed]
109. De Valdez, M.R.W.; Nimmo, D.; Betz, J.; Gong, H.F.; James, A.A.; Alphey, L. Genetic elimination of dengue vector mosquitoes. *Proc. Natl. Acad. Sci. USA* **2011**, *108*, 4772–4775. [CrossRef] [PubMed]
110. Van Veen, S. The worldbank and pest management. In *Integrated Pest Management in the Global Arena*; Maredia, K.M., Dakouo, D., Mota-Sanchez, D., Eds.; Commonwealth Agricultural Bureau International (CABI): Wallingford, UK, 2003.
111. Gillson, L.; Dawson, T.P.; Jack, S.; McGeoch, M.A. Accommodating climate change contingencies in conservation strategy. *Trends Ecol. Evol.* **2013**, *28*, 135–142. [CrossRef] [PubMed]
112. Cook, S.M.; Khan, Z.R.; Pickett, J.A. The use of push-pull strategies in Integrated Pest Management. *Ann. Rev. Entomol.* **2007**, *52*, 375–400. [CrossRef] [PubMed]
113. Khan, R.Z.; Midega, C.A.O.; Bruce, T.J.A.; Hooper, A.M.; Pickett, J.A. Exploiting phyto-chemicals for developing a "push-pull" crop protection strategy for cereal farmers in Africa. *J. Exp. Bot.* **2010**, *10*, 1–12.

MDPI AG

St. Alban-Anlage 66

4052 Basel, Switzerland

Tel. +41 61 683 77 34

Fax +41 61 302 89 18

http://www.mdpi.com

Sustainability Editorial Office

E-mail: sustainability@mdpi.com

http://www.mdpi.com/journal/sustainability

www.ingramcontent.com/pod-product-compliance
Lightning Source LLC
Chambersburg PA
CBHW051730210326
41597CB00032B/5669